Stranded at the Top of the World

Mary R. Tahan · Cornelia Lüdecke

# Stranded at the Top of the World

## A Story of Exploration and Heroic Rescue in the Arctic

 Springer

Mary R. Tahan
Vancouver, BC, Canada

Cornelia Lüdecke
Munich, Germany

ISBN 978-3-031-56287-7        ISBN 978-3-031-56288-4  (eBook)
https://doi.org/10.1007/978-3-031-56288-4

This Springer imprint is published by the registered company Springer Nature Switzerland AG
The registered company address is: Gewerbestrasse 11, 6330 Cham, Switzerland

Paper in this product is recyclable.

*In memory of Olga Rosette Karam Tahan.*

*To K.A. and Catherine.*

*In memory of Bernhard Fritscher.*

*And in memory of the seven German expeditioners who disappeared and the one Norwegian expeditioner who died in Svalbard in 1913.*

*And finally, for all the animals who have aided humans—at home and on the ice.*

# Preface: A Tale of Ambitious Exploration and Heroic Rescue in the Arctic

This book tells the story of the Norwegian Captain Arve Staxrud's Rescue Expedition of 1913 that searched for and successfully rescued members of the German Lieutenant Herbert Schröder-Stranz's expedition of 1912 to Spitsbergen (today called Svalbard). It is the first book to portray the Staxrud expedition in detail and combine the chronological events of both the Staxrud and Schröder-Stranz expeditions as well as the concurrent rescue expeditions that failed. The story follows the overland ice, fjord, and glacier crossings of Staxrud and his expedition members to reach the German expedition vessel *Herzog Ernst* where it lay locked in the ice in Treurenberg Bay and to find the stranded members who had gone separate ways across the ice in West Spitsbergen and who had disappeared in Northeast Land. It recounts the adversities and adventures of both the rescuers and those who hoped to be rescued. And it examines the controversial personalities and methodologies involved in both enterprises, including the ambitious and unprepared Schröder-Stranz, and the organized and at times criticized Staxrud.

A brief summary of this story had been featured in author Mary R. Tahan's previous book *Roald Amundsen's Sled Dogs: The Sledge Dogs Who Helped Discover the South Pole*, and the events were further documented in her sequel book *The Return of the South Pole Sled Dogs: With Amundsen's and Mawson's Antarctic Expeditions*, including the tale of the two sled dogs Lussi and Storm from Roald Amundsen's Norwegian Antarctic Expedition of 1910–1912 (the South Pole Expedition) who participated in the Staxrud Expedition dispatched to search for and rescue Schröder-Stranz. This new account features a wider documentation and expanded version of the story, incorporating details written by Tahan and co-author Cornelia Lüdecke, who has also written about Schröder-Stranz in previous publications, including *Wissenschaft, Abenteuer und Prestige—Deutscher Brennpunkt in Spitzbergen in der Vorkriegszeit 1910–1914*.

Recent research conducted for this book has resulted in the discovery of interesting new information that sheds light on the Staxrud Rescue Expedition's impact on later search-and-rescue expeditions in Spitsbergen/Svalbard. The new information also provides further insight about the inner workings of the Schröder-Stranz expedition.

The Staxrud rescue expedition with the *Hertha*, under the leadership of trail-blazing topographer and surveyor Arve Staxrud, was composed of Norwegian officers and crew, local overwintering experts and trappers, Sámi reindeer herders (reindeer walkers), trained and experienced sledge dogs, and sledge-pulling draft reindeer, all of whom conducted an extensive overland and sea search-and-rescue operation among the frozen icefields, fjords, and glaciers of Spitsbergen, to find and save the German officers, scientists, and marine artist, as well as the Norwegian sailors and crewmembers, of the Schröder-Stranz expedition aboard the *Herzog Ernst*. The Staxrud rescue expedition crossed the length of West Spitsbergen four times, in two separate journeys, using sled dog teams during both journeys, as well as sledge-pulling reindeer during the first journey, to reach the iced-in Schröder-Stranz ship and the Swedish Arc-of-Meridian hut in Treurenberg Bay, where some of the expeditioners were stranded, and to endeavor to reach Northeast Land where Schröder-Stranz himself and three of his companions, along with their dogs, had attempted to sledge across the ice and had disappeared.

Information incorporated into the book includes how the Schröder-Stranz expedition materialized and came to be in Spitsbergen at such an unusually late time of year and how the Staxrud expedition was formed, planned, and expertly carried out. It also includes the attempts and failures of other rescue expeditions to find and save the Schröder-Stranz expedition members—failures which sometimes endangered the lives of the very people whom the rescuers were seeking to save, and failures which also necessitated the rescue of the would-be rescuers. And it includes how previous German expeditions had conducted forays into the Arctic, how the Germans had used Spitsbergen as a staging ground and as preparatory fieldwork for launching expeditions to Antarctica, and how subsequent rescue missions and expeditions in Svalbard were strengthened and better-strategized due to the successes and results of the Staxrud rescue mission and based on the Staxrud expedition's effective performance.

A description of the use of animals in both the Norwegian and German expeditions and in the search-and-rescue efforts conducted by Staxrud is included, demonstrating the crucial role that sled dogs and draft reindeer performed in preserving human lives and in aiding those expedition members who sought to rescue the stranded expeditioners. The role of animals as life-sustainers and as nutrition for the marooned expeditioners is analyzed as well in terms of the Polar bears, Arctic foxes, and wild reindeer that were hunted and consumed by the stranded men as well as by the rescue expedition members.

The activities and presence of Sámi in the rescue efforts also are featured in the story, depicting the important part they played in acting as guides and reindeer walkers and in providing expert assistance that was vital to both the rescuers and the rescued expedition members.

This is the story of a group of people who were stranded on a group of islands in the Arctic Ocean—islands that themselves seem isolated in the middle of the ocean at the top of the world, north of Norway, and south of the North Pole. It is the story of how the men and dogs of an Arctic expedition became victims of the uninformed ambitions of their leaders and how some of them were saved by a humble and unassuming interspecies rescue mission made up of men, dogs, and reindeer, who found some of the stranded expeditioners and brought them home.

Most of the documentation contained in this book is based on original source material including letters, telegrams, and reports written by Staxrud, newspaper interviews with Staxrud, diaries and recollections written by members of Staxrud's expedition and Schröder-Stranz's expedition, first-hand account books and reports written by members of Schröder-Stranz's expedition, official reports from Norway and Germany, and correspondence written by influential personalities affiliated with the organization of the rescue expedition. These letters, diaries, articles, and reports have been translated from their respective Norwegian and German languages into English.

The sources researched and utilized include original unpublished documents and original photographs housed in archives at several institutions, including the National Library of Norway in Oslo; the Norwegian Polar Institute in Tromsø; the Svalbard Museum in Svalbard; the Gazert Estate, Volkert Gazert, in Partenkirchen; the Herzog Ernst II estate in Mauritianum, Altenburg; the Drygalski estate, Thomas Mörder, in Feldkirchen-Westerham; the Schröder-Stranz Estate, Leibniz Institute for Regional Geography, in Leipzig; the Ritscher estate, Leibniz Institute for Regional Geography, in Leipzig; and the Ritscher Estate, Cornelia Lüdecke, in Munich. The sources also include original research conducted by Mary R. Tahan regarding the history and biographies of the South Pole Expedition sled dogs.

The Staxrud rescue mission of 1913 brought together humans and animals in an organized, engaged, and well-prepared expedition that sought not to seek fame but to seek and save lives who were at risk in the Arctic. Though lesser known in the annals of history, this diverse expedition exemplified cooperation and concern for life—worthy traits and actions to study and spotlight, especially in light of how world events continued to unfold after the expedition and through today. It is hoped that this documentation and telling of the true tale, as featured in this book, will bring more recognition to this portion of history and will inspire a closer look at the remarkable actions of these human and animal actors in the Arctic.

Vancouver, BC, Canada                                                 Mary R. Tahan
Munich, Germany                                                      Cornelia Lüdecke
July 2023

# Acknowledgements

The authors wish to thank the National Library of Norway (Nasjonalbiblioteket) in Oslo for providing us with access to important historical documents and archival material related to Arve Staxrud, Spitsbergen, Leon Amundsen, Roald Amundsen, and the South Pole expedition, for use in this book, including letters of correspondence, diaries, original photographs, and historical publications preserved in the NB Manuscripts Collection, and for granting us use of rare photographs in the Picture Collection. We extend a very special thank you to Research Librarian Anne Melgård for her wonderful assistance, thoroughness, and helpful translations of letters, diaries, newspaper articles, and documents. We also recognize Librarian Nina Korbu for her significant role in the Library, and we express many thanks to Research Librarians and Curators Guro Tangvald and Jens Petter Kollhøj for their kind assistance with photographs in the NB Picture Collection.

We also very sincerely thank the Norwegian Polar Institute (Norsk Polarinstitutt) in Tromsø for their great assistance and for providing us with crucial historical documents including diaries, letters of correspondence, reports, original photographs, and rare publications related to Staxrud's expedition and expedition members, Herbert Schröder-Stranz's expedition and expedition members, and Spitsbergen and Green Harbour, including material related to Arve Staxrud, Daniel Nøis, Hilmar Nøis, and Einar Rotvold, for use in this book. Thank you very much to Harald Dag Jølle, Polar Historian at the NPI, and a truly special thank you to Petr Masat, Librarian at the NPI Library (Bibliotek). Thanks also go to Head Librarian Ivar Stokkeland and Photo Librarian Ann Kristin Balto.

Our sincere gratitude as well goes to the Svalbard Museum in Svalbard for providing us with the authorized use of historical photographs and with the granted access to diaries and archival documents from Staxrud, Schröder-Stranz, Hilmar Nøis, and Spitsbergen, for inclusion in our book. We especially thank Trygve Sikveland Røysland, Collection Leader; Sander Solnes, Head of Collection; and Anita Utsi, Office Manager.

We thank the Leibniz Institute of Regional Geography, Leipzig, for the generous authorization to use archival material and photographs from the Ritscher and Schröder-Stranz estates, as well as Volkert Gazert for Gazert's letter to Drygalski from Gazert's estate. We also received authorization from Dietrich Reimer Verlag for using photos from Miethe (1914).

Additional photographs were given to us from the Mauritianum, Altenburg, as well as from Brigitte Zeinel Abidine, Frank Berger, and Thomas Mörder. We also thank Dietmar Kade for providing a copy of the advertising leaflet of the polar voyage of North German Lloyd in 1911 and Karsten Piepjohn for preparing the English version of the German map of Northern Svalbard.

We wish to also thank Anders Bache, Curator at Roald Amundsen's Home (Roald Amundsens Hjem), Follo Museum, Museene i Akershus, for information regarding Helmer Hanssen's letters.

A special and significant thank you goes to Debbie Archeck DVM, Doctor of Veterinary Medicine, for her expert information on dogs' gestation and weaning times and for her professional opinion on the age of the puppies featured in the historical newspaper photograph of Lussi's puppies.

Thank you to professional translator Elin Melgård for translations of the expedition diaries, letters of correspondence, reports, newspapers, and articles. And thank you also to our other translators. We would also like to thank our publisher and the reviewer.

Mary R. Tahan would like to express her profound thanks to K.A. Colorado, Catherine Tahan-Corpus, and Elena Tahan for their ever-present support and patience during her intense work. She would also like to especially acknowledge the support of Olga "Rosette" Tahan, who, to our deep sadness, has passed away, but who will always remain with us.

In addition, Cornelia Lüdecke would like to thank her husband Gerhard for his patience while she was working for a long time on the manuscript.

The authors would like to state that this has been a very positive, productive, and special collaboration, working together and using Norwegian and German sources to tell the little-told tale about a rescue mission comprised of humans and animals who cooperated to rescue members of an ambitious but poorly-planned expedition and who persevered in order to save and preserve lives and thus honor the importance of life.

# Contents

# About the Authors

**Mary R. Tahan** is a writer, producer, and documentarian, with a professional background in journalism and marketing. She has authored journal articles and historical books and produced and directed documentary films. Her articles, lectures, and presentations focus on Antarctica, the "Heroic Age of Antarctic Exploration," and the mid-twentieth century, as well as the Arctic region, and cover many countries. As part of her research on Polar exploration, Mary R. Tahan traveled to Antarctica by invitation of the Dirección Nacional del Antártico (Instituto Antártico Argentino), where she was awarded an art residency to live and work on the Antarctic continent. There she performed on-site photography and videography of the Antarctic landscape, wildlife, animals, and historical sites, as well as conducted interviews with scientists and curators. Her research has also taken her to Argentina, Norway, France, Russia, the USA, Australia, Canada, and England, where she has visited significant sites and also interviewed descendants of the early explorers of the Arctic and Antarctic. She has presented her work at The Scientific Committee on Antarctic Research (SCAR) and the Encuentro de Historiadores Antárticos Latinoamericanos (Meeting of Latin American Antarctic Historians) conferences, as well as at Centro Austral de Investigaciones Científicas del Consejo Nacional de Investigaciones Científicas y Técnicas (CADIC-CONICET). She was a featured speaker at the *Visions on Antarctica* Fundación PROAntártida International Symposium, is a member of the SCAR Standing Committee on the Humanities and Social Sciences (SC-HASS), and was nominated as Honorary Historian for the Argentine Navy.

Mary R. Tahan's previous books include *Roald Amundsen's Sled Dogs: The Sledge Dogs Who Helped Discover the South Pole*; *The Return of the South Pole Sled Dogs: With Amundsen's and Mawson's Antarctic Expeditions*; and *The Life of José María Sobral: Scientist, Diarist, and Pioneer in Antarctica*.

**Cornelia Lüdecke** is a meteorologist, an expert on German polar history, and a retired professor of history of natural sciences from the University of Hamburg. She has written 22 books and 195 papers focusing on the history of geography, polar research, meteorology, and oceanography. She gained practical experience during meteorological expeditions on aircraft in the Alps and the Northern Territories (Australia) as well as on the research vessels *Gauss III* and *Meteor II* in the German Bight and the Atlantic. Later, she accompanied an expedition to Svalbard on board the *Maria S. Merina* as polar historian, where she visited the remains of Polhem, Treurenberg Bay, and Schröder-Stranz's last camp. Currently she is lecturing about polar expeditions on board cruise ships to the Arctic and Antarctic. Since 1991, she has been leading the German History of Polar Research Working Group. She also chaired the German History of Meteorology Specialist Group (1995–2018). In addition, she was Vice President (2001–2006) and President (2006–2009) of the International Commission on History of Meteorology. In 2004 she founded the Action Group on History of Antarctic Research for The Scientific Committee on Antarctic Research, which was transformed into the SCAR Standing Committee on the Humanities and Social Sciences (SC-HASS), which she co-led until 2021. She also served as Vice President of the International Commission on the History of Oceanography (2012–2021). She received the Reinhard Süring Medal in 2010 and the Paulus Award in 2019, both from the German Meteorological Society. In 2012 she became a corresponding member of the International Academy of the History of Science in Paris.

Cornelia Lüdecke's previous books include *Germans in the Antarctic*, *The Third Reich in Antarctica*, *Erkundung der Arktis im Dienste der Wettervorhersage*, and *Amundsen—ein biographisches Portrait*.

# Chapter 1
# Introduction: Norwegians and Germans in Spitsbergen (Svalbard)

## Prologue: The Meeting in Tromsø—Fridtjof Nansen and Alfred Ritscher (August 5, 1913)

The esteemed Northern explorer Fridtjof Nansen, formulator of the ice drifting method of ship travel near the North Pole, mentor to South Pole explorer Roald Amundsen, designer of the rounded vessel *Fram*, and trusted global diplomat with stellar scientific background and reputation, stepped off the steamer in Tromsø, Norway and went about completing the final tasks of preparation prior to undertaking the major portion of his travel northward (Nansen, 1914; Tahan, 2019, 2021).

He had left Christiania (Kristiania, today: Oslo) a few days prior, on August 2, 1913, traveling by train to Trondhjem (Trondheim), then boarding a transit steamer to Tromsø, where he arrived on a rainy midday on the 5th of August (Nansen, 1914, 1–8). The famous Norwegian figure was due to depart again that very evening, on the steamer *Correct*, embarking on a journey to Siberia to establish a trade route between Europe and Russia—specifically North Norway and the interior of Siberia—by way of the Arctic Ocean's Kara Sea, through which the great explorer himself had once sailed, and by way of the mouth of the Yenisei River.

In the midst of his preparations in Tromsø, a message was delivered to him, requesting that the respected explorer visit Alfred Ritscher, the young German captain of the ill-fated Schröder-Stranz expedition who had survived an unimaginable ordeal in Spitsbergen (today: Svalbard) and who now lay convalescing at the Catholic hospital in Tromsø, having lost his foot as a result of the tragic events (Nansen, 1914, 8–10). Ritscher's injuries had led to the amputation of half of his right foot as well as the large toe of his left foot and the two leading joints of his small finger on his right hand (Ritscher, 1916, 28).

Herbert Schröder-Stranz's expedition, in which Ritscher had participated, had ventured north to Spitsbergen the previous summer of 1912 and, by autumn, had found itself in dire straits (Nansen, 1914, 8–10). Nansen himself had helped to coordinate a rescue operation for the German expedition, resulting in a relief expedition

© The Author(s), under exclusive license to Springer Nature Switzerland AG 2024
M. R. Tahan and C. Lüdecke, *Stranded at the Top of the World*,
https://doi.org/10.1007/978-3-031-56288-4_1

organized and led by the Norwegian captain Arve Staxrud and dispatched the ensuing spring. Now, on this day, Nansen graciously accepted the invitation to meet with this survivor, and visited him at his bedside. He marveled at the young individual's perseverance in survival and his forthrightness about the lack of experience which had led the Schröder-Stranz enterprise to the disaster it had encountered. For the young captain candidly admitted to his visiting guest that the German expeditioners' inexperience and lack of knowledge were two factors that had led to decisions which later proved to be unwise and that were the source of the predicament in which Ritscher now found himself. All of the expeditioners had suffered. Eight men had died, and only seven had survived. The young captain now reclined in his hospital bed, injured, thoughtful, and wishing to return to Spitsbergen to retrieve his rescued ship, the ship that had been imprisoned in the ice as a result of the hasty and ill-timed voyage.

It was the draw of the North, the draw of the ice that drew some people to their misfortune and to their deaths, thought the Norwegian explorer Nansen after his meeting with the young German (Nansen, 1914, 10). Inexperience and lack of knowledge only compound the already adverse conditions that await a traveler in the Arctic. The older gentleman walked away from the young captain pondering *what if*—What if he, Nansen, had been able to stop the German expedition in time, to have helped them avoid the tragedy that eventually befell them. After all, he had been sailing in his small yacht *Veslemøy* on a scientific cruise in that very same area north of Spitsbergen and Hinlopen Strait during the very same time that the German expedition had been there—in August of 1912. Maybe a fortuitous encounter would have changed the course of events? If only he had had an opportunity to advise the expedition men on the *Herzog Ernst*, maybe they would have all lived to tell the tale? But fog had been prevalent in that area at that moment of time, recalled Nansen, and the two vessels surely must have passed one another unknowingly.

They must have passed one another literally like two ships proverbially passing in the night.

And so it was. It was too late to go back, impossible to travel back in time. What had happened, had already happened. It had occurred in the far North, on the group of islands that reside in the Arctic Ocean, situated at the top of the world, nearly halfway between northern Norway and the North Pole. It had happened here, in the extreme Arctic archipelago that was Spitsbergen.

## The Archipelago in the Arctic Ocean: "No Man's Land" (June 1596)

The first documented sighting of the archipelago in the Arctic Ocean was made in June 1596 by the Dutch sailor Willem Barentsz (ca. 1550–1597), a navigator who led an expedition from Amsterdam in search of the Northeast Passage (Rabot, 1919; Rudmose Brown, 1919; Svalbard Museum/Willem Barentsz; Berg, 2020). The voyage was Barentsz's third attempt to establish a trade route in the North Sea area for the Dutch, having previously traveled in 1594 and 1595.

This third expedition comprised two ships, and Barentsz sailed on the ship captained by Jacob van Heemskerck, while the second ship was captained by Jan Cornelisz Rijp. Rather than finding a sea route along the north, the expedition came upon land—first, Bear Island (Bjørnøya), at the southern vicinity of the group of islands, and then, further north, up to nearly 80° N, the mass of land that was part of the Spitsbergen (Svalbard) archipelago. This land, described as mountainous and covered with sharp peaks, was given the name Spitsbergen. And the first mapping of it was created and subsequently introduced to The Netherlands in 1597 by the returning members of the expedition led by Barentsz, who himself had died on the return journey while overwintering in Novaja Zemlja.

Situated approximately 600 miles south of the North Pole and 400 miles north of the northern coast of Norway, and comprising two large islands—West Spitsbergen and Northeast Land—as well as many smaller islands, the Spitsbergen archipelago covers approximately 25,000 square miles along the Arctic Ocean; as far as was known then and today, there were no indigenous peoples who resided there (Rudmose Brown, 1919).

It is as though, in modern terms, the archipelago were a veritable miniature Antarctica existing in the vast Arctic Ocean, in that there were no inhabitants. Indeed, it would subsequently become known as *terra nullius* (no one's land) or "no man's land" – but that designation was not just for the lack of people but rather for the lack of sovereign rule or territorial claim or imposed laws (Berg, 2020; Rudmose Brown, 1919). The Arctic was connected, however, by the group of countries and territories surrounding the Arctic Circle, including Norway, Greenland, and Russia, and the knowledge that Barentsz's expedition brought back about Spitsbergen's presence elicited further investigation and an excited rush for living resources. Soon humans were attracted to the abundant wildlife in the Spitsbergen area—not for observation, but for consumption. The 1600s and 1700s were filled with waves of hunters and trappers who sailed from Europe and Scandinavia and Russia, hunting and depleting the walrus, whale, and seal populations on and around the islands, as well as hunting polar bears, foxes, and reindeer in the area (Rabot, 1919; Rudmose Brown, 1919; Svalbard Museum/The Pomors). These hunters and fur trappers were meeting the brisk demands of an economy back home and supplying popular animal products as well as the initial products of a burgeoning whale oil industry.

## The Run for Resources: Whaling, Hunting, Touring, and Mining (Seventeenth Century–Twentieth Century)

Serious whaling ensued and became an international enterprise in Spitsbergen during the early seventeenth century and the eighteenth century, beginning in the fjords among the islands and along the coast and spreading out to the sea around the archipelago (Svalbard Museum/Whaling; Berg, 2020). As of the end of the eighteenth century and the beginning of the nineteenth century, a near-extinction of the whales in the area had unfortunately resulted. The ceaseless and merciless massive large-scale hunting of the whale, which had destroyed the population and nearly consumed the entire existence of this mammal, was finally at an end in this region.

But the hunt for animals did not cease. In the twentieth century, tourists from different parts of the world, armed with money and more, now joined the fray, flocking to the archipelago and taking cruises along the coast and the fjords, sailing on ships from where some of these summertime vacationers would leisurely shoot at polar bears as they spotted them along the way, or hiking through the valleys where they would happily shoot a multitude of reindeer for gleeful sport, leaving behind their carcasses strewn across the ground (Berg, 2020; Rudmose Brown, 1919). Perhaps this, too, was an early form of trophy hunting.

Hunting as a means of living and as a vocation was taken up by certain individuals—hunters and trappers—who would overwinter on the archipelago, spending 10 to 11 months on the islands hunting, shooting, capturing, and trapping as many animals as they could to take back home (Svalbard Museum/Trapper Life). The targets of their long stays and physical endeavors were the fur and/or meat or body parts of foxes, polar bears, walruses, seals, birds, and reindeer. Norwegian overwintering hunting and trapping expeditions began in 1795 and continued in the early 1800s—the first of which, in 1795/1796, included Russians who helped and trained the Norwegian expedition members. Over the years the number of hunters increased, and a system of essential hunter's cabins and hunter's huts was established, as well as a network of animal traps devised. The type of prey dictated the method used to incapacitate and kill the animals. The round-the-clock daylight of the summer constituted the majority of the hunting activity timeframe. Many of the overwinterers who faced danger, winter's darkness, isolation, and extreme elements in order to hunt, came from Norway's mainland, including Hammerfest, Tromsø, and the northern coast.

The hunting of whales, walruses, and Arctic animals, which had begun at the beginning of the 1600s, had caused the convergence of English, Dutch, Norwegians, Danish, Spaniards, French, Germans, and Russians, in the Far North. They would be joined by Americans.

At the end of the nineteenth century, coal was found on Spitsbergen and brought renewed attraction from all parts of the globe. In 1899, the Norwegian ice pilot and captain Søren Zachariassen (also spelled Zakariassen)[1] (1837–1915) shipped a major load of coal from Spitsbergen to the mainland of Norway—it was the first such commercial shipment and was sold successfully (Svalbard Museum/Mining; Stenersen, 1988; Norsk Polarhistorie.no/Søren Zachariassen). The pursuit of natural mineral resources accelerated at this time. Just as whale oil had attracted sailors, now fossil fuel drew entrepreneurs. John Munro Longyear (1850–1922), an American mining businessman originally from Lansing, Michigan, visited Spitsbergen in 1901, and established the Arctic Coal Company in 1906. The mining company,

---

[1] Zachariassen is the spelling used in the text throughout this book.

located in Advent Bay in the settlement Longyear City, later would be purchased by the Norwegian company Store Norske Spitsbergen Kulkompani—Great Norwegian Spitsbergen Coal Company—in 1916 (Svalbard Museum/Mining; Rudmose Brown, 1919; Longyear Museum; National Mining Hall of Fame; Berg, 2020). Other mineral mining companies founded on the archipelago included the Englishman Ernest Mansfield's Northern Exploration Company. The establishment of the American coal company and similar companies, with their civilian infrastructures and Norwegian labor forces, marked the advent of mineral mining enterprises on Spitsbergen during the twentieth century and the rise of mining activity. It was a veritable coal rush in the wild, wild north.

## A Taste for Scientific Exploration: Norwegians, Swedes, and Britons (During the 1800s)

Meanwhile, scientific interest in Spitsbergen was sparked among many, including notable British explorers and expeditions, and was pursued quite strongly and intensely by several Swedish explorers and expeditions.

It is reported that one of the first Norwegian expeditions to Spitsbergen for scientific purposes was undertaken by Professor B. M. Keilhau (Balthazar Mathias Keilhau [1797–1858]) of the University of Christiania, in 1827, during which he studied the geology, paleontology, and botany of the islands (Rabot, 1919; Committee to Norske Geografiske Selskap 18 March 1913). It is also reported that Norwegian hunters stalking walrus and seal populations geographically explored the land and waters of the islands during the mid and latter part of the 1800s, venturing out from western Spitsbergen—where the numbers of their prey had dramatically decreased due to their extreme hunting—and traveling further eastward and northward to Northeast Land, which is the second and the northernmost of the two largest islands within the archipelago.

When the Swedish scientist and explorer Adolf Erik Nordenskiöld (1832–1901) planned his expedition from Spitsbergen (Svalbard) toward the North Pole to take place in 1872–1873, it would be the first overwintering in this area (Nordenskjöld, 1880, 154–241). Due to an infectious canine disease at the time, there were not enough dogs from Greenland available for the sledge trip from his planned wintering station on Parryøya (Parry Island), one of the Seven Islands (Sjuøyane) in northern Svalbard. Alternatively, he now wanted to use reindeer as draft animals, for the first time. The reindeer also presented the added advantage of serving as an excellent meat reserve for the men in case of food shortage.

Nordenskiöld set out from Tromsø on July 21, 1872, with two ships carrying the wintering crew of 22 individuals as well as the provisions and wood for the wintering station. Dense drift ice prevented them from advancing as far as Parryøya. When they turned back, they met a Norwegian ship whose crew informed them of very unfavorable ice conditions in northern and eastern Spitsbergen. In consequence of this situation, Nordenskiöld's expedition retreated to Fair Haven in northeast Spitsbergen. On August 13, their third ship arrived there with 40 reindeer accompanied by four Sámi to care for the reindeer,[2] and with 3000 sacks of reindeer moss and coal. On August 17, a Norwegian fishing steamer passed by, having previously been trapped by ice in Liefde Bay for three weeks. Ice conditions seemed to be very bad in that year.

Later the Englishman Benjamin Leigh Smith (1828–1913) visited them on his way back from a hunting trip, having previously been stranded in Wijde Bay for five weeks (Nordenskjöld, 1880, 166; Capelotti, 2013, 82–83). Before Leigh Smith continued on his trip, he assured Nordenskiöld "that he wanted to be one of the first to find them the next year" (Nordenskjöld, 1880, 166). He already suspected that they would get into trouble.

Nordenskiöld turned a deaf ear to all warnings and continued his expedition north with two ships, but encountered drift ice already at a latitude of 80° 5′ N, so they had to turn back without success. Eventually, they reached Mosselbukta (Mossel Bay), where they set up their winter quarters and named their dwelling hut after their expedition ship Polhem. Three days later, the third expedition ship arrived with the reindeer moss and reindeer, who were now released to forage on land under the care of the Sámi. During a violent storm, all three ships were caught in the bay and later were trapped by the ice. Now 67 persons had to spend the winter together. By reducing the available provisions to 2/3rd portions, the food supplies would last just until April 1873, when new provisions were to arrive from Sweden.

During another strong storm, the 40 reindeer ran away to seek shelter in the mountains. Only one reindeer returned later. As a result, the expedition plan for a long sledge trip north had to be abandoned and, in addition, an important food source was lost. Due to insufficient food and the absence of hunting game, more and more cases of scurvy occurred during the winter. Nevertheless, a variety of scientific measurements were performed during the wintering.

On April 24, 1873, Nordenskiöld finally set out on his journey north with ten companions and a reindeer sledge and two other manhauled sledges. After running out of reindeer moss, they slaughtered their reindeer, who had proved to be a very useful draft animal. The reindeer's meat was a welcomed change to the daily pemmican. On May 16, the men finally reached Parryøya, their originally planned starting point. So that the "effort and work already used would not be entirely lost," Nordenskiöld chose for the return trip a route around the east coast of Northeast Land, which he wanted to map for the first time. When it became apparent that there was open water along the east coast, they could not continue their exploratory tour with the sledges. Nordenskiöld rescheduled again and steered his way south on June 1, where they came ashore Northeast Land at about 28°E and became the first to successfully cross the inland ice to Wahlenberg Bay.

---

[2] A. E. Nordenskiöld used the at-that-time contemporary term "Laplanders"; this term is no longer in use and will not be used in this book, as the proper name is Sámi, and therefore Sámi is the name used throughout this book text.

In the meantime, those who stayed behind fought scurvy but were not able to obtain much fresh meat by hunting. On June 13,[3] a steamer unexpectedly appeared in Mosselbukta. It was Benjamin Leigh Smith, who had guessed that he would find Nordenskiöld's expedition at that place (Capelotti, 2013, 103–106). As announced earlier, he brought plentiful provisions, such as "fresh potatoes, preserved vegetables, soups and preserved meats of various kinds, … citron juice, wine, tobacco, etc." (Nordenskjöld, 1880, 234) for the emaciated and sick men. The expedition doctor expressed it in these words: "To this help, which came just in time, we owe the salvation of several lives" (Nordenskjöld, 1880, 234).

On June 29, Nordenskiöld's group returned safely to Polhem. On the same day, the two supply ships left their anchorage for the voyage home. After further explorations, *Polhem* returned to Tromsø on August 6. Had Nordenskiöld taken seriously the warnings about the poor ice conditions, the expedition might have been more successful. However, he was able to demonstrate the usefulness of reindeer for sledging in the far north, and he was the first to describe the inner ice cover of Northeast Land. He also left his wintering hut Polhem and the unused bags of reindeer moss for later expeditions. In addition, his expedition served as important preparation for Nordenskiöld's great Northeast Passage navigation of 1878–1879.

Following Nordenskiöld's expedition on Spitsbergen, a cooperative scientific expedition to measure an arc of meridian in high latitudes was initiated at the end of the nineteenth century. The arc of meridian measurement expeditions were conducted jointly by Swedes and Russians, simultaneously in the north and south of Spitsbergen, respectively, and took place from 1898 to 1902 (Rudmose Brown, 1919). Both expeditions established overwintering huts: the Swedish in Treurenberg Bay and the Russian at the entrance of Hornsund. The facilities were not demolished after the return of the expeditions and therefore could be used later on.

Nordenskiöld's expedition employing reindeer, and the Swedish-Russian arc of meridian measurement expedition, would later have a direct and unforeseen effect on stranded members of a German expedition in 1912, and on the Norwegian rescue expeditioners who sought to save them in 1913. These events will unfold later in this narrative.

## The Time for Topography: Surveying and Geology in the Far North (1906)

Up until the beginning of the twentieth century, the archipelago that was considered *terra nullius* also contained land that was *terra incognita*, as little was known about the archipelago's interior (Rabot, 1919).

Serious undertaking of topographical and geological surveying of Spitsbergen by Norwegians began in 1906 and continued beyond that year.

---

[3] Nordenskiöld's report incorrectly states July 12 (Nordenskjöld 1880, 233).

Prince Albert I of Monaco (1848–1922) financed and conducted the first such survey in 1906, enabling the Norwegian army cavalry captain Gunnar Isachsen (1868–1939), via an oceanographic cruise on the prince's yacht *Princess Alice*, to lead an expedition, consisting of a group of Norwegian researchers, to survey the land along West Spitsbergen over two consecutive summers—in 1906 and 1907 (Committee to Norske Geografiske Selskap 18 March 1913; Rudmose Brown, 1919; Rabot, 1919; Hoel, 1933; Stenersen, 1988). Working with Isachsen on this first expedition was a Norwegian lieutenant named Arve Staxrud (1881–1933), who employed a new method of photogrammetric surveying, and who led sledging journeys across the ice (Hoel, 1933; Stenersen, 1988) (Fig. 1.1). It was the first detailed mapping, geographical exploration, and geological study of a contiguous area in the interior of Spitsbergen, and the results were significant and received considerable response.

Norway's government took up the cause in 1909 and initiated a grant that funded Isachsen to conduct another expedition in 1909–1910 (Committee to Norske

**Fig. 1.1**  Captain Arve Staxrud, in a portrait by Anders Beer Wilse, date given 1911. *Photographer* Anders Beer Wilse; Source/Owner: Norsk Folkemuseum/National Library of Norway

Geografiske Selskap 18 March 1913; Norwegian Polar Institute/History; Rabot, 1919). It was an expanded affair that included additional scientists making a detailed survey of the northwest area of Spitsbergen, again led by Isachsen, and it incorporated travel by dog sledge (Rabot, 1919). The 1910 expedition again included the topographer Staxrud (Hoel, 1933; Stenersen, 1988). In 1911 and again in 1912, as part of what now became Norway's new state-supported Spitsbergen surveying and research program, the Norwegian geologist Adolf Hoel (1879–1964) and Captain Arve Staxrud led surveying expeditions to record the topography and geology of a now extended area of the western side of Spitsbergen (Committee to Norske Geografiske Selskap 18 March 1913; Rabot, 1919; Hoel, 1933). At the recommendation of Hoel and Staxrud, a committee was formed by the Norwegian Geographical Society in 1911 to administer the Spitsbergen expedition accounts (Stenersen, 1988).

Thus, detailed mapping of Spitsbergen's interior, specifically the inner districts of West Spitsbergen, was performed at this time by Norwegians, under the leadership of Adolf Hoel and Arve Staxrud, and under the auspices of state-financed surveys authorized by Norway's parliament (Committee to Norske Geografiske Selskap 18 March 1913; Rudmose Brown, 1919). It was the dawn of a new age of scientific exploration in the rugged Arctic frontier (Stenersen, 1988). These investigative surveying missions would be carried on annually, and would eventually help position Norway as the likely sovereign of these Arctic islands—a feat later accomplished in 1920 with the signing of the Svalbard Treaty (Norwegian Polar Institute/History).

## The Telegraph Station: Communicating with the Outer World (1911–1912)

When the American journalist Walter Wellman (1859–1934) tried to fly to the North Pole with a dirigible from Danskøya in 1906, he sent the first wireless message from Spitsbergen (Svalbard) to *The Chicago Record-Herald* on July 24, 1906 (Capelotti, 1999, 71).

In the year 1911, Norway established the first permanent means of wireless communication between Spitsbergen and the outside world, by constructing a radio telegraph station in Green Harbour under the auspices and administration of the Norwegian state (Rudmose Brown, 1919; Norsk Polarhistorie.no/Spitsbergen Radio). Thus, for the very first time, overwinterers, explorers, and workers could communicate with the mainland of Norway and therefore send and receive news to and from the rest of the world south of the Arctic. The telegraph station in Green Harbour became a center of operations in Spitsbergen and would factor into the events of 1912 and 1913 quite significantly. By August 1912 it would be able to receive telegrams from Advent Bay, where a smaller telegraph station was erected by the Arctic Coal Company (Dole, 1922, vol. 2: 136–137), thus enabling intra-island communication. The Green Harbour telegraph station's existence also further strengthened

Norway's candidacy for sovereignty over the islands—sovereignty which would take effect when the Svalbard Treaty went into force in 1925.

But for now, as of 1912, the Spitsbergen land still belonged to no one, and continually attracted a small group of mining entrepreneurs, coal laborers, hunters/trappers, and scientific researchers who came to work on the islands, and whose activities came under no law or rule, and no jurisdiction.

With the stage set thusly at the end of the nineteenth century and the beginning of the twentieth century, into this timeframe entered the German exploration expeditions onto the Spitsbergen scene, seeking to combine science, sport, and surveillance through their own personal interactions with the icy Arctic archipelago—the archipelago that presented boundless potential, and that was known as Spitsbergen.

## Unpublished Sources and Original Material

Committee Letter to Norske Geografiske Selskap, 18 March 1913, Letter from Committee to Norske Geografiske Selskap board, 1913_03_18_Brev_til_Norske_Geografiske_Selskap, Library Archives, Norwegian Polar Institute, Tromsø. Received 20 July 2021.

## References

Berg, R. (2020). The Svalbard "Channel", 1920–2020—A geopolitical sketch. In Weber, J. (Eds.), *Handbook on geopolitics and security in the Arctic. Frontiers in international relations* (pp. 303–321). Cham: Springer. https://doi.org/10.1007/978-3-030-45005-2_18

Capelotti, P. J. (1999). *By airship to the North Pole: An archaeology of human exploration.* New Brunswick, New Jersey, and London: Rutgers University Press.

Capelotti, P. J. (2013). *Shipwreck at Cape Flora: The expeditions of Benjamin Leigh Smith, England's forgotten Arctic explorer.* Calgary: University of Calgary Press.

Dole, N. H. (1922). *America in Spitsbergen: The Romance of an Arctic Coal-Mine.* In Two Volumes. Boston: Marshall Jones Company. Volume II. Facsimile reprint by Scholar Select.

Hoel, A. (1933). Arve Staxrud. In *Norsk Geografisk Tidsskrift [Norwegian Journal of Geography], 4*(6), 345–346. https://doi.org/10.1080/00291953308622018

*Longyear Museum.* John Munro Longyear. https://www.longyear.org/learn/pioneer-index/longyear-john-m/ and https://www.longyear.org/learn/research-archive/john-longyear-landlooker-from-michigan/. Accessed April 13, 2022.

Nansen, F. (1914). *Through Siberia, the land of the future,* (A. G. Chater, Trans.). New York: Frederick A. Stokes Company; London: William Heinemann. Digitized by Google. https://books.google.com. Accessed August 4, 2014.

*National Mining Hall of Fame & Museum.* John Munro Longyear. https://www.mininghalloffame.org/hall-of-fame/john-munro-longyear. Accessed April 13, 2022.

Nordenskjöld, A. E. (1880). *Die Nordpolarreisen Adolf Erik Nordenskjöld's 1858 bis 1879.* Leipzig: F.A. Brockhaus.

*Norsk Polarhistorie—Polarhistorie.no.* Søren Zachariassen. "Søren Zachariassen: 1837–1915, Ishavsskipper fra Tromsø som innledet kulldriften på Spitsbergen." (Norwegian Polar Institute, University of Tromsø, Troms County Municipality.) https://polarhistorie.no/personer/Zachariassen,%20Soren.html. Accessed May 12, 2022.

*Norsk Polarhistorie—Polarhistorie.no.* Spitsbergen Radio. "Etablering av Spitsbergen Radio i 1911." Oddvar Ulvang. (Norwegian Polar Institute, University of Tromsø, Troms County Municipality.) https://www.polarhistorie.no/artikler/2014/etablering-spitsbergen-radio.html. Accessed May 12, 2022.

*Norwegian Polar Institute.* The History of the Norwegian Polar Institute. https://www.npolar.no/en/history/. Accessed July 21, 2021.

Rabot, C. (1919). The Norwegians in Spitsbergen. *The Geographical Review*, 8(4/5), 209–226. https://doi.org/10.2307/207837. Accessed July 21, 2021.

Ritscher, A. (1916). Wanderung in Spitzbergen im Winter 1912. *Zeitschrift der Gesellschaft für Erdkunde zu Berlin*, pp. 16–34.

Rudmose Brown, R. N. (1919). Spitsbergen, Terra Nullius. *The Geographical Review*, 7(5), 311–321. https://doi.org/10.2307/207588. Accessed March 29, 2022.

Stenersen, H. (editor/publisher). (1988). Tingelstad-brødrene som kartla Svalbard: Arve og Olav Staxrud har satt varige spor etter seg. In: *Brandbu'stikka: Lokalhistorisk Kvartalskrift for Brandbu. 1. KV. 1988 Mars NR. 11* (March 1988). (Svalbard-forskerne Arve og Olav Staxrud.) pp. 1–26. Biography Archive/Library Archives, Norwegian Polar Institute, Tromsø. Received July 20, 2021.

*Svalbard Museum.* Mining Communities. Gerd Johanne Valen/Bjørg Evjen. https://svalbardmuseum.no/en/kultur-og-historie/gruvesamfunn/. Accessed January 26, 2022.

*Svalbard Museum.* The Pomors Enter the Area. Per Kyrre Reymert/Christian Lydersen/Kit Kovacs. https://svalbardmuseum.no/en/kultur-og-historie/pomorene/. Accessed January 26, 2022.

*Svalbard Museum.* Trapper Life/Hunting Life. Marit Anne Hauan/Gerd Johanne Valen. https://svalbardmuseum.no/en/kultur-og-historie/pelsjegerliv/. Accessed January 26, 2022.

*Svalbard Museum.* Whaling in the Arctic. Kristin Prestvold. https://svalbardmuseum.no/en/kultur-og-historie/hvalfangst/. Accessed January 26, 2022.

Svalbard Museum. Willem Barentsz and the Discovery of Svalbard. Thor Bjørn Arlov. https://svalbardmuseum.no/en/kultur-og-historie/oppdagelsen/. Accessed January 26, 2022.

Tahan, M. R. (2019). *Roald Amundsen's sled dogs: The sledge dogs who helped discover the South Pole.* Cham: Springer International Publishing.

Tahan, M. R. (2021). *The return of the South Pole sled dogs: With Amundsen's and Mawson's Antarctic expeditions.* Cham: Springer International Publishing.

# Part I
# Visions of Northeast Land: The Failed Herbert Schröder-Stranz Expedition

# Chapter 2
# German Interest in Spitsbergen

## Previous German Expeditions

"When Spitsbergen appears from afar in the sea, it looks like a cloud; so that whoever has no idea of this landscape can hardly distinguish it from the air, for the mountains give such a reappearance in the sea that one is in doubt whether one is seeing clouds or land." (Quoted from Zorgdrager, 1723 [1975, 115]). There were still no scientific studies about Spitsbergen in the eighteenth century. The ship's doctor Friderich Martens (1635–1699) from Hamburg published the first detailed travel report about this fascinating landscape in the far north, after he had accompanied a whaling ship to Spitsbergen in 1671 (Martens, 1675). This report was a standard work until well into the nineteenth century, when German whalers and sealers came mainly from Schleswig–Holstein and Hamburg (Oesau, 1937, 1955).

Exploration of Spitsbergen did not occur until 1868 by the first German North Polar Expedition, which scientifically explored the waters around Spitsbergen (Koldewey & Petermann, 1871). Koldewey's travel report attracted a lot of attention at that time. Spitsbergen was not yet developed for tourism, but the travels of well-known personalities and spectacular ventures gradually boosted tourism.

Over time, this region came more and more into the German focus. On the one hand, Wilhelm Bade (1843–1903), who had participated as 2nd officer in the 2nd German North Polar Expedition (1869–1870), organized the first cruises to the Arctic since the beginning of the 1890s (Wikipedia/Wilhelm Bade). On the other hand, Spitsbergen became the destination of research voyages or an easily accessible training area for larger expeditions to the Arctic and Antarctic.

M. R. Tahan and C. Lüdecke, *Stranded at the Top of the World*,
https://doi.org/10.1007/978-3-031-56288-4_2

## Tourism: Holidaymaking in the North

In the last decade before the turn of the twentieth century, Norway and the North Cape, in Germany known as Nordland (Northland), became more and more interesting for German holidaymakers. Last, but not least, Kaiser Wilhelm II's so-called "Nordlandfahrten" (voyages to the northland, i.e., Norway), which took him to Norway year after year on his yacht MS *Hohenzollern*, continuously aroused interest through the newspaper reports published about them. The public interest was further intensified by the cinema film *Die Nordlandfahrt Kaiser Wilhelms II (Kaiser Wilhelm II's Voyage to the Northland)* in 1907 (Marschall, 1991, 200–201).

Soon tourist cruises were extended to Spitsbergen. The unique landscape, the wildlife, and also the special attractions, such as the attempts of the Swedish engineer Salomon August Andrée (1854–1897) to fly to the North Pole in a balloon from his launch site in Virgohamna on the Danish Island in the northwest of Spitsbergen in 1896 and 1897, triggered a boom in which mainly the Hamburg-Amerikanische-Paketfahrt-Actien-Gesellschaft (HAPAG) and, from 1911 onward, the Norddeutscher Lloyd (NL), participated (Wegener, 1897; Marschall, 1991, 186–188) (Fig. 2.1).

The attempts of the American journalist Walter Wellman (1858–1934), who unsuccessfully launched an airship for a flight towards the North Pole in 1906, 1907, and 1909, provided the basis for a further boom (Capelotti, 1999) (Fig. 2.2).

In any case, the launch site in Virgohamna thus became a popular destination for German voyages to the North (Fig. 2.3) at a time when newspaper headlines announced that Cook and Peary had discovered the North Pole in 1908 and 1909, respectively (Norddeutscher Lloyd, 1911; Sieberg, 1913).

## Count Zeppelin: Airships in the Arctic

Not only tourists and hunting parties looking for polar bear trophies, but also scientifically oriented expeditions, headed for Spitsbergen at the end of the imperial era. During the summer of 1910, Count Ferdinand von Zeppelin (1838–1917) organized the German Arctic Zeppelin-Expedition to Spitsbergen together with Hugo Hergesell (1859–1938), director of the Meteorological Service of Alsace-Lorraine and President of the International Commission for Scientific Aeronautics, and Prince Heinrich of Prussia (1862–1929), brother of the German Emperor (Miethe, 1911; Capelotti, 1999, XVIII, 78). They chartered the NL steamer *Mainz* in addition to two other ships to explore whether airships could be used for the investigation of the unknown parts of the Arctic (Fig. 2.4) (Hergesell, 1911a).

**Fig. 2.1** Flyer of the cruise to Spitsbergen on *Großer Kurfürst* of Norddeutscher Lloyd from Bremen in 1911. *Source* Norddeutscher Lloyd (1911), Title page, private collection Dieter Kade, Munich

One of the additional ships, the small wooden steamer *Fönix*, was converted in Tromsø for voyages into polar waters around Spitsbergen according to plans by the renowned shipyard owner Max Oertz (1871–1929). *Mainz* and *Fönix* were equipped with a radio-telegraphic system from the Telefunken-Gesellschaft, which allowed radio communication with the ships over a distance of 150 km (Fig. 2.5).

**Fig. 2.2** Wellman's airship hangar visited in 1910. *Source* Miethe and Hergesell (1911, 139)

From 16 July to 10 August 1910, they sailed along the west coast of Spitsbergen. In Advent Fjord, the first anchoring tests for airships were carried out with an ice anchor (Miethe, 1911). Further north, in Kongs Fjord, the first aerological ascents were made with a tethered balloon and pilot balloons, to which registering instruments for measuring air pressure, temperature, and humidity were attached (Hergesell, 1911b, 268–269; Hergesell, 1933; Lüdecke, 2012). With these devices the meteorological conditions of the upper air could be investigated. They revealed that in fine weather the local winds in the fjords, which extended only a few hundred meters upward, were often caused by the difference in temperature over land, ice, and sea, but were completely absent in fog (Hergesell, 1933, 268). However, there was a lack of meaningful information about cloud heights and precipitation to be expected at the flight level of airships between 300 and 1000 m height, which could lead to dangerous icing of the airship hull. Besides these meteorological measurements and successful anchoring tests in ice and on land, Count Zeppelin was looking for suitable places to land and launch airships (Hergesell, 1911b, 234–235; Miethe, 1911, 77). Finally, good conditions were found for an airship harbor in Kings Bay, where they discovered a little bay that was excellently suited for landing and launching airships. This bay is still listed under the name Zeppelinhamna near Ny-Ålesund in the Handbook for the European North Sea (Fig. 2.6).

Signe Harbour (Signehamna) at the northwestern end of Lilliehöök Fjord was also a good place (Miethe, 1911, 119). However, they considered the topographical conditions at Red Bay in Raud Fjord in northern Spitsbergen most suitable for an airship harbor (Hergesell, 1911b, 259, 261). Because of the great distance from the last European landing site, Signe Harbour was finally favored as the central station, while research flights were then to start from Red Bay (Zeppelin, 1911, 289).

**Fig. 2.3** Map of the western part of Spitsbergen with the planned itinerary of the steamer *Großer Kurfürst*. *Source* Norddeutscher Lloyd (1911, 5), private collection Dieter Kade, Munich

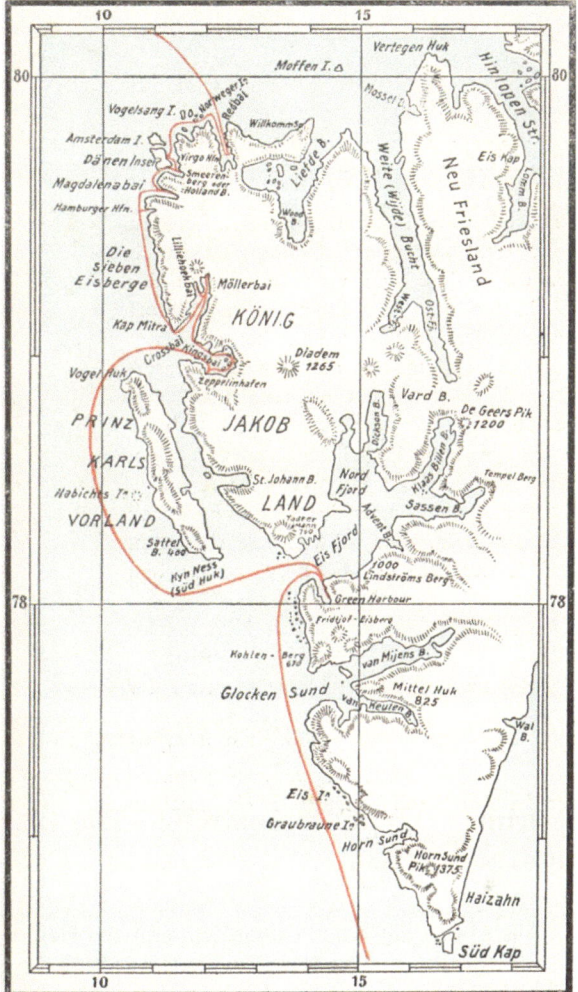

Before research flights with an airship could be thought of, however, the zeppelin had to be further developed in order to "cover long distances over land and sea without much recourse to outside assistance" (Hergesell, 1911a, 6).

**Fig. 2.4** *Mainz* in Advent Bay. *Source* Miethe (1911, 56)

## Filchner: Arctic for Antarctic—The Training Ground for the South

At the same time, when Count Zeppelin and Hergesell were planning their expedition to Spitsbergen, the Bavarian lieutenant Wilhelm Filchner (1877–1957) presented his plan for a German Antarctic expedition to the Berlin Geographical Society (Gesellschaft für Erdkunde zu Berlin) (Filchner, 1922, XI, 1994, VII). Zeppelin's project was already strongly supported by the highest circles. To prevent competition, it was suggested to Filchner that he abandon his own expedition plan in favor of Zeppelin's expedition. He was even invited to take part in this expedition, but he declined.

Instead, Filchner wanted to investigate the connection between West Antarctica and East Antarctica, both of which appeared to be separated from each other by the inlets discovered by James Weddell (1787–1834) and by James Clark Ross (1800–1862) (Filchner, 1994, VII, 1–6). Filchner was supported by the chairman of the Berlin Geographical Society, Albrecht Penck (1858–1945), who also helped him to draw up the scientific plan for the expedition, which was enthusiastically approved by the society. Although the Bavarian lieutenant had already gained plenty of expedition experience in the Pamir Mountains and in Tibet, neither he nor any of the

**Fig. 2.5** Radio communication aboard *Mainz. Source* Miethe and Hergesell (1911, 64)

four scientists who were in prospect to accompany him to Antarctica had any polar experience.

In order to remedy this shortcoming, a training expedition was first to be carried out to the nearest polar region, i.e., to Spitsbergen, in the summer to test the equipment of the Antarctic expedition planned for 1911 and to familiarize the expedition members, meteorologist Erich Barkow (1882–1923), geographer Heinrich Seelheim (1884–1964), astronomer Erich Przybyllok (1880–1954), and geologist Hans Philipp (1878–1944), with traveling and working in polar environments under the guidance of the Austrian alpinist and physician Karl Potpeschnigg (born 1875) (Filchner & Seelheim, 1911). A first transport of the equipment and the two ponies to Spitsbergen failed due to heavy ice conditions along the way. Luckily the expedition was invited onboard the *Äolus*, the ship of the excursion to the west coast of Spitsbergen organized by the

**Fig. 2.6** Zeppelin Harbour near Ny-Ålesund. *Source* Hergesell (1911b, 251)

XIth International Geological Congress, which had taken place under the presidency of Gerard de Geer (1858–1943) in Stockholm previously. Unfortunately, the ponies had to stay behind for reasons of space (Filchner & Seelheim, 1911, 33, 35f). This meant that the men had to pull the sledges themselves. It was a great misfortune that the equipment had not been designed accordingly for light weight. Due to this, the crossing to Spitsbergen was used to sort out everything superfluous and to make for themselves new pulling harnesses for the sledges. After a discussion with de Geer, who was considered the best expert on Spitsbergen at the time, Filchner decided onboard *Äolus* to focus on the working area in the central part of West Spitsbergen east of the Ice Fjord (Filchner & Seelheim, 1911, 39ff) (Fig. 2.7).

On 4 August, Filchner's expedition was disembarked with the help of the geologists in the interior of Temple Bay at von Post Glacier (Von Postbreen) (Filchner & Seelheim, 1911, 40ff). First, they brought all the expedition material from the coast over the stony moraine walls to the bare glacier ice. Then the men had to harness themselves in front of the heavily loaded sledges and pull and heave everything up the glacier, which took a lot of strength and time. A maze of crevasses forced tedious detours (Fig. 2.8).

**Fig. 2.7** Filchner's planned expedition route from Temple Bay to Stor Fjord (see dots in the square) in the east of Ice Fjord (Central Spitsbergen). *Source* Philipp (1914, Tf. XIV)

**Fig. 2.8** Exhausting way up the von Post Glacier. *Source* Filchner and Seelheim (1911, 48)

The equipment and its arrangement on the sledges were adapted to the daily requirements on the way by skillful improvisation. Nevertheless, there was "nothing more unpleasant than having to cover such distances with self-drawn sledges" loaded with heavy instruments and apparatus to be used at the station during the wintering in Antarctica (Potpeschnigg, 1914, 4, 6).

During this unaccustomed strenuous work, the men were plagued by hunger and above all thirst (Filchner & Seelheim, 1911, 50ff). Scouting the way, setting up and breaking down the camps, cooking, and the necessary repairs were time-consuming, so that hardly any other investigations could be carried out apart from the topographical survey of the surroundings.

The expedition had started in Temple Bay in the south-east of the Ice Fjord and passed by the northern side of Hayes Glacier on Stor Fjord about 50 km to the east via the von Post Glacier. On 11 August 1910, after hurricane-like storms, they finally set foot on the highest point of the von Post Glacier, where they set up their central camp at the watershed. Filchner and his companions reached their easternmost point on the ridge north of the Hayes Glacier, where they had a wonderful view of the ice-covered Stor Fjord (Fig. 2.9).

They returned to the central camp on 14 August only after a total of 53 ½ h of marching in driving snow with few interruptions. On 18 August, they reached the shore at the Temple Bay, and on 27 August, they set off for home from Longyear City [today: Longyearbyen] on the Arctic Coal Company's steamer *Munroe*

**Fig. 2.9**   First view of the east coast of Sabine Land. *Source* Filchner and Seelheim (1911, 81)

(Potpeschnigg, 1914, 10ff). The result of this expedition was a detailed map of an up to then unexplored part of central Spitsbergen (Brunner & Lüdecke, 2012).

## Duke Ernst II of Saxony Altenburg

After the successful training expedition, Filchner asked Duke Ernst II of Saxony Altenburg (1871–1955) in the winter of 1910/11 to join the honorary board of his committee for the promotion of his Antarctic expedition, of which Prince Heinrich of Prussia was already a member (Herzog Ernst, 1943, 240–269). Inspired by Filchner's reports on his Spitsbergen expedition, Duke Ernst wanted to undertake an expedition to Spitsbergen himself and planned to be the second to cross Northeast Land. The first crossing of this remote region had been carried out by Adolf Erik Nordenskiöld (1832–1901) in June 1873 from the northeast corner to Wahlenberg Bay (Fig. 2.10).

This time, Duke Ernst wanted to cross the island between Rijp Bay in the north and Wahlenberg Bay and, in addition to magnetic measurements, examine the glaciation of this region. The well-known German ship builder Max Oertz (1871–1929), an old friend of Duke Ernst, arranged for a 22 m yacht, which was christened *Senta*. Duke Ernst (Fig. 2.11) was accompanied by his valet, court marshal von Breitenbuch, a

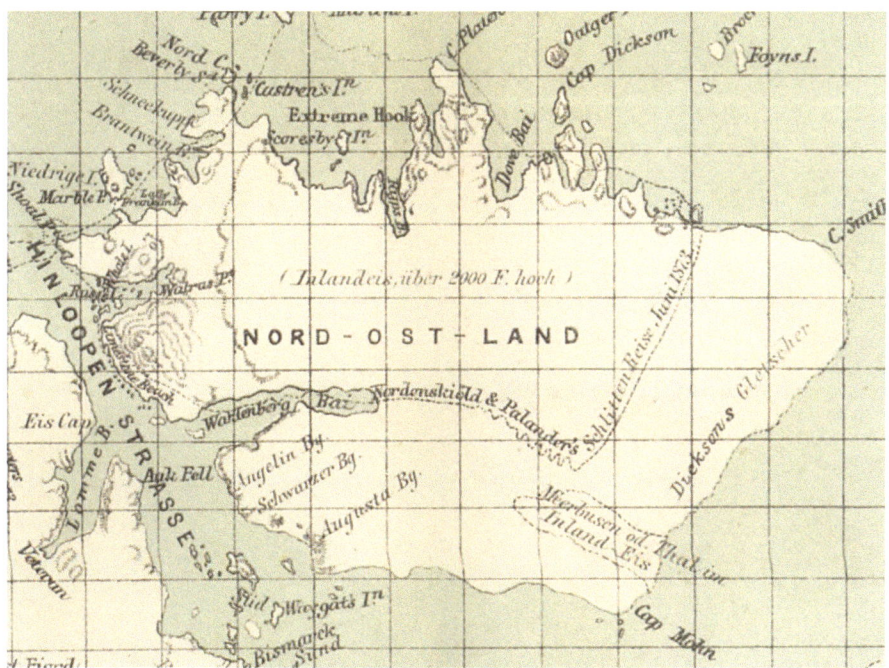

**Fig. 2.10** Detail of a map showing Northeast Land and the route of Adolf Erik Nordenskiöld's crossing in June 1873. *Source* Nordenskjöld (1880)

doctor, and Karl Potpeschnigg, the doctor and mountaineer recommended by Filchner as a proven ice expert, while the captain was provided by HAPAG. In Tromsø, they were also joined by the Norwegian ice pilot August Stenersen (who, as shall be seen, would later work on the Herbert Schröder-Stranz expedition in 1912).

Shortly before they reached Spitsbergen, they received their first mail from the steamer *Blücher*. After taking on more provisions and fuel in the harbor of the coal mine in Advent Bay, they departed on 23 July. Sometime later they discovered that *Senta*'s fuel tank was leaking, but luckily the crew of *Großer Kurfürst*, which was on its first cruise through Arctic waters, could repair the fuel tank in Magdalenen Bay, and delivered further mail to Duke Ernst. Like all tourists in the Far North, the Germans also visited the relics of early aviation in Virgohamna. When they finally encountered dense drift ice north of Moffen Island, they had to abandon their destination of Northeast Land and turned back.

**Fig. 2.11** Duke (Herzog) Ernst II of Saxony Altenburg. *Source* Mauritianum, Altenburg

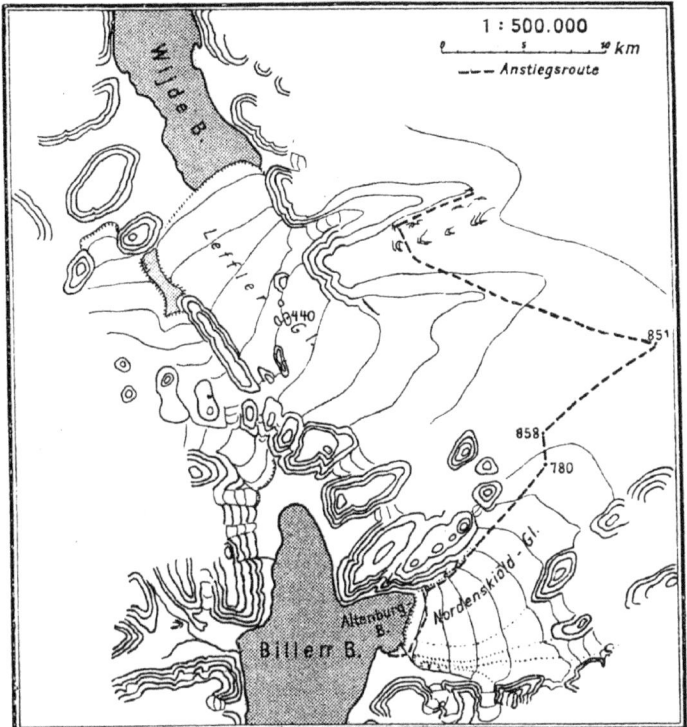

**Fig. 2.12** Route of Duke Ernst's expedition in August 1911. *Source* Penck (1912, 792)

After some landings in Möller Bay and Signehamna and hunting for the zoological collection at Zeppelinhamna, they reached the Ice Fjord and anchored in a small bay at the end of the Nordenskiöld Glacier, which they called Senta Bay (Penck, 1912, 792). Starting from there they conducted a 14-day sledging expedition across Nordenskiöld Glacier towards Wijde Bay, climbing De Geer's Peak (1023 m) on the way (Pluntke, 2008, 97) (Fig. 2.12).

After their return to Tromsø on 7 September, all expedition members said goodbye to each other, while helmsman Demelius brought *Senta* back to Hamburg on 15 October. *Senta* was "the first German sailing yacht" that had "travelled almost 3000 nautical miles to the 80th parallel" (Pluntke, 2008, 98).

## The German Geophysical Observatory in Ebeltofthamna

When the German Arctic Zeppelin Expedition had shown that there was still far too little known about the meteorological conditions for flights in the Arctic, Hergesell campaigned for a German Geophysical Observatory to be set up on Spitsbergen for

**Fig. 2.13** Aerological measurements with a registering instrument fixed on a captive balloon at Advent Bay in May 1912. *Source* Rempp and Wagner (1914, Table 2)

meteorological, magnetic, and seismic measurements, and to be manned all year round (Hergesell, 1914; Rempp & Wagner, 1914). In the summer of 1911, two meteorologists were able to move into a house belonging to the U.S. firm Arctic Coal Company on the outskirts of Longyear City (where the airport of Longyearbyen is located today). As often as possible, they carried out aerological ascents, determining the temperature and humidity up to 3300 m above sea level (Fig. 2.13).

When the first wintering crew was picked up by the Reichsforschungsdampfer (Reich Research Vessel) *Poseidon* in the summer of 1912, the station was to be moved from Advent Bay to Ebeltofthamna in Cross Bay (Cross Fjord) in the northwestern part of Spitsbergen (Hergesell, 1914; Wegener, 1914). There, two houses prefabricated in Norway were to be built for the meteorologists Kurt Wegener (1878–1964), who also acted as station manager, and Max Robitzsch (1887–1952), as well as two assistants (Wichmann, 1912) (Fig. 2.14).

**Fig. 2.14** The German Geophysical Observatory in Ebeltofthamna (Cross Bay). *Source* Steinhagen (2008, 51)

Wegener had been observer at the German Samoa Observatory of the Royal Society of Sciences at Göttingen in Apia (German Samoa) from 1909 to 1910 (Angenheister, 1974, 48). He was a very experienced meteorologist, well versed in measurements and improvisations far from home, but, like all his companions, completely inexperienced in polar regions. While he was arriving in Spitsbergen, his brother Alfred (1880–1930), as a participant of the four-man Danish expedition led by Captain Johann Peter Koch (1870–1928), was preparing their wintering on the east coast of Greenland at 76° N to cross Greenland from east to west the following year.

When *Poseidon* arrived in Ebeltofthamna, Captain Ferdinand Gluud (1875–1913), who had been first officer of the steamer *Mainz* during Count Zeppelin's expedition to Spitsbergen, was to "take possession on behalf of Count Zeppelin and prof. Hergesell of a strip of land suitable for the station installation" (Wichmann, 1912, 94). This terrain seemed well suited for the airship expedition planned by Zeppelin. Experience had shown that such an expedition would generate high tourist traffic, so that Norddeutscher Lloyd had already taken possession of some sites. In particular, a tourist hotel should be built, where researchers could also stay for a longer period of time.

The station in Ebeltofthamna became a popular destination for cruise ships on their way north. In addition, the cruise ships were of great help, such as when the NL steamer *Großer Kurfürst* brought more food in August (Sieberg, 1913, 10–11). In the same month, Hergesell delivered a radio station for the observatory on board the HAPAG steamer *Victoria Luise* (Wegener, 1914, 22–23; Steinhagen, 2008, 53–58). As no one had been able to radio before, Hergesell sent Robitzsch to the Spitsbergen radio telegraph station in Green Harbour to train as a radio telegraph operator. The meteorologist was very talented and within a few days he was able to receive 60 signals per minute.

**Fig. 2.15**  Whaling station at Green Harbour in 1910. *Source* Miethe and Hergesell (1911, 48)

For his return trip to Ebeltofthamna, Robitzsch had to wait for the next mail cutter due to bad weather conditions in the European Arctic Ocean. This situation gave him plenty of time to explore the whaling station and whale processing (Fig. 2.15).

Sometime later, he happened to meet the Spitsbergen expert Arve Staxrud (1881–1933) and the famous Norwegian polar explorer Fridtjof Nansen (1861–1930), who was on a little expedition to the waters of Spitsbergen (Fig. 2.16).

After his return to Ebeltofthamna, Robitzsch was busy setting up the radio telegraph station. A lot of improvisation had to be done before it was fully functional in April 1913 (Fig. 2.17).

During the winter night, when the sun disappeared behind the horizon from 14 October 1912 to 21 February 1913, the period of additional auroral observations began, which was a splendid interruption of the daily routine (Wegener, 1914, 23–24). In the following summer of 1913, the wintering crew was replaced again.

**Fig. 2.16** Robitzsch meets Arve Staxrud (left), Fridtjof Nansen (middle), and telegraph station manager Olaf Henriksen (right) in Green Harbour. *Source* Steinhagen (2008, 55)

**Fig. 2.17**  Radio Telegraph
station of the observatory in
Ebeltofthamna. *Source*
Steinhagen (2008, 57)

## Photographs

Herzog Ernst II estate, Mauritianum, Altenburg.

## References

Angenheister, G. G. (1974). Geschichte des Samoa-Observatoriums von 1902–1921. In H. Birett, K.
    Helbig, W. Kertz & U. Schmucker: *Zur Geschichte der Geophysik. Festschrift zur 50-jährigen
    Wiederkehr der Gründung der Deutschen Geophysikalischen Gesellschaft* (pp. 43–66). Berlin,
    Heidelberg, New York: Springer.
Brunner, K., & C. Lüdecke. (2012). Übung für die Antarktis: Wilhelm Filchners Vorexpedition
    nach Spitzbergen im Jahr 1910. Ein Beitrag zur Expeditionskartographie. In C. Lüdecke & K.
    Brunner (Eds.), Von A(ltenburg) bis Z(eppelin). Deutsche Forschung auf Spitzbergen bis 1914.
    100 Jahre Expedition des Herzogs Ernst II. von Sachsen-Altenburg. *Schriftenreihe des Instituts
    für Geodäsie der Universität der Bundeswehr München*, Neubiberg, Heft 88, pp. 69–76.

Capelotti, P. J. (1999). *By Airship to the North Pole: An Archaeology of Human Exploration.* New Brunswick, New Jersey, and London: Rutgers University Press.

Filchner, W., & Seelheim, H. (1911). *Quer durch Spitzbergen. Eine deutsche Übungsexpedition im Zentralgebiet östlich des Eisfjords.* Berlin: E.S. Mittler und Sohn.

Filchner, W. (1922). *Zum sechsten Erdteil.* Berlin: Ullstein.

Filchner, W. (1994). *To the Sixth Continent: The Second German South Polar Expedition, 1911–1913.* Translated and edited by W. Barr. Eccles: Erskine Press and Bluntisham: Bluntisham Books.

Hergesell, H. (1911a). Das arktische Luftschiffunternehmen und der Zweck unserer Studienreise. In A. Miethe & H. Hergesell (Eds.), *Mit Zeppelin nach Spitzbergen* (pp. 4–16). Berlin: Deutsches Verlagshaus Bong & Co.

Hergesell, H. (1911b). Die Fahrten des „Fönix". In A. Miethe & H. Hergesell (Eds.), *Mit Zeppelin nach Spitzbergen* (pp. 227–282). Berlin: Deutsches Verlagshaus Bong & Co.

Hergesell, H. (1914). Die Deutsche wissenschaftliche Station in Spitzbergen. In H. Hergesell (Ed.), Das Deutsche Observatorium in Spitzbergen. Beobachtungen und Ergebnisse I, *Schriften der Wissenschaftlichen Gesellschaft in Straßburg* 21, pp. 1–5.

Hergesell, H. (1933). Der erste Registrierballonaufstieg in hohen arktischen Breiten. *Beiträge Zur Physik Der Atmosphäre, 20,* 261–268.

Herzog Ernst. (1943). Beschreibung der Spitzbergenexpedition des „Herzog Ernst" II. von Sachsen-Altenburg (1943) In: U. Gillmeister. (2009). *Vom Thron auf den Hund.* Borna: Südraum, 3. Auflage, pp. 240–269.

Koldewey, K., & Petermann, A. (1871). Die erste Deutsche Nordpolar-Expedition im Jahre 1868. *Petermanns Geographische Mitteilungen.* Ergänzungsheft 28.

Lüdecke, C. (2012). Die Zeppelin-Studienexpedition nach Spitzbergen (1910). In C. Lüdecke, & K. Brunner (Eds.), Von A(ltenburg) bis Z(eppelin). Deutsche Forschung auf Spitzbergen bis 1914. 100 Jahre Expedition des Herzogs Ernst II. von Sachsen-Altenburg. *Schriftenreihe des Instituts für Geodäsie der Universität der Bundeswehr München,* Neubiberg, Heft 88, pp. 99–107.

Marschall, B. (1991). Reisen und Regieren. Die Nordlandfahrten Kaiser Wilhelms II. *Schriften des Deutschen Schiffahrtsmuseums* 27. Bremen: Schiffahrtsmuseum, Hamburg: Ernst Kabel.

Martens, F. (1675). *Friderich Martens vom Hamburg Spitzbergische oder Groenlandische Reise Beschreibung gethan 1671. Aus eigner Erfahrunge beschrieben/die dazu erforderte Figuren nach dem Leben selbst abgerissen/(so hierbey in Kupffer zu sehen) und jetzo durch den Druck mitgetheilet.* Hamburg.

Miethe, A. (1911). Die Reise der „Mainz". In A. Miethe & H. Hergesell (Eds.), *Mit Zeppelin nach Spitzbergen* (pp. 17–164). Berlin: Deutsches Verlagshaus Bong & Co.

Miethe, A., & Hergesell, H. (Eds.). (1911). *Mit Zeppelin nach Spitzbergen.* Berlin: Deutsches Verlagshaus Bong & Co.

Norddeutscher Lloyd. (1911). *Polarfahrt 1911 mit dem großen Schraubendampfer „Großer Kurfürst".* Bremen: Norddeutscher Lloyd.

Nordenskjöld, A. E. (1880). *Die Nordpolarreisen Adolf Erik Nordenskjöld's 1858 bis 1879.* Leipzig: F.A. Brockhaus.

Oesau, W. (1937). *Schleswig-Holsteins Grönlandfahrt auf Walfischfang und Robbenschlag vom 17.-19. Jahrhundert.* Glückstadt: Augustin.

Oesau, W. (1955). *Hamburgs Grönlandfahrt: auf Walfischfang und Robbenschlag vom 17.-19. Jahrhundert.* Glückstadt: Augustin.

Penck, A. (1912). Polargebiete. *Zeitschrift der Gesellschaft für Erdkunde zu Berlin,* 91–93.

Philipp, H. (Ed.). (1914). Ergebnisse der W. Filchnerschen Vorexpedition nach Spitzbergen 1910. *Petermanns Geographische Mitteilungen, Ergänzungsheft,179.*

Pluntke, M. (2008). Expedition des Herzogs Ernst II. von Sachsen-Altenburg im Sommer 1911 nach Spitzbergen. *Altenburger Geschichts- und Hauskalender für den Kreis Altenburger Land* (Vol. 17 NF, pp. 93–98).

Potpeschnigg, K. (1914). Verlauf und Ausrüstung der Expedition. In H. Philipp (Ed.), Ergebnisse der W. Filchnerschen Vorexpedition nach Spitzbergen 1910. *Petermanns Geographische Mitteilungen*, Ergänzungsheft (Vol. 179, pp. 1–13).

Rempp, G., & Wagner, A. (1914). Die Station in der Adventbai. In H. Hergesell (Ed.), Das Deutsche Observatorium in Spitzbergen. Beobachtungen und Ergebnisse I. *Schriften der Wissenschaftlichen Gesellschaft in Straßburg* (Vol. 21, pp. 6–20).

Sieberg, A. (1913). *Im Nordpolareis des nordwestlichen Spitzbergen.* Straßburg: M. DuMont.

Steinhagen, H. (2008). *Max Robitzsch. Polarforscher und Meteorologe.* Jakobsdorf: Die Furt.

Wegener, G. (1897). *Zum Ewigen Eise. Eine Sommerfahrt in nördliche Polarmeer und Begegnung mit Andrée und Nansen.* Allgemeiner Verein für Deutsche Litteratur.

Wegener, K. (1914). Das Observatorium in der Crossbai 1912/13. In H. Hergesell (Ed.), Das Deutsche Observatorium in Spitzbergen, Beobachtungen und Ergebnisse. I. *Schriften der Wissenschaftlichen Gesellschaft in Straßburg* (Vol. 21, pp. 21–29).

Wichmann, H. (1912). Deutsche Nordostdurchfahrt. *Petermanns Geographische Mitteilungen, 58/II*(7), 94.

*Wikipedia/*Wilhelm Bade. https://de.wikipedia.org/wiki/Wilhelm_Bade. Accessed October, 6 2021

Zeppelin, F. G. (1911). Hat unsere Expedition die Zweckmäßigkeit der Verwendung meiner Luftschiffe zur Erforschung der Arktis ergeben? In A. Miethe & H. Hergesell (Eds.), *Mit Zeppelin nach Spitzbergen* (pp. 283–291). Berlin: Deutsches Verlagshaus Bong & Co.

Zorgdrager, C. G. (1723 [1975]). *Alte und neue Grönländische Fischerei und Wallfischfang.* P. C. Monath. Reprint: Kassel: H. Hamecher.

# Chapter 3
# Herbert Schröder-Stranz's Training Expedition for His Planned Northeast Passage (1912)

## Schröder-Stranz and His Ambitious Plan

In the German scientific journal of geographical matters, *Petermann's Geographische Mitteilungen*, a short report on the plan of a German expedition to the Taimyr Peninsula appeared in the second volume of 1911 under the heading "Asia" (Wichmann, 1911). A certain Lieutenant Schröder of the Colberg Grenadier Regiment, who had traveled through Russian Lapland the previous winter, wanted to take advantage of his experience and connections with locals and Russian authorities and to start a larger journey in January 1912. To do this, he had set for himself the following tasks:

> Improvement of the unreliable map of those areas, zoological, anthropological and ethnological research, meteorological observations. He intends to complete the first part of his journey by April 1912; then he will follow the course of whales from the White Sea to the Northern Arctic Ocean, in order to participate in the hunts of the Russians and Samoyeds. In summer, the expedition is to undertake an advance through the Kara Sea and attempt to reach the Taimyr Peninsula, into the interior of which he wants to penetrate on reindeer sledges; for this journey, eight months are envisaged; under favorable conditions and with sufficient means, this research is to be continued for another year. /.../ In order to reproduce everything that has been seen and explored as true to nature as possible, not only for processing, but also for visualization for further circles, all means of modern technology, cinematograph, color and stereo photography, telephotography and flash light, phonograph, are to be used.

(Wichmann, 1911, 26).

Who was this Lieutenant Schröder, who apparently planned a major polar expedition for the first time? Only little was known about him (Rüdiger, 1913, 208–209). Herbert Schröder (1884–1912) was born on the manor of his parents in Stranz near Deutsch-Krone in West Pomerania (today in Poland). Since the name Schröder was widely used in Germany, he appended the place of birth to his name and called himself Schröder-Stranz thereafter (Fig. 3.1).

At the age of 19, he volunteered for the Prussian Army and joined the 4th Guards Regiment on foot for a year. When he heard about the uprising of the Herero and

**Fig. 3.1** Castel Stranz near Deutsch-Krone. *Source* Drygalski estate, Mörder, Feldkirchen-Westerham

Nama in the colony of German Southwest Africa in 1904, he was able to join the Protection Force (Schutztruppe) as a volunteer with the rank of lieutenant and travel to Africa (Schröder-Stranz, 1910). Here, his "tradition as a Prussian nobleman was paired with a certain spirit of adventure" that contrasted with his aversion to higher education (Heidbrink, 2015, 152). As a youth, he preferred to spend time in nature rather than studying for school. During the suppression of the uprising in Africa, however, he was unable to prove himself militarily, first falling ill with dysentery and later with typhus (Heidbrink, 2015, 153). Finally, he was sent back to Germany to recover without being able to participate in the decisive battle at Waterberg. German naval historian Ingo Heidbrink interprets this interruption of his military career as a "personal defeat /.../ that was to shape his entire later life" (Heidbrink, 2015, 153). However, his stay in Africa aroused his interest in German overseas policy, which was being pursued very forcefully at the time.

At home, he joined the hunters in Kulm (West Prussia) as an officer cadet in 1905, but had to quit again due to illness (Rüdiger, 1913, 208–209). His doctor advised him to take an extended sea voyage to recover, which took him to Scandinavia, England, and then to the North and South American continents. Finally, he tested his regained health during a long ride through Argentina. In 1907, he joined the 9th Colberg Grenadier Regiment, which was stationed in Stargard. Finally, in the winter of 1910/11, he travelled across Russian Lapland, crossing the Kola Peninsula and Karelia, accompanied only by his dog Tell (Schröder-Stranz, 1911) (Fig. 3.2).

When he returned from his journey, he had the idea "to make the Northeast Passage usable and thus to open up the treasures of Northern Siberia for Europe" (Rüdiger, 1913, 209). After his first announcement in 1911, Schröder-Stranz expanded his expedition plan a year later to include the navigation of the entire Northeast Passage, which had not been repeated for more than 30 years since the crossing by Adolf Erik Nordenskiöld on the *Vega* (Wichmann, 1912a) (Fig. 3.3).

The expedition was to set out in June 1913 and within three to four years to carry out oceanographic, cartographic, geographic, geological, meteorological, and

**Fig. 3.2** Schröder-Stranz, in the clothing of the locals of Russian Lapland, and his dog Tell. *Source* Ritscher estate, Leibniz Institute for Regional Geography, Leipzig, Germany

magnetic measurements, as well as to create zoological and botanical collections. Names of the participants were also already mentioned (Wichmann, 1912b). Dr. Max Mayr from Munich would be responsible for geography and geology, Dr. Hermann Rüdiger from Rostock for oceanography, Dr. E. Detmers from Hannover for zoology, and Dr. Wenke from Berlin for botany, while the ship's officers Captain Ritscher and Captain Lieutenant Sandleben would take over the meteorological, geomagnetic, and physical work.

In addition, Ritscher was not only to act as first officer but also as pilot of a flying machine, because it would be much better to explore a suitable driving crack from a higher altitude (Wedemeyer, 1914, VII). Ritscher did not have much time to get his pilot's license at Johannisthal Airport near Berlin (now a district in the Treptow-Köpenick borough of Berlin) in addition to his job. When he was seriously injured in a plane crash in June, this plan had to be abandoned for the time being.

Captain Berg, who had already gained a lot of experience in northern Siberian waters, was designated for the ship's command (Wichmann, 1912a). In addition to the Northeast Passage, the interior of the Taimyr Peninsula would also be investigated during the winter. Perhaps a small group would also spend the summer there, while oceanographic investigations and coastal surveys would be made from the ship. The eastern part of the Northeast Passage was to be explored in the third summer. The return voyage was planned via the Great Ocean (Pacific) and the Atlantic Ocean (Rave, 1913, 5–6). During the expedition, both an airplane and radio telegraphy were

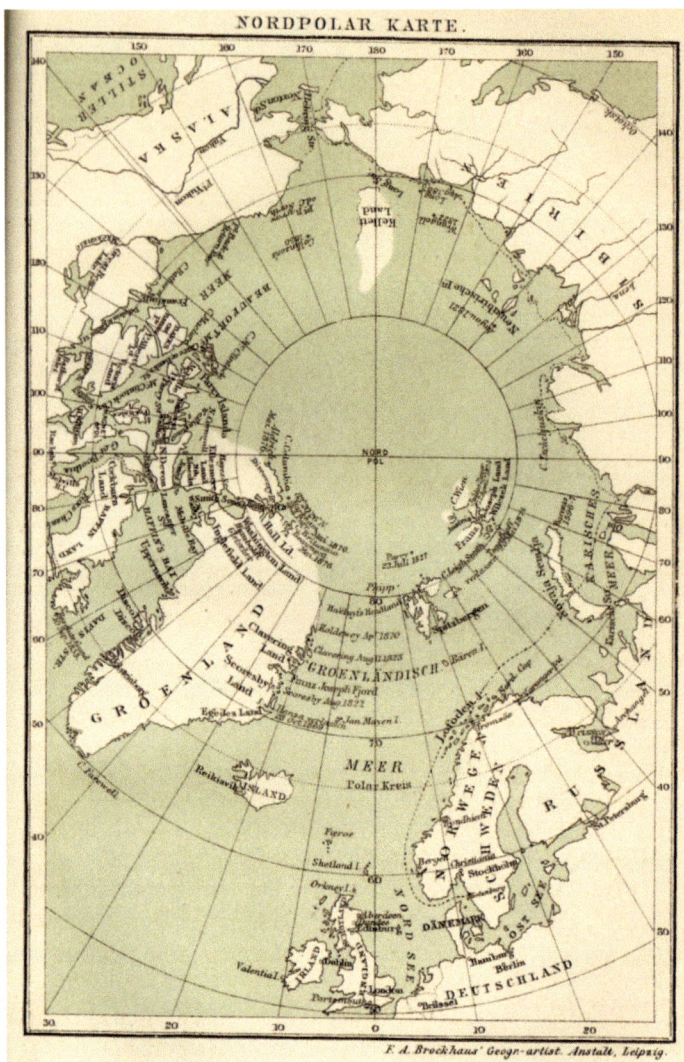

**Fig. 3.3** Route of Adolf Erik Nordenskiöld's Northeast Passage along the Siberian coast in 1878–1879. *Source* Nordenskjöld, 1880

to be used (Behm, 1912). The expedition would probably cost 2,000,000 Marks, so Schröder-Stranz had to rely on donations. The German Maritime Observatory in Hamburg therefore published an appeal for donations in its journal *Annalen der Hydrographie und Maritimen Meteorologie*.

Schröder-Stranz's expedition seemed politically desirable after the German Empire had occupied Schantung (Shandong) Province in China, including the Bay of Kiaut-Schou (today: Jiaozhou Bay) with the fortress of Tsingtau (today: Qingdao),

**Fig. 3.4** Schantung (Shandong) Province in northeast China. Tsu Schima (Tsushima) is visible in the Korea Strait east of Schantung. *Source* Andree, 1893, 107

as a protectorate in China in 1898 (Westphal, 1984, 204–216, 287–288, 306–308) (Fig. 3.4).

Two years later, Germany participated in the suppression of the Boxer Rebellion, which was directed against the colonial rulers in China. German troops had to be moved to China as quickly as possible in order to deal with what at that time was referred to as the "yellow peril." The journey from Germany around Africa to the site of operations in northern China took almost two months. The outdated Russian Baltic Fleet took even about seven months to reach Port Arthur during the Russo-Japanese War (1904–1905), en route relying on coal supplies from chartered HAPAG cargo ships (Heidbrink, 2015, 154–156). When the fleet finally arrived at Tsushima, it was already so weakened as to be easily defeated by the Japanese fleet. The route via the Northeast Passage would have been much shorter, but it was still largely unexplored. The opening of the Northeast Passage would be also useful to German overseas interests throughout the Pacific, especially since Tsingtau was developed as a "harbor colony" and an important naval base by the Reichsmarineamt (National Naval Office) (Kneitz, 2016, 421) (Fig. 3.5).

For further preparation, Schröder-Stranz oriented himself to Filchner's German Antarctic Expedition Association and also set up a committee for his German Arctic Expedition, for whose honorary presidium he was able to recruit both Therese Princess of Bavaria and Duke Ernst II von Sachsen-Altenburg, who was very inter-ested in his venture (Lüdecke, 1995, 137, A 18). There was a remarkably large

**Fig. 3.5** Kiau-Tschou Region at Schantung Province. *Source* Andree 1903, 140

overlap in membership in both associations. A total number of 14 members supported Filchner's as well as Schröder-Stranz's organizations at the same time.

Since Nordenskiöld had to overwinter close to the eastern exit of the Northeast Passage during his crossing (1878/79), Schröder-Stranz planned a scientific expedition lasting several years. For this purpose, he needed a polar research vessel to be built by the company Stocks & Kolbe according to plans prepared by the well-known Max Oerzt in June 1912 (Behm, 1912; Rave, 1913, 5–6) (Fig. 3.6).

The three-master of the German Arctic expedition was to have a length of 50 m, a breadth of 12.5 m, and a draft of 5 m, as well as an ice strengthened hull and an ice ton at the middle mast. There was a large workroom that ran the full width of the ship, a laboratory, a darkroom, and a radiotelegraphy and instrument room. Individual cabins were provided for the captain, the expedition leader, the first and second officers, and the 12 scheduled scientists (Oertz, 1912). The two machinists and the helmsman and ice pilot shared double cabins. Finally, two cabins accommodated six seamen and six stokers, respectively. The large saloon was to be furnished with 16 armchairs, while the crew mess consisted of two u-shaped corner benches. A smoking room with a library was not to be missed for the long voyage. In total, the polar research ship was to accommodate 32 men.

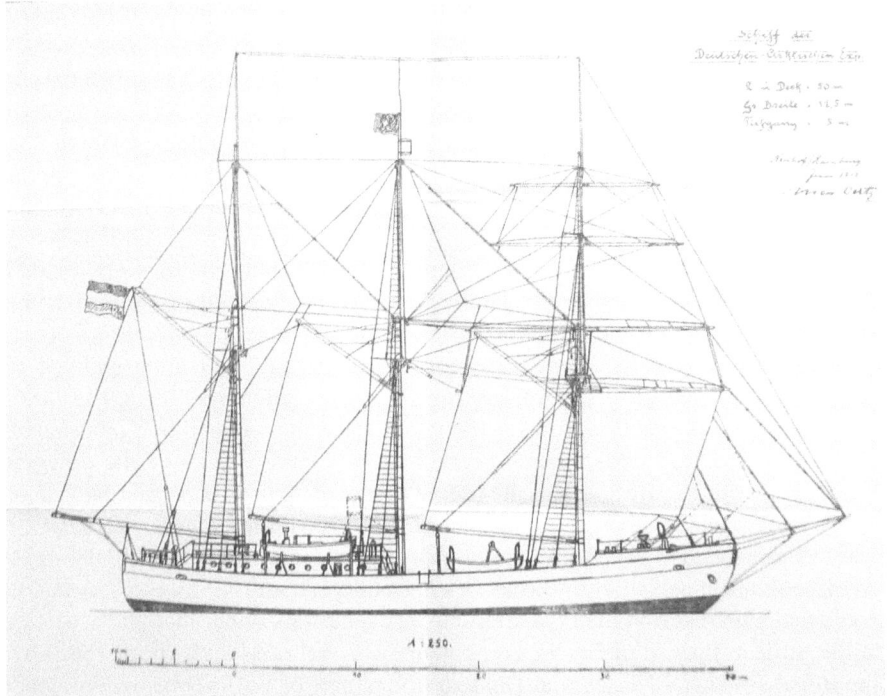

**Fig. 3.6** Outline of the ship of the German Arctic expedition, planned by Max Oertz in 1912. *Source* Schröder-Stranz estate, Leibniz Institute for Regional Geography, Leipzig, Germany

## Schröder-Stranz's Trial Expedition to Spitsbergen

Before the main expedition could set out for the Northeast Passage, the provisions and equipment had to be tested and the participants had to be familiarized with the movement on snow and ice during a pre-expedition (Rüdiger, 1913, 2ff). Like Filchner, Schröder-Stranz also chose Spitsbergen as his training area. There, he wanted to clarify "whether or not the interior of the Northeast Land is covered by a uniform ice sheet" (Rüdiger, 1913, 33). Since Nordenskiöld's crossing in 1873, no further information had been added to the interior cartography of Northeast Land. This endeavor corresponded to "the energetic personality of this man, who wanted to accomplish something whole at once, and who had in mind as a goal a penetration into the almost completely unknown interior of the Northeast Land and possibly its crossing" (Rüdiger, 1913, 3).

Plans changed a bit, because on the training expedition Schröder-Stranz was accompanied by his private secretary Richard Schmidt (born 1887), sergeant of the 9th Colberg Grenadier Regiment, who would become the taxidermist of the main expedition (Rave, 1913; Rüdiger, 1913). As Captain Berg had stepped back, Alfred Ritscher (1879–1963), second officer of the Hamburg-Amerika-Linie, became captain (Fig. 3.7), while commander lieutenant of the Imperial Navy, August Sandleben (born 1882), became first officer.

Wilhelm Eberhard (born 1886) was a technician, chauffeur, and aeronautical engineer, who would have been in charge of the motor-sledge as well as the flying machine; however, neither motor-sledge nor flying machine could be taken to Spitsbergen. Instead, Eberhard took care of the dogs and later acted as machine operator on the expedition ship. Four Norwegian sailors—Knut Stave (1878–1913), Einar Rotvold (1881–1934), Julius Jensen (born 1877), and his brother Jørgen Jensen (born 1892)—and the Norwegian ice pilot August Stenersen (born 1876), were joined by four scientists (Fig. 3.8). The five Norwegian crewmembers all came from Tromsø (Nøis, 1929, 4).

Max Mayr (born 1885) studied geography, geology, and anthropology, and received his doctorate in geography from Professor Erich von Drygalski (1865–1949), and was perfectly suited for the tasks of the main expedition (Rave, 1913; Rüdiger, 1913). After his studies he became a reserve officer of the Telegraph Battalion in Munich. Besides this, he was an excellent mountaineer and linguistically very talented, which allowed him to become friends with foreigners very easily. He was recommended to Schröder-Stranz by Drygalski. Hermann Rüdiger (1889–1946) was a geographer and polar historian. Erwin Detmers (born 1888) was a very talented zoologist, who received his PhD after only six semesters rather than the usual eight at the earliest. Instead of Dr. Wenke, botanist Dr. Walter Moeser (born 1885) from the Royal Botanical Museum at Dahlem (today part of Berlin) was recommended by his supervisor privy councilor professor Adolf Engler (1844–1930) to take part

**Fig. 3.7** Captain Alfred Ritscher (left) and expedition leader Herbert Schröder-Stranz (right). *Source* Rüdiger, 1913, 9

**Fig. 3.8** The nautical crew of *Herzog Ernst*. First row from left to right: Captain Alfred Ritscher, ice pilot August Stenersen, cook Knut Stave; second row from left to right: the sailors Julius Jensen, Jørgen Jensen, Einar Rotvold. *Source* Rüdiger, 1913, 68

in the training expedition. Finally, marine painter and cameraman Christopher Rave (1881–1933) was added to the participants to document the events of the expedition. They all met aboard the former fishing vessel *Sterling*, a two-master with a 45 hp engine, which was converted for the expedition in Norway (Fig. 3.9).

Shortly before departure, the sailing ship was christened *Herzog Ernst* after the protector of the expedition Duke Ernst II of Saxony-Altenburg.

According to a newspaper report, 16 dogs were employed to work with sledges as part of the Schröder-Stranz expedition, and according to Daniel Nøis, who would serve as second in command of the rescue mission led by Arve Staxrud, 24 dogs had originally been included—all of various breeds, of large size, and without proper training for sledge-pulling (*Dominion*, 1913, 5; Tahan, 2021, 294; Nøis, 1929, 5).

Originally, Schröder-Stranz only planned to test equipment and instruments and to familiarize expedition members with Polar regions and by no means spend the winter in Spitsbergen, because there was a lot to prepare in Berlin for the main expedition (Wedemeyer, 1914, VIII-XI). Dr. August Wedemeyer (born 1867), from the nautical department of the Reichsmarineamt in Berlin, and Dr. Ludwig Kohl (since 1913 Kohl-Larsen) (1884–1969), the physician of Filchner's Antarctic expedition, which

**Fig. 3.9**  The expedition members on the *Herzog Ernst*. First row from left to right (f.l.t.r.): August Sandleben, Herbert Schröder-Stranz with his dog Tell, Prof. Wedemeyer (as guest before departure), Alfred Ritscher, Christopher Rave; second row f.l.t.r.: Walter Moeser, Hermann Rüdiger, Erwin Detmers, Max Mayr; third row f.l.t.r.: Richard Schmidt, Wilhelm Eberhard; fourth row f.l.t.r.: the two brothers Julius and Jørgen Jensen, Einar Rotvold, ice pilot August Stenersen, behind them the cook Knut Stave. *Source* Leibniz Institute for Regional Geography, Leipzig, Germany

he had to leave early due to an appendectomy previously performed in South Georgia before they entered Antarctic waters, arrived in Tromsø shortly before the expedition leader. Schröder-Stranz then announced his final plans on August 1st, stating that he wanted to accomplish something extraordinary by crossing Northeast Land, which might be done within 40 days. A success of this crossing would promote further financial support which was desperately needed. Under these new conditions, Kohl resigned, because it seemed doubtful to him that he would be back to Germany in November. Wedemeyer also resigned, because he shied away from possible dangers. In addition to making this surprise announcement, Schröder-Stranz did not inform the expedition's honorary presidium about his new plan.

It was already August 5, when the expedition set sail (Rüdiger, 1913, 10–35). Due to the advanced season, an overwintering on Spitsbergen was not excluded. Since the east coast of Spitsbergen was often difficult to access due to ice, the first attempt should be to advance north along the west coast, which is usually free of ice in summer (Fig. 3.10).

In Magdalen Bay, the expedition met the HAPAG tourist steamer *Victoria Luise* with the former captain of the *Senta*, Paul Vollrath, as officer on board. Letters were

**Fig. 3.10** *Herzog Ernst* close to Adams Glacier south of Magdalen Fjord. *Source* Rüdiger, 1913, 22

quickly written to the families back home, which the *Victoria Luise* would take along to Germany. Coincidentally, professor Hergesell was also on board, on his way to the German observatory in Cross Bay to bring the material for wireless telegraphy. "This information was of the greatest interest to us", recalled Rüdiger (Rüdiger, 1913, 23). It would be the first time that a radio telegraph station would be available in northwestern Spitsbergen.

Further along the way, the expeditioners took every opportunity to hunt polar bears and walruses in which not only the biologist had some interest (Fig. 3.11).

## Disembarkation on the Northeastern Ice for the Attempted March Westward: Schröder-Stranz, Richard Schmidt, August Sandleben, and Max Mayr

On August 15, *Herzog Ernst* reached "about midway between the North Cape and Cape Platen – 80° 25′ N and 21° 15′ E – where the unbroken pack ice called for an irrevocable halt" (Rüdiger, 1913, 34). The eastern part of Northeast Land was inaccessible because of the pack ice cover. Thereupon, Schröder-Stranz decided to set out on the ice with Schmidt, Sandleben, and Mayr as far east as possible, in order to be able to reach the ice with "the large dinghy, ... three kayaks, two sledges, the eight best dogs, two tents, the scientific instruments, weapons, skis, sleeping bags, etc., etc. To this was added, as the most important, the provisions calculated for

**Fig. 3.11**  The first polar bear killed. *Source* Rüdiger, 1913, 32

about two to three months" (Rüdiger, 1913, 34–35). (It was also reported that 12 dogs disembarked with Schröder-Stranz, per the newspaper *Dominion* [*Dominion*, 1913, 5; Tahan, 2021, 294]). The dogs had come from Berlin and Hamburg and were, however, completely untrained as pulling forces, although they were expected to pull the sledges on the crossing of Northeast Land to the Hinlopen Strait (Fig. 3.12). (It is not known for certain if Schröder-Stranz's companion dog Tell disembarked with him onto the ice.)

Schröder-Stranz planned to then cross Hinlopen Strait with the kayaks and continue the way via Treurenberg Bay, Wijde Bay, and Liefde Bay to Cross Bay in the west. Meanwhile, the *Herzog Ernst* was to set up a depot with provisions for several months in Treurenberg Bay and then wait in Cross Bay until December 15 for the sledge expedition. Should the expedition not arrive in time, Schröder-Stranz ordered them to leave further provisions for wintering at the German station in Ebeltofthamna and to start the voyage home to Germany immediately, as far as the ice conditions allowed. After the expedition leader had presented his plan to the men, they spent two hours unloading all the material needed from the ship onto the ice (Fig. 3.13).

Finally, Schröder-Stranz and his three companions said goodbye to the men who stayed behind on board. He was very keen on proving his abilities as a polar explorer. Apparently, he had completely misjudged the situation due to a lack of sufficient prior knowledge, because his sledge group disappeared and tragically was never to be seen again (Fig. 3.14).

**Fig. 3.12** Dogs on the ice before the first attempt to pull a sledge. *Source* Ritscher estate, Lüdecke, Munich

**Fig. 3.13** One hour before the departure of the sledge group north of Northeast Land on 15 August 1912. *Source* Rüdiger, 1913, 36

**Fig. 3.14** Map section indicating the starting point of Schröder-Stranz's crossing of Northeast Land. *Source* Schmidt, 1913

## The Meeting that Never Was

On the same day that Schröder-Stranz and his three companions had disembarked at Northeast Land, Fridtjof Nansen was in the general vicinity, carrying out an investigation of the water bodies and ocean currents at Spitsbergen as well as a study of the currents in the drift ice, undertaken with his English yacht the *Veslemøy* (Nansen, 1920, Foreword, 184–186). On that day, August 15th, Nansen was drifting in the fog, likely off of Verleegen Huk (Verlegen Hook, Verlegenhuken), in the same waters where the *Herzog Ernst* evidently had passed two days prior, unseen by Nansen and his people while Nansen had been a relatively short distance away—"a few quarter-miles." Although he never encountered the Schröder-Stranz expeditioners in those waters at that time, he would later wonder if a chance meeting with them might have possibly prevented the tragic disaster that ensued. The main source of the problem, maintained Nansen, was Schröder-Stranz's utter lack of experience—most exemplified by his mistaken notion and instructions that the ship should wait for him until December 15, during the height of winter and darkness and cold and ice in the North, and during the basically impossible time to travel by either sledge or vessel.

And yet, those were the instructions received by the Schröder-Stranz crew and expeditioners, and those were the orders that were followed that day as the *Herzog Ernst* made its way back westward from the northern waters of Northeast Land.

# The Ice Imprisonment of the *Herzog Ernst*

Drift ice and fog delayed the journey of *Herzog Ernst* to Treurenberg Bay, also called Sorge Bay (a name whose meaning conveys mourning), so that they did not arrive there until August 21 (Rüdiger, 1913, 34–35). As arranged, they set up the first depot in the house of the Swedish Arc-of-Meridian expedition of 1899, which was located on the east coast inside Treurenberg Bay (Fig. 3.15).

Due to the prevalent ice conditions, they were not able to leave the bay, because they were "locked in there for a full eight days" (Rave, 1913, 9). Three men availed themselves of the opportunity and settled into the best chamber of the Swedish house, while the others remained on board the ship. To use the time wisely, Detmers wanted to ring petrels. Rave willingly caught some for him with a fishing rod to which a piece of seal meat was attached. With outside temperatures of − 5 °C, this was no particular pleasure for him. (Fig. 3.16).

Eventually, they were able to leave Treurenberg Bay, only to be pinned again by the ice shortly thereafter. Finally, they ended up back in Treurenberg Bay, where they would stay to overwinter. Due to there being four less people on board by now, they at least enjoyed having much more space at this time.

It was already early September when they made themselves at home on *Herzog Ernst*. On September 3, some took the skis and went on an excursion. Rüdiger and Moeser climbed Magdalenen-Berg (Magdalenafjellet), while Detmers and Rave explored the neighboring valley (Fig. 3.17), where Rave found a nice motif to paint.

**Fig. 3.15** The house of the Swedish Arc-of-Meridian expedition of 1899 in Treurenberg Bay, September 1912. *Source* Rüdiger, 1913, 51

**Fig. 3.16**  The zoologist Detmers with ptarmigans for his collection. *Source* Rüdiger, 1913, 59

Unfortunately, he could not carry out his project because of the cold and strong wind (Rave, 1913: 14ff).

During a small kayak trip, Rüdiger and Rave wanted to go sealing on the ice edge at the entrance of the bay. When Rave left his kayak and pushed forward with his shotgun over the ice floes, a floe turned over, and he fell into the icy water. Fortunately, Rüdiger was able to help him to get out onto the solid ice. After Rave was totally soaked, they returned aboard *Herzog Ernst*. Later, Rave was able to accompany the sailor Julius Jensen on the hunt in the rowboat. To the men, the steaks of the hunted seal were a nice change from the Knorr and Maggi soups.

After an ice field pressed *Herzog Ernst* closely, they could only save themselves with effort and engine power by going into the interior of the bay. Now they were inevitably trapped for the next months. On the evening of September 8, a conference was held. Stenersen had not been able to see any more open water on his mountain tour, so that "the moment has come for us to think in all seriousness of abandoning ship to take our way to return overland" (Rave, 1913, 19) (Fig. 3.18). The provisions would easily suffice for wintering. However, since they consisted mainly of pemmican, many of the men were afraid that they would contract scurvy.

Rave and Rüdiger prepared the provisions for the eleven-man crew, which were to be sufficient for the sledge trip estimated to last four weeks. In addition, warm clothing, heating material, sleeping bags, and weapons were added, while other valuable items such as the expensive photographic and film equipment had to be left behind. Once again, *Herzog Ernst* had to be moved to a safe place and was now moored opposite the Swedish house of Arc-of-Meridian expedition. On September

**Fig. 3.17** Detmers, Rüdiger, and Moeser on a mountain tour. *Source* Ritscher estate: Cornelia Lüdecke, Munich

**Fig. 3.18** *Herzog Ernst* beset by ice. *Source* Rüdiger, 1913, 66

12, there was another consultation. The two biologists Detmers and Moeser, together with Rüdiger, wanted to leave for Advent Bay as early as September 15 in order to reach Norway with the last steamer of the year. But this plan was contradicted by all the others. So they had no other choice but to join the others, who wanted to set off on September 23 at the latest, if the ice situation around the ship did not improve. They planned for an approximately 300 km long sledge trip, and, diligently, the necessary preparations were made.

## Leaving the Ship, and the Departure of Dr. Erwin Detmers and Dr. Walter Moeser from the Group

A last attempt to break out of Treurenberg Bay on September 19 ultimately failed, so that everything was now unloaded for the sledge trip to Advent Bay. However, the crew had changed its mind in the meantime and preferred to spend the winter on the ship first and then set off in January. Finally, the entire group left the *Herzog Ernst* on September 21 (Fig. 3.19).

The skilled Norwegian skiers were greatly admired by the Germans, since for the Germans skiing was completely unfamiliar. After the first night in the open, the Norwegian crew decided to return to the ship. Now the Germans were on their own.

**Fig. 3.19** On the march west of Wijde Bay on October 3, 1912, as depicted by the artist and expeditioner Christopher Rave. *Source* Rüdiger, 1913, 89

It took five days and a very exhausting march to reach Nordenskiöld's hut Polhem ("Pole Home" in Swedish, referring to his attempt at the North Pole) located in Mossel Bay, which he had built in 1872 for wintering before crossing Northeast Land (Fig. 3.20).

**Fig. 3.20** Map of Svalbard with routes of the *Herzog Ernst* and the different groups of expedition members. (1) Point where Schröder-Stranz and companions left the ship, (2) Winter harbor of the *Herzog Ernst*, (3) Hunter's hut in Second Valley (Rüdiger and Rave), (4) Russian hut, (5) Cross Point hut (at Cape Petermann), (6) Hunter's hut at West Fjord, (7) Hunter's hut at North Fjord, (8) Longyear City, Advent Bay, (9) Telegraph Station, Green Harbour, (10) German Observatory, Cross Bay. *Source* Rüdiger, 1913

Meanwhile, Detmers, who already had an ear frostbitten, had taken the opportunity to separate from the group together with Moeser to make faster progress. In order for the remaining men to have more provisions for the sledge trip, on the following day, four of them returned to the *Herzog Ernst* on skis—a trip which took them only six hours instead of five days.

Also on September 28, another tour to the ship was undertaken. The return trip to Polhem was joined this time by the ice pilot Stenersen, the sailor named Rotvold, and the cook Stave, who, however, was much too slow for the group because of rheumatism and therefore soon returned to the ship. The provisions should now be enough for the next 20 days to get to Advent Bay.

With this last movement of people, the three remaining crew members on board the ship were Stave and the brothers Julius and Jørgen Jensen. The two biologists, Detmers and Moeser, had unfortunately disappeared. The remaining six expedition members—Ritscher, Rave, Rüdiger, Eberhard, Rotvold, and Stenersen—now embarked on their southward journey.

## Unpublished Sources and Original Material

Oertz, M., (1912). Schröder-Stranz estate, Box 26, Leibniz Institute for Regional Geography, Leipzig, Germany.

## Photographs

Drygalski estate, Mörder, Feldkirchen-Westerham.

Ritscher estate, Leibniz Institute for Regional Geography, Leipzig, Germany.

Ritscher estate, Lüdecke, Munich.

## References

Andree, R. (1893). *Andrees allgemeiner Handatlas*. 3., völlig neubearb. und verm. Aufl. Bielefeld: Velhagen & Klasing.

Behm. (1912). Die deutsche Arktische Expedition Schröder-Stranz. *Annalen der Hydrographie und Maritimen Meteorologie, 40*, 449.

Dominion. (1913, October 31). Arctic Tragedy. *Dominion*, Volume 7, Issue 1894, 31 October 1913, p. 5. National Library of New Zealand. https://paperspast.natlib.govt.nz/newspapers/DOM191 31031.2.34. Accessed 4 August 2014.

Heidbrink, I. (2015). Vom Scheitern vor dem Beginnen. Überlegungen zu den Motiven der gescheiterten Schroeder-Stranz-Expedition nach Spitzbergen. *Jahrbuch für Europäische Überseegeschichte* 13. Wiesbaden: Harrassowitz, pp. 147–166.

Kneitz, A. (2016). German Water Infrastructure in China: Colonial Qingdao 1898–1914). *Zeitschrift für Geschichte der Wissenschaften, Technik und Medizin* (N.T.M.) 24, pp. 421–450.

Lüdecke, C. (1995). Die deutsche Polarforschung seit der Jahrhundertwende und der Einfluß Erich von Drygalskis. PhD Thesis. *Berichte zur Polarforschung* 158. Bremerhaven: Alfred-Wegener-Institut für Polar- und Meeresforschung.

Nansen, F. (1920). *En ferd til Spitsbergen.* Kristiania: Jacob Dybwads Forlag. Digital Archive, National Library of Norway, Oslo, online archives, https://www.nb.no. Accessed 28 September 2023.

Nøis, D. (1929). Med kaptein Staxrud på leiting etter Schrøder-Stranz og hans folk [With Captain Staxrud on the search for Schröder-Stranz and his people]. In: *And-Ungen*, July 1929, (pp. 4–12). Andenes: Andøyposten. Library Archives, Norwegian Polar Institute, Tromsø. Received 20 July 2021.

Nordenskjöld, A. E. (1880). *Die Nordpolarreisen Adolf Erik Nordenskjöld's 1858 bis 1879.* Leipzig: F.A. Brockhaus.

Rave, C. (1913). *Tagebuch von der verunglückten Expedition Schröder-Stranz.* Schaffsteins Gründe Bändchen, vol. 49. Cöln: Schaffstein.

Rüdiger, H. (1913). *Die Sorge Bay. Aus den Schicksalstagen der Schröder-Stranz-Expedition.* Berlin: Georg Reimer.

Schmidt, C. (1913). Übersichtskarte von Spitzbergen zur Veranschaulichung des Verlaufs der Expedition Schröder-Stranz und der Hilfsexpeditioin zu deren Rettung. *Petermanns Geographische Nachrichten, 59*(II), table 29.

Schröder Stranz, H. (1910). *"Süd-West". Kriegs- und Jagdfahrten.* Berlin: Wilhelm Süsserott.

Schröder Stranz, H. (1911). Quer durch Russisch-Lappland. *Die Woche, Nr., 26*, 1090–1095.

Tahan, M. R. (2021). *The return of the South Pole sled dogs: with Amundsen's and Mawson's Antarctic expeditions.* Cham: Springer International Publishing.

Wedemeyer, Dr. (1914). Die Spitzbergen-Expedition des Leutnant Schröder-Stranz. In A. Miethe (ed.), *Die Expeditionen zur Rettung von Schröder-Stranz und seinen Begleitern geschildert von ihren Führern Hauptmann A. Staxrud und Dr. K. Wegener* (pp. VI–XII). Berlin: Dietrich Reimer.

Westphal, W. (1984). *Geschichte der deutschen Kolonien.* München: C. Bertelsmann.

Wichmann, H. (1911). Plan einer deutschen Expedition nach der Taimyrhalbinsel. *Petermanns Geographische Mitteilungen, 57*(2), 26.

Wichmann, H. (1912a). Deutsche Nordostdurchfahrt. *Petermanns Geographische Mitteilungen* 58/II (7), p. 34.

Wichmann, H. (1912b). Deutsche Nordostdurchfahrt. *Petermanns Geographische Mitteilungen,* 58/II(7), 94.

# Chapter 4
# Three Long Christmas Treks, Wilhelm Eberhard's Disappearance, and Captain Alfred Ritscher's Stressful Journey Across the Ice

## From Polhem to Wijde Bay

When Alfred Ritscher, Wilhelm Eberhard, Christopher Rave, Hermann Rüdiger, Einar Rotvold, and August Stenersen left Polhem, the temperatures were between −20 °C and −30 °C, which did not exactly make it easier to move forward. In addition, they left behind more and more luggage, including finally the methylated spirits and the cooking apparatus, although now they would get neither a warm meal nor warm tea. In the morning "there was a piece of frozen cheese. Afterwards one could nibble a piece of Plasmon chocolate. In the evening there was uncooked hard pemmican, which is not much of a treat, while it tastes excellent when cooked in pea soup" (Rave, 1913, 28). Instead of leaving the Nansen sledge behind as well, Rave shortened the frame by half and was thus able to pull his luggage. In front of the sledge, of the three dogs they had, only "the big yellow Caesar proved itself excellently as a pulling dog" (Rave, 1913, 28). Then, leaving a large sleeping bag behind, only four men could sleep at a time, while the other two stood guard outside for three hours and twenty minutes. Finally, they reached a wooden hut in Second Valley (Wijde Bay), where the Norwegians who had hurried ahead had already lit a warming fire and where hot tea was waiting for them (Fig. 4.1). They arrived there on October 4 (Aftenposten, 1913, evening: 1–2; Tahan, 2021).

After Rüdiger experienced pain in his feet, two men had to help him undress. "In the process it turned out that one foot was literally frozen and frostbitten in the boot" (Rave, 1913, 30). With this, Rüdiger could go on no further, so Rave decided to stay behind in the hut with his comrade to take care of the foot. Fortunately, there were extra provisions in the hunter's hut. Unfortunately, their sick dog Caesar had to be shot here, as he had probably inadvertently poisoned himself by consuming polar fox bait. Now Jule was the only dog left to take care of polar bears. Described as having a "dark yellow color with black spots, white chest, and white paws," her markings, according to Rüdiger, did "not suggest any breed" (Rüdiger, 1913, 104); she would prove to be a reliable companion to the two men. Meanwhile, the third

**Fig. 4.1** Arrival at the hut on the west side of Wijde Bay. *Source* Rave (1913, 29)

dog Bella accompanied Ritscher's group, which departed on October 8 to organize a relief expedition with dog sledges (Ritscher, 1916).

## Alfred Ritscher's Group: The Captain, the Machinist, and the Norwegians

On the date that the captain Alfred Ritscher, the machinist Wilhelm Eberhard, the Norwegian sailor Einar Rotvold, and the Norwegian ice pilot August Stenersen, along with Ritscher's dog Bella, left Rave and Rüdiger and the dog Jule at the hut, the time was already so far advanced that the sun shone only for two to three hours during the day. Also, the time of the twilight shortened daily, so that Ritscher's group set up the night camp behind a fast established snow wall, due to the fact that darkness broke in as early as 5 p.m. They still found enough driftwood for a campfire on their way at the western shore of Wijde Bay, and they adored Stenersen, who was very skilled in making a fire out of a log (Fig. 4.2).

Provisions were unfortunately quite scarce at half a pound per man per day, so they were very pleased when they were able to kill an old reindeer along the way. However, after a substantial dinner, they could not take much meat with them for the onward march. Each now had to carry 60–70 pounds on his back. This made the men

**Fig. 4.2** Map showing
Ritscher's route from Wijde
Bay to Advent
Bay–..–Attempted advance
October 21–26,
1912,– –Route from Second
Valley to Advent Bay
(October 8–16 and
December 18–27,
1912),—boundary of open
water. *Source* Ritscher
(1916, 19)

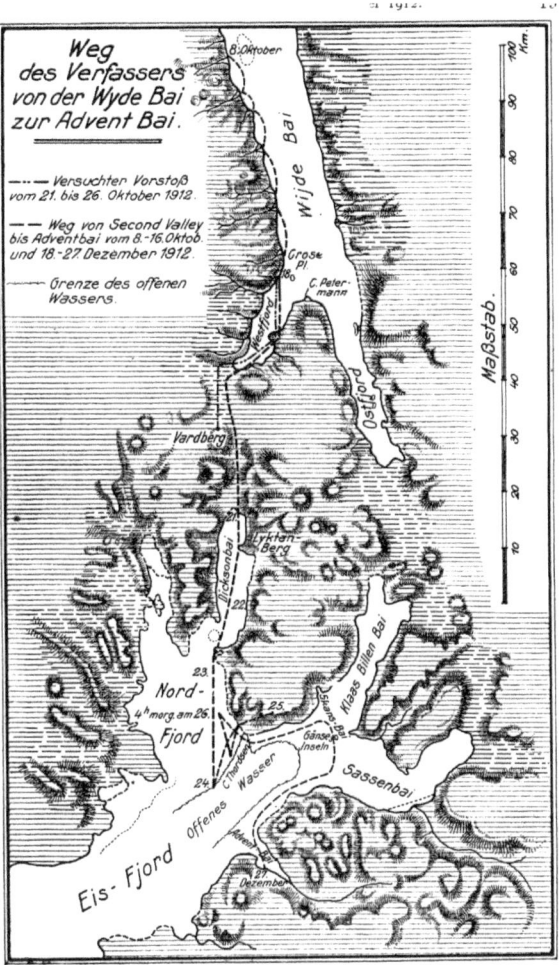

so heavy that they sometimes sank into the soft snow up to their hips and therefore
did not reach the entrance to West Fjord until October 16. Opposite Cape Petermann,
they were able to stop in a simple hut built from driftwood logs. Here they could rest
and build sledges from the available material, mend the equipment, and replenish the
provisions a little bit. When they were on the move again on October 21 and crossed
the West Fjord, they discovered a hut made of driftwood on a small promontory
southeast of Cape Petermann, but it was so small that the four of them could hardly
lie in their sleeping bags at night. When they did not find driftwood for the evening
campfire as expected for the next camp after persistent snow flurries and cutting
cold, they spent the night on a moraine and reached the northern side of Vardberg the
following day. Now they had only four to five hours of twilight left in the day and
"fog, southwesterly storm, and driving snow, cutting cold and crevasses and finally

the approaching darkness" put them under a lot of strain. When the weather did not improve the next day, they had to abandon their goal of Advent Bay and returned to the hut at Cape Petermann on October 26.

Here they wanted to wait for the weather to improve until the next moonlight period, which would begin on November 19. But the sky remained overcast, it was still cold, and they could not leave because of heavy snowfalls. Although they still had managed to kill four reindeer in October, they now consumed food only once a day. They hoped for a good moonlight period in December, otherwise they would be close to starvation. After the petroleum ran out, they improvised some light using reindeer tallow as a candle and a cable yarn as a wick. As soon as the weather permitted, they were busy gathering driftwood to make a fire. "The hard physical work, the constant darkness, and the mental strain partly caused by it made us await the next moonrise with ever greater longing", Ritscher would later recall (Ritscher, 1916, 21).

## Christopher Rave and Hermann Rüdiger: The Artist and the Geographer

Back at the Wijde Bay hut, after Rüdiger could no longer walk, Rave was responsible for everything: Finding and chopping wood, heating, baking, cooking, and sewing pants out of fur or canvas (Rave, 1913). In the meantime, the others were on their way again under Ritscher's leadership to the next settlement (Longyear City), which was a good 150 km further south. On October 20, the sun showed itself for the last time.

Again and again, Rave cleaned Rüdiger's feet and re-bandaged them, which took a total of two hours each time. Then he baked fresh bread every three or four days from "five parts oatmeal, one part wheat flour and some pemmican" (Rave, 1913, 31). Otherwise, they had salted reindeer meat with rice or with peas. Having no change of clothing with them, they also had to sleep in their same clothes at night.

After six weeks of waiting, they had given up hope of rescue. It was dark, food was only enough for a week as of November 15, and Rüdiger's left foot, unlike his right, was still not healed. So, to avoid starvation, they decided to return to the ship. For this march, Rave constructed a special shoe for Rüdiger to help him walk (Figs. 4.3 and 4.4).

On the night of November 23, the weather had improved for them—in contrast to Ritscher's group, which was about 45 km to the south as the crow flies. The clouds had cleared, the moon came out, and stars were visible in the sky. Rave quickly prepared the farewell meal of hot pemmican cake and warm water. Then he loaded the two-person sleeping bag and the luggage onto the sledge, put the specially made shoe onto Rüdiger's frostbitten foot, and finally wrote a note indicating when they had left for the north (Fig. 4.5, see also map in Fig. 3.20).

After four hours, they had reached the other shore of Wijde Bay, and after another hour, they took a rest behind an iceberg. Rave unrolled the sleeping bag, took

**Fig. 4.3** The shoe made by Rave for Rüdiger. *Source* Rüdiger (1913, 113)

**Fig. 4.4** Cross-section of Rüdiger's shoe, which Rave made for his companion's frostbitten left foot. *Source* Rave (1913, 38)

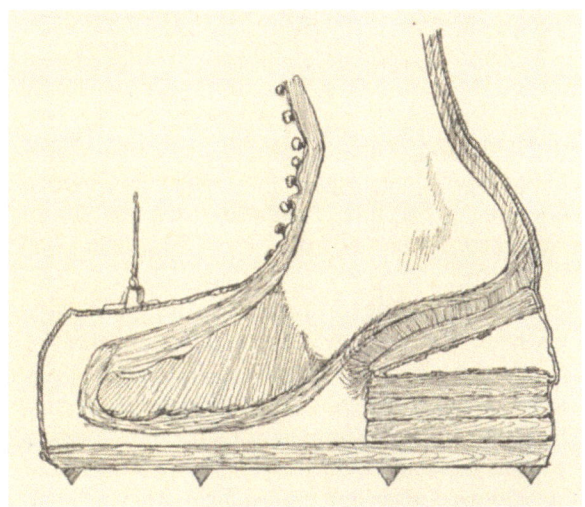

Rüdiger's shoes off of him, and made an effort to get him into the sleeping bag as quickly as possible. Nevertheless, Rüdiger froze the big toe on his right foot again during this procedure. They were so exhausted from marching that they did not wake up until 1:30 p.m. Since they were neither hungry nor thirsty, they left immediately. They only sucked some freshwater ice and ate some snow. At times, clouds obscured the moonlight. Also on this day, after five hours of hard work in the pack ice, they again allowed themselves a few hours of rest. On the way to Mossel Bay, they still had to cross two mountain ranges. Their dog Jule pulled tirelessly, and she was rewarded for it in each case with a larger piece of precooked reindeer meat than the two men had. At night, she was able to curl up on a piece of reindeer skin in the windbreak next to their sleeping bag.

**Fig. 4.5** Rüdiger (left) drags himself up the last ascent, while Rave (right) and Jule pull the sledge, as portrayed in artwork by Christopher Rave. *Source* Rüdiger (1913, 127)

On the following day, which was the third day of travel, the moon had already disappeared almost completely behind the clouds, and the north wind came on. Only after 34 h of waiting could they leave the sleeping bag and continue the march. Now they still had to cross Mossel Bay, which had taken 3 ½ h during the day two months prior. Every now and then, when the trail permitted, Rave pulled Rüdiger on the sledge. They had not eaten for 94 h and were at the end of their strength. Only the vision of the dark black spot in front of them, just visible in the sparse moonlight, which they realized was the longed-for hut (Polhem), mobilized their last remaining energy. Finally, they entered the hut at 1:30 a.m. on the morning of November 27, and were saved—for the time being. Rave's shoe for Rüdiger had worked perfectly. Immediately, a fire was made, and snow was melted in an empty sugar bowl. Otherwise, unfortunately, they had only flour left in the hut, from before they had continued their march southward at the end of September. Rave processed some flour with salt and water to make the first warm soup. The added pemmican greatly improved the second serving.

They slept more poorly than not until 2 p.m. Meanwhile, the weather had continued to deteriorate. An inspection of the provisions left behind by various ships, some of which oozed out of destroyed or opened barrels, revealed only one barrel of barley and one barrel of peas that were still edible. The hard bread was scattered on the ground and most of it was barely usable because of mold. A hearty meal of the reindeer meat, barley, and pemmican brought everyone back to strength, including Jule.

With the tools and material from the hut, Rave was able to convert Rüdiger's boot to ski for the last stretch to the ship. Rave also had to bandage Rüdiger's hands in

**Fig. 4.6** Rave bandages Rüdiger in Polhem at the end of November 1912. *Source* Rave (1913, 65)

addition to his feet, since they had suffered badly from the strain of the ski poles and were partially frostbitten (Fig. 4.6).

They did not want to stay long in the hut but rather intended to continue to the ship. During the middle of the night of 29/30 November, at midnight, Rave observed and assessed the weather. The wind had died down in the meantime, and the moon now shone unclouded, but it no longer shone as brightly as it had when the moon had been full. At 4 a.m., after a refreshment from the still edible pieces of hard bread, Rüdiger and Rave were ready to leave again. Now they went directly to the east, Arcturus and Vega serving as their guiding stars. After one hour, they had to go uphill again; however, Rüdiger came along quite well. When the ascent became steeper and steeper, they had to use their hands as well as their feet to move forward with the sledge. Although it was so cold, Rave was sweating with exertion. Thanks to the hearty food and rest, Jule again pulled along tirelessly. After a good three hours of marching, they reached the high plateau. Then the moon disappeared behind clouds, and it became so dark that they could see only a few steps away. After they could no longer make the last climb, they stopped and waited in their sleeping bags for better light conditions.

When the moon appeared as a hazy spot in the sky 22 h later, Rave decided to march on. Fortunately, the sky cleared. The highest altitude was reached, and on the way down, Rüdiger could sit on the sledge again. Luckily, they had found the right gorge in the direction of Treurenberg Bay. Now they only had to find the *Herzog Ernst*, but they were on the right track and were soon in front of the ship. When nothing moved, Rave fired three shots and called out loudly. It wasn't until the second call from close by that someone came on deck and asked who was there. Quickly the sailor Julius Jensen lowered the ladder and came to meet them. Only when he saw with his own eyes who was standing in front of him did he greet his comrades. His younger brother Jørgen and the cook Knut Stave had appeared on deck in the meantime and were very happy about the arrivals. Rüdiger and Rave were led into the crew lodgings where a warming fire was burning. Soon there was plenty of food to eat and loads of cocoa to drink and their cold feet were put into warm woolen boots. Here they had plenty of fresh meat, because Julius had killed a polar bear eight days ago. There was much to talk about before the two Germans retired to the aft cabin, which had been tidied up and heated by Jørgen. It was December 1, and they were finally safe.

Now Rave was able to amputate the first half of Rüdiger's foot. He would also take care of the sick cook Stave, who suffered from tuberculosis and received 1% morphine from him every three hours with bitter almond water on a piece of sugar. In addition, Rave would also take over the duties of the cook, while the two brothers took care of snow for melting and driftwood for heating. Varied food was enough, they thought, so they could spend the winter without worries.

Following their arrival at the *Herzog Ernst*, Rave also proceeded to perform several operations on Rüdiger's feet and both hands to remove the dead and already rotten flesh and expose the bones. He had to amputate the front half of the left foot and individual finger limbs, which was not easy for him, but which helped Rüdiger substantially (Rave, 1913, 52–55).

## Separation of Ritscher's Group, the March of Einar Rotvold and August Stenersen, and the Disappearance of Wilhelm Eberhard

Meanwhile, dramatic turns of events had been continuing to occur with Ritscher's group during the nearly three months that those men had been attempting to reach Advent Bay to retrieve assistance. Based on a report later written by Rüdiger, the following events took place.

After parting from the two Germans Rüdiger and Rave on October 8, the two Norwegians Rotvold and Stenersen, together with Ritscher and Eberhard, had spent almost two months in the hut at the end of Wijde Bay opposite Cape Petermann (Rüdiger, 1913, 138–140). There they had enough provisions: "about 40 kg of flour, 10 kg of large barley, 5 kg of peas, 4 kg of rice, also 40 rolls of tobacco, 1 rifle, 500 cartridges, medicine and kerosene" (Rüdiger, 1913, 138). With this, they killed several reindeer, which thus ensured their survival for the next few weeks. On October 21, they had set out again and hiked to the small refuge that lay 8 km south of Cape Petermann. Fog and snow had made the group turn back south of West Fjord on October 24, although Eberhard, the machinist, was the only one who insisted on continuing. The next day, they had reached a hut north of Cape Petermann. Now they found that they possessed the provisions of two huts; they also killed ten reindeer in the course of time. They planned to leave at the coming moonlight in November "via the East Fjord, Snow Valley, Klaas-, Billen and Sassen Bay to Advent Bay" (Rüdiger, 1913, 139).

Based on a report later written by Ritscher, on December 18, Ritscher's group at Cape Petermann lay ready (Ritscher, 1916, 22). They had previously packed the sledges in anticipation of the coming moonlight, so that they could set out in the evening of the same day. The next evening, when they had reached the terminal moraine in West Fjord, machinist Eberhard was already so weakened after the idle waiting time that he could not continue. Thereupon, Ritscher decided to go alone to Advent Bay to get help for Rüdiger and Rave, and sent the two strong Norwegians, Rotvold and Stenersen, as escorts for Eberhard back to the *Herzog Ernst*, which was about 100 km to the north. Ritscher estimated that his way south would be about twice as far. They separated at 8 p.m.

As would later be told to Rüdiger, and as he would understand and report it, supposedly nothing came of the Ritscher group's plan to proceed due to the bad weather that subsequently occurred. When they finally set out in December, the two Norwegians (at the insistence of Ritscher) had turned back together with Eberhard, while Ritscher went on for help (Rüdiger, 1913, 139–140). On December 23, the three reached the hut where Rüdiger and Rave previously had stayed for seven weeks (at Wijde Bay). They found that Rüdiger and Rave had left the hut. Here, Eberhard's behavior became increasingly strange, so that "the Norwegians believed in a disturbance of his mind" (Rüdiger, 1913, 139). In the evening, when they were singing some Christmas carols, Eberhard complained profusely of a headache (Rüdiger, 1913, 139–140). To speed up the march back to Treurenberg Bay, in order to bring the worsening Eberhard back to the ship, Rotvold and Stenersen, with Eberhard, left their sleeping bags behind, safely stored in the hut, after spending the night there, and crossed Wijde Bay as quickly as possible. Eberhard was not keeping up so well. The two Norwegians turned around several times, noticing his ashen face color, but when they asked him how he was, he replied that he was fine so far. By the time they reached the frozen Mossel Bay, Eberhard had disappeared from the face of the earth. In Nordenskiöld's hut Polhem, Rotvold and Stenersen waited together for four hours, during which time they went out again and again, calling out and firing signal shots, but unfortunately unsuccessfully. Eberhard never showed up. Therefore, Rotvold and Stenersen proceeded toward the ship at Treurenberg Bay, without the machinist (Fig. 4.7).

It would later be reported by the Norwegians to the rescue expeditioners, according to one of the Norwegian rescuers, Daniel Nøis (Nøis, 1929, 7), that Rotvold and Stenersen searched and waited in desperation for their missing colleague Eberhard, who had suddenly disappeared in the darkness on the Bangenhook Peninsula. But finally, in despair, with the threatening weather and the lack of strength and lack of food, they were forced to trek on ahead toward Treurenberg Bay, crossing over the mountain in an exhausting march that Stenersen later stated nearly caused them to resign themselves to their dangerous situation. After almost losing hope, however, they managed to make it to the vicinity of the vessel and to the frozen ship itself, where much-needed food, sustenance, and shelter awaited them.

**Fig. 4.7** Machinist Eberhard measuring wind velocity with an anemometer on board the *Herzog Ernst*. *Source* Rüdiger (1913, 141)

In the accounts Rotvold and Stenersen themselves would later give, as reported by a newspaper (Aftenposten: Ugens Nyt, 1913, 2–3),[1] which both seemed to agree with this sequence of events, they stated that, at the Wijde Bay hut,[2] Eberhard ate well but seemed discouraged and was holding his forehead. He often said how eager he was to reach the ship, and Rotvold could not detect anything wrong with him other than Eberhard's melancholy. When they had made their way across Wijde Bay and reached Steil Hook, Eberhard gave two cheers of "hurra," as he recognized the familiar location. But soon his face took on an alarming yellow shade, although he spoke well and, when asked if he was cold or if something was wrong with him, he answered no and seemed to indicate he was willing to go on—which the men absolutely had to do in order not to freeze to death due to the amount of sweat from physical exertion—and, very talkative, he looked forward to re-boarding the ship. The three men continued together, Eberhard always with the other two. Coming upon an unfamiliar river valley, Eberhard exclaimed (presumably in English), "I can not more," thus apparently signifying his feeling that he could not make it any further. The other two continued to encourage him, urging him to do his best along with them to

---

[1] The article's title, which includes "A Christmas-night tragedy", refers to the tragic occurrences on the night between Christmas Eve (December 24) and Christmas Day (December 25).

[2] The report seems to indicate that they reached the Wijde Bay hut in Second Valley in the late afternoon of December 22.

cover the final little remaining distance to Mossel Bay, to which Eberhard responded very positively, perking up and promising to make his best effort, and he continued on with the other men, although he now sometimes lagged behind, and they would stop and wait for him. A few times, at a hill or a ridge, they briefly could not see him but could hear his ski poles hitting against the hard snow crust, and they waited until he was in their line of vision again. When the three men approached an isolated mound, with Eberhard fully visible further behind the other two men, Rotvold and Stenersen took the higher side, nearer the land, while Stenersen turned and called to Eberhard to do the same, skiing closer toward him and shouting directions, completely within seeing and hearing distance, but Eberhard did not reply and instead took the lower side ahead, the west side, and disappeared—he never emerged from the other side of the hill. They rushed down the hill to the other side but could not see any trace of him, not even ski tracks, and so they went searching, rounding the hill yet again, calling out to him, and searching further. After the fervent and fruitless attempts to find him, and with their own survival now in question, the two conjectured that he might have possibly sat down behind a rock in order to rest or to adjust something, and had fallen asleep there – Stenersen had previously observed in him what he believed to be signs of some type of sleeping sickness from which he thought Eberhard suffered. And so, as the men continued to search for him for a long time, on the road and at Mossel Bay, and as Christmas Eve had turned into Christmas Day, the lost Eberhard sadly must have frozen to death. The two remaining men had to go on, lest they meet the same fate, weakened and stiffening and starving as they were, and with a storm coming, and so they approached the ship exhausted and hoping that healthy individuals on board would be able to continue the search for Eberhard.

Meanwhile, aboard the ship, on Christmas Eve, the Germans and the remainder of the Norwegian crew were biding their time as best they could (Rave, 1913, 53–55). Rave was preparing special surprises for his companions: In addition to a large quantity of small cakes he had baked and a ginger liqueur he had prepared, there were cigars, which everyone was very happy about. According to Rave, on Christmas Day (December 25), they heard voices outside shouting "Happy Christmas." It was Rotvold and Stenersen, who arrived with deeply sad news. The exact sequence of events was at first not so easy to grasp for the Germans, as the two returning crew members spoke no German and could only report in a mixture of English and Norwegian that was not completely understood by their German colleagues. But the Germans generally understood what the Norwegians were attempting to explain. And the sad news was digested by all.

## Ritscher's Travels with the Dog Bella, the Case of the Christmas Candle, and the Advance to Advent Bay

After taking leave of Rotvold, Stenersen, and Eberhard at West Fjord on December 19, Ritscher took his dog Bella with him on his long journey (Ritscher, 1916, 22–27). He also took three pieces of cooked and already rock-hard frozen reindeer meat, which were no bigger than a fist and which had to be enough for the both of them. He also had four kilos of uncooked pearl barley with him, which he planned to cook on the way in Dickson Bay, if he found a hut.

In order to move forward as quickly as possible, Ritscher had left behind all superfluous luggage, including his heavy fur sleeping bag. At the first rest, however, he already missed it very much. In order not to freeze to death during the rest breaks, he rolled up on the snow on the ground, as he later recalled, "and covered the legs with snow. I put my pocket alarm clock in my inner palm under my gloves, put my hand to my ear, pulled the canvas backpack over my head and tied it tightly over my chest" (Ritscher, 1916, 22). Ritscher set the alarm clock so that it would wake him after 10–15 min. After this short rest, he set off again, and continued marching and resting in this manner. Now the crevasses encountered earlier near Vardberg had been snowed over, and the cover was frozen. At 5 a.m. on December 20, he passed the mountain on the east side and went down the valley toward Dickson Bay (see Fig. 4.2). The descent from the pass was extremely strenuous, as he repeatedly sank into snow holes. It took him seven hours to cover four kilometers in this way. And it took him a total of 16 h to reach the lowlands after 18 km. Fortunately, on the plains, he continued quickly, so that he could now cover a distance of 15–20 km a day. According to measurements in Advent Bay, the temperature at this time fluctuated between − 27 °C and − 39 °C. Unfortunately, Ritscher could not find a hut in Dickson Bay, although black rocks sometimes made him think he had seen one and tempted him to take detours. But he could not afford that anymore, because the moonshine period would soon be over. Bella was even more exhausted than Ritscher.

The way became more and more unpleasant, because the ice in Dickson Bay had been broken up by the tides, so that the open crevasses were only lightly covered with snow. Restful short breaks were out of the question because of the wetness of the ground. Ritscher now continued across the pack ice in the North Fjord. In the distance he could already recognize the rocky peaks of Advent Bay in the moonlight. On the evening of December 24, he was only 14 km north of Advent Bay, but, lamentably, his advance was stopped by open water in the Ice Fjord. His courage was almost taken away by his disappointment, but he pushed forward mechanically on his skis along the ice edge. First, he tried to reach a hut at Cape Thordsen 6 km inland for a rest and a meal, but he was already too exhausted to tackle the steep face 300 m from his destination. All that remained for him was to hope that the ice would soon freeze sufficiently solid. When he tried to advance south again from Cape Thordsen on December 25, he needed to pass through the young ice twice. The air temperature now dropped to − 39 °C and clouds were gathering. For more than 5 ½ days he had eaten nothing except a handful of uncooked pearl barley and some reindeer tallow, which he actually had with him for lighting, for this had been the candle he had manufactured at the Cape Petermann hut.

A further devastating detail describing Ritscher's desperation at this time is revealed in the diary kept by Olaf Henriksen (Olaf Henriksen Diary), manager of the Spitsbergen Radio telegraph station in Green Harbour (Spitsbergen Radio), who reports, as he had been informed, that while on Cape Thordsen, from where he could see the lights of Advent Bay, on Christmas Eve, Ritscher ate his last meal, which consisted of the crafted candle. On the following day, Christmas Day, evidently feeling the sting of hunger, Ritscher made a desperate attempt to seek nourishment by trying to pierce the skin of his dog Bella so as to draw blood from his canine companion and to suck blood from her for nutrition. She would have none of it, however, and—apparently despite any exhaustion she might have felt—she managed to escape his attempt to puncture her.[1]

Fortunately, during the brief rest that Ritscher took, the clouds moved again, and he had gained so much energy that he made a last attempt with the courage of despair (Ritscher, 1916, 26–27). First, at 4 a.m. on December 26, he directed his skis to Cape Thordsen, going from floe to floe until he reached the young ice, which was bent by his weight. Bella could hardly follow him. Then suddenly Ritscher broke through the ice. It took him ten minutes to pull himself up from the frigid water onto solid ice. To keep his feet from freezing, he tried to wring out his wet stockings. His clothes froze in a flash, as did his right hand. There was nothing left for him to do but to throw off all ballast and run as fast as he could around the open water hole in a wide arc across Goose Islands and the entrance to Sassen Bay. After ten hours, he finally reached Advent Bay, at 2:30 a.m. on December 27, and he was saved.

As Ritscher would later tell Fridtjof Nansen, as recounted by Nansen after his meeting with him seven months later, the "terrible walk" from Treurenberg Bay to Advent Bay, "over the mountain and across the Ice Fjord," had nearly ended when he "fell through the new ice and was nearly drowned," and so he had arrived at Advent Bay "at the last gasp," his foot critically injured from frostbite (Nansen, 1914, 8–9).

At Advent Bay, the Norwegian machinist on watch, Lovset, discovered Ritscher in the settlement and took care of him in the nearest apartment building (Ritscher, 1916, 27). A doctor attended to his frostbite, while Lovset's wife gave him copious amounts to drink. Then Ritscher sank into a deep sleep for two days. Only then could he report on what had happened and try to organize an immediate relief expedition. He also sent a telegram to Berlin. The news about the failed expedition attracted a lot of attention back home.

Ritscher summarized his accomplishment on this forced march in his report, which he did not write until he was in World War I. Within 7 ½ days, he had managed to cover about 210 km as the crow flies, including 55 km in the final 22 h. On average, he had covered 28 km over mountains, glaciers, and pack ice in 24 h.

In contrast, Rüdiger would later state that he wondered, however, why Ritscher had not also taken advantage of the moonlight in November, which had been sufficient for Rüdiger and Rave to return to the ship (Rüdiger, 1913, 139). The weather had probably been bad in some cases, but never long-lasting. Ritscher's group could well have gone to Advent Bay for help in October or in November, thought Rüdiger. He also criticized the fact that no one had returned to the hut within the agreed upon four to six week timeframe that he and Rave had remained there.

Olaf Henriksen (ca. 1886–after 19 August 1968) (Norwegian Polar Institute) made it a point to report in his diary that Ritscher arrived with only his one dog (Bella), and that he had had to leave his guns—his rifle and revolver—as well as his skis behind due to having fallen twice through the young ice (Olaf Henriksen Diary). What stood out to Henriksen from Ritscher's report was that Rüdiger remained in the Wijde Bay hut in a life-threatening state, with no food and with severe frostbite, and that there was no hope for him. The impression also was that the two Norwegians Rotvold and Stenersen remained with Rüdiger while awaiting rescue, that other expeditioners had headed Northward, and that the supply of provisions was not sufficient for beyond a month, with Ritscher fearing that some of the expeditioners would succumb to scurvy. Ritscher's own health was by now vastly compromised, with Henriksen reporting about the seriousness of the frostbite injury and the likeliness of his foot being removed.[2]

In fact, Ritscher reported that he expected the "amputation of half of the right foot, the big toe of the left foot and the two foremost phalanges of the little finger of the right hand" (Ritscher, 1916, 28).

Ritscher had fought for his life, journeying with little sleep and little food. His sojourn would ultimately result in severe physical loss. Yet he had arrived in Advent Bay. His telegram finally notified the world of the Schröder-Stranz Expedition's predicament, and his call for help mobilized forces. It was the long-hoped-for call that the stranded expeditioners desperately needed. As a result, the first forays into rescue efforts and feverish flurries of activity began.

## Notes on Original Material and Unpublished Sources

Olaf Henriksen's diary is in the Library Archives at the Norwegian Polar Institute (NPI) in Tromsø.

1. O. Henriksen diary, 2 January 1913, (NPI), D-305/D00305_0001
2. O. Henriksen diary, 2 January 1913, (NPI), D-305/D00305_0001

## Unpublished Sources and Original Material

Henriksen, O., Diary, Olaf Henriksen's diary written while he was the Green Harbour Telegraph Station Manager, "Dagbok for Olaf Henriksen, 1911–13" ["Diary for Olaf Henriksen, 1911–13"], dated 14 August 1911 to 2 May 1913, D-305, D00305_0001. Library Archives, Norwegian Polar Institute, Tromsø, https://brage.npolar.no/npolar-xmlui/handle/11250/2600678. Accessed 24 July 2021.

Norwegian Polar Institute, Tromsø. Olaf Henriksen correspondence in Library Archives. Approximate date-of-birth and date-of-death dates per Fred I. Presteng in memo dated 7 February 2008. Received 22 July 2021.

Spitsbergen Radio, Personnel List, "Personalfortegnelse", handwritten Green Harbour telegraph station staff record 1 July 1911–17 August 1914, courtesy of Per Kyrre Reymert, from the National Archives in Tromsø (Arkivverket Statsarkivet i Tromsø), File Spi radio065, received from Petr Masat, Library, Norwegian Polar Institute, Tromsø (Bibliotek, Norsk Polarinstitutt). Received 31 March 2022.

# References

Aftenposten. (1913, May 15). Hjælpeexpeditionen paa Spitsbergen. Kaptein Staxruds beretning til "Aftenposten". [The rescue expedition on Spitsbergen. Captain Staxrud's account to "Aftenposten".]. *Aftenposten* (Kristiania [Oslo]), 15 May 1913, evening edition, pp. 1–2. National Library of Norway, Oslo, online archives, www.nb.no. Accessed August 28, 2012.

Aftenposten: Ugens Nyt. (1913, September 2). En julenats-tragedie paa Spitsbergen. [A Christmas-night tragedy on Spitsbergen.]. *Ugens Nyt* published by *Aftenposten* (Kristiania [Oslo]), 2 September 1913, pp. 2–3. National Library of Norway, Oslo, online archives, https://www.nb.no. Retrieved April 2, 2022.

Nansen, F. (1914). *Through Siberia, the land of the future*, (A. G. Chater, Trans.). New York: Frederick A. Stokes Company; London: William Heinemann. Digitized by Google. https://books.google.com. Accessed August 4, 2014.

Nøis, D. (1929). Med kaptein Staxrud på leiting etter Schrøder-Stranz og hans folk [With Captain Staxrud on the search for Schröder-Stranz and his people]. In: *And-Ungen*, July 1929, pp. 4–12. Andenes: Andøyposten. Library Archives, Norwegian Polar Institute, Tromsø. Received July 20, 2021.

Rave, C. (1913). *Tagebuch von der verunglückten Expedition Schröder-Stranz.* Schaffsteins Gründe Bändchen 49. Cöln: Schaffstein.

Ritscher, A. (1916). Wanderung in Spitzbergen im Winter 1912. *Zeitschrift der Gesellschaft für Erdkunde zu Berlin*, pp. 16–34.

Rüdiger, H. (1913). *Die Sorge Bay. Aus den Schicksalstagen der Schröder-Stranz-Expedition.* Berlin: Georg Reimer.

Tahan, M. R. (2021). *The return of the South Pole sled dogs: With Amundsen's and Mawson's Antarctic expeditions*. Cham: Springer International Publishing.

# Part II
# The Stranded: Seven Men Marooned on an Iced-In Ship and in Hunter's Huts

# Chapter 5
# The Remaining Expeditioners: Hermann Rüdiger, Christopher Rave, Einar Rotvold, Knut Stave, August Stenersen, and Julius and Jørgen Jensen

## Close Relations on the Ship

Christmas 1912 had seen the reunion of Einar Rotvold and August Stenersen with Hermann Rüdiger, Christopher Rave, Knut Stave, and Julius and Jørgen Jensen on board the iced-in ship *Herzog Ernst* in Treurenberg Bay. The disappearance of Wilhelm Eberhard as well as Dr. Erwin Detmers and Dr. Walter Moeser on the attempted trek south was confounding and heavy news. And the mystery of whether Herbert Schröder-Stranz, Richard Schmidt, August Sandleben, and Max Mayr, with their team of eight (or 12) dogs, had survived on Northeast Land, was still an unfathomable question. Equally excruciating was the question of whether or not Captain Alfred Ritscher had reached Advent Bay—for the remaining men on the ship had no way of knowing that their captain and his dog Bella had successfully made the trip, at great cost, and had notified the world of their situation. No, the remaining men were caught in a grip of frozen ice and a state of frozen knowledge, wherein they did not know what the outside world knew, if there would be a rescue, and if they would survive. They greeted the New Year 1913 with much uncertainty.

Now these seven stranded men had to come to terms with one another. Rave was very bothered by the fact that the sailors spat all over the place (Rave, 1913, 58). He described Julius Jensen, the smaller and older of the two brothers, as honest and reliable, while Jörgen, to Rave, stood out for nothing. Rotvold, too, appeared to Rave to be of an honest nature but came across to him as an "inseparable shadow of the ice pilot [Stenersen], who often gave cause for trouble by his moodiness" (Rave, 1913, 58). Little did Rave know at this moment how instrumental Stenersen would be later in the rescue efforts to save Rave and his German colleague Rüdiger. The two Germans at this time, however, did not see much of the Norwegians, who chopped and prepared wood at noon for warmth and cooking, fetching snow for water production and driftwood for the fire.

Rotvold, meanwhile, recorded in his diary, nearly every day, his descriptions of the long and empty days aboard the ship, the men's state of mind and their desperate

© The Author(s), under exclusive license to Springer Nature Switzerland AG 2024
M. R. Tahan and C. Lüdecke, *Stranded at the Top of the World*,
https://doi.org/10.1007/978-3-031-56288-4_5

longing to go home, and his own profound sadness and concern (Einar Rotvold Diary). "It is getting sad now, even though we are all well, except for Knut [Stave]," he wrote on January 12. "We have nothing to read to pass the time with. It was an unexpected over-wintering for us. We had hoped to be at home in October, but in that we were disappointed. God with us. All well."[1] The final five words of his entry were his signature sign-off at the end of each and every diary entry but two in his diary of January 1913 to June 1913, during the immediate aftermath of the expedition's failure. His struggles and sadness, as well as sympathy and concern for his mates, pour out of each page.

The severe cold conditions caused the men to remain on the ship, with the Norwegian crew members alternating in carrying out chores, and all looking out for one another. Rotvold wrote on the 13th (Einar Rotvold Diary):

> We are still by necessity forced to stay put onboard. These are long sad days. Knut is better now, he says. He drinks tar-water for the chest and the cough, and gets massage in the hip each evening. Julius has been in good moods and obliging since last Sunday. We take turns, one day each, with chopping wood and fetching snow to thaw for water, one day Julius and Jørgen, the other Stenersen and I. Knut cooks for us. God with us. All well.[2]

The routine of the men continued on the next day, with the only break being Julius and Jørgen checking fox traps, to no avail.[3] On the 17th, the Jensen brothers managed to catch a Polar fox. On this day also, Knut Stave took a turn for the worse.[4] As of January 18, Stave became bedridden, remaining in his bunk and experiencing hip pain, coughing, and little appetite, while the severe cold outside the ship hovered at $-37\ ^{\circ}\text{C}$ to $-38\ ^{\circ}\text{C}$ temperature.[5]

In mid-January, Rave and some of the crew members had made a trip across the bay to the Swedish Arc-of-Meridian house to check on coal supplies and to revise the list of existing provisions (Rave, 1913, 59–61). Some days later, Rave began to finally accomplish some major cleaning and washing after months of necessity. During the process, enough food was provided. There was tea and cocoa and something like a varied weekly schedule of menus:

Sunday: soup, meat, vegetables and sometimes pudding,

Monday: soup, meatballs,

Tuesday: porridge,

Wednesday: soup and cooked together,

Thursday: fish or dumplings,

Friday: soup, meat, vegetables,

Saturday: soup, pancakes.

(Rave, 1913, 61).

Although they had enough ready-made soups from Maggi and Knorr, which tasted very good mixed with pemmican, Rave reported, the Norwegians preferred to eat sweet soups as long as sugar was available.

Meanwhile, on the ship, Rave had fetched the typewriter from Herbert Schröder-Stranz's cabin so that Rüdiger could begin to write his book about the expedition (Rave, 1913, 61–62).

But the time spent onboard was a trying one for the men, as Rotvold reported on January 28 (Einar Rotvold Diary): "It is creaking and groaning in the ship's joints. The vessel has come higher up out of the ice. It is a sad and poor existence to be here onboard, only wish that the time had passed, so that we were at home again, but there are many long, sad days until that time yet. God with us. All well."[6]

It was getting colder and colder, and the men had to economize on heating. One day, a large piece of coal landed in the stove, causing the iron to burn red hot (Rave, 1913, 64). Shortly thereafter, heavy smoke developed. Everyone was awakened and all helped to find the cause. Fortunately, the dynamite to blast the ship free had already been stored outside. Now they only had to carry the detonating cord and explosive caps out of the ship. They found the fire site behind the iron plate on the stove, which was exposed with an axe. Plenty of water finally extinguished the fire. That could have gone wrong, if Rave had not noticed the smoke when he had gone to bed.

Rotvold described the harrowing details of the alarming fire thusly, in his January 29th diary entry (Einar Rotvold Diary):

> At 5 o'clock pm Mr. Rave came forward and said that we had to come aft, as there was such a lot of smoke. When we came thereto, there was fire behind the stove in between the paneling. Dr. Rüdiger hopped [limped] forward in the cabin and sat down. We broke away an iron sheet, and had to break away enough of the paneling so that one could get to where it was burning, and could pour water on it. One got it pretty soon extinguished, and without anything getting ruined. We could nearly have been both foodless and homeless in such a tremendous weather, as we have everything onboard, amongst this a lot of paraffin.[7]

Thus, the fire that had broken out on board and been discovered by Rave was extinguished by Rotvold and Stenersen—a fortunate outcome that did not add to the plethora of painful woes for the expeditioners.

And, so, the month of January ended for the stranded men on the ship, with extreme cold and extinguished fires, menus and medicines, and "intensely whirling snow... . Severe cold. ... [and] weather... so terrible that one can just about not get up on deck."[8]

# The Arctic Coal Company Workers' Rescue Attempt

When Alfred Ritscher arrived in Advent Bay, the moon had set and it was again a dark winter night. Moreover, the ice had broken up again in Advent Bay and also in Ice Fjord. When the next moonlight period began on January 19, 1913, and it had become cold again, the first relief expedition could be organized by the Arctic Coal Company in Advent Bay on behalf of the Norwegian government, which felt also responsible for the Norwegian participants (Ritscher, 1916, 28; Rüdiger, 1913, 169–171). It consisted of leader Ingvar Jenssen and workers Einar Pedersen, Einar Tessem, and Jakob Rognli, as well as four sledges with provisions and clothing pulled by 13 dogs, including Ritscher's dog Bella, who ran back with them. However, after finding that the ice situation in the Ice Fjord was still too unstable, the men had to turn back.

The rescue wheels had actually already been put into motion immediately following Ritscher's arrival. As of January 2, Green Harbour telegraph station manager Olaf Henriksen was reporting on Ritscher's stranded expedition after the news had reached him via a visitor from Advent Bay (Olaf Henriksen Diary).[9] As the manager of Spitsbergen Radio, Henriksen sat at the crossroads of communication, and was a witness to historic events and figures. Just several months prior, in July of 1912, he had been invited onboard the *Veslemøy* by none other than Fridtjof Nansen himself, who inquired about news from Roald Amundsen (who was returning from his South Pole expedition [Tahan, 2021, 233–234]) and who regaled his guests with dinner and fine examples of his abilities as a raconteur (Olaf Henriksen Diary).[10] [This was the cruise that Nansen would later wish had resulted in his crossing paths with the *Herzog Ernst* and thus stopping Schröder-Stranz from proceeding into oblivion (Chaps. 1 and 3)]. As Henriksen reported, Nansen also spoke with members of the Nøis family, who were packing seal-skins when Nansen arrived (the Nøis family members would later factor substantially into the rescue expedition, as shall be seen in this narrative). A month later, in August of 1912, Henriksen was again enjoying the company of both Nansen and Arve Staxrud, who was working on Spitsbergen in his survey-related activities.[11] Nansen himself would later write about his initial meeting with Henriksen on July 16 and the true hospitality shown by the manager at the telegraph station during his stay (Nansen, 1920, 53, 229), as well as document his dinner meeting with Staxrud and Henriksen on board the *Veslemøy* on August 25, and sing Staxrud's praises regarding mapping Spitsbergen (Nansen, 1920, 229–230).

And so Henriksen was in a unique position to see this new history unfold pertaining to the Schröder-Stranz Expedition and the ensuing rescue operations. In fact, to some degree, he participated in helping organize some of those rescue efforts.

On January 3, 1913, in Green Harbour, Henriksen (Olaf Henriksen Diary) reported on the survival of only one puppy from the five puppies born to a dog named Letta—and the fact that the surviving puppy had had to be saved.[12] This seemingly unrelated piece of information will factor quite significantly into the story of the relief efforts, for indeed there was a shortage of dogs on Spitsbergen. On the same day, he reported people leaving from Green Harbour to Advent Bay, and on the 5th reported that the expedition had arrived there safely on the previous day.[13] On January 12, Henriksen recorded in his diary that the rescue expedition in Advent Bay still had not yet departed—presumably this was the Arctic Coal Company relief expedition trying to leave for its pursuit of the stranded Schröder-Stranz members.[14] And on January 21, Henriksen reported that Jenssen—most likely Ingvar Jenssen who was leading the relief expedition—had begun the excursion on the 20th but had stopped due to encountering open water midway through the fjord, hence turning back.[15] Jenssen telegrammed to Henriksen that he needed dogs—most likely finding that the excursion would be impossible without the help of dependable sledge-pulling dogs. And so Henriksen immediately joined with Tessem—possibly Einar Tessem, also of the relief expedition—and traveled that day from Green Harbour toward Advent Bay with Tessem and three dogs, trekking over land as the ice was too wet. Arriving the following day in Advent Bay, he met with Ritscher himself on January 23, and heard directly from the captain the entire tale of the Schröder-Stranz expedition—up to, of course, Ritscher's arrival in Advent Bay.[16] On January 24, according to Henriksen's diary entry, the Arctic Coal Company's relief expedition departed at 1:00 a.m. with Jenssen, three men, 13 dogs, and three sledges.[17]

Thus, the second and final departure of the four men was delayed until January 24, 1913, and is documented by the participants as follows, in a report of the relief expedition sent out by Captain Ritscher: (The travel route is given in Fig. 10.2a, b).

24/1.13. The expedition departed Advent Bay at noon, assisted by a horse and 3 men. The horse had to turn back at Advent City because the ice was too weak; the 3 men continued to Rævnes. About 3 km before Rævnes we encountered screw ice in strips of 2–300 meters. Between these strips was new ice with heavily watery snow that made it very difficult to drive the sledges. Here the runner of one sledge broke. Since the damage could not be repaired, the load was distributed to the other two sledges. We camped for the night at Cape Thordsen. Distance covered about 26 km.

25/1.13. Went [along] the North Fjord further towards Dickson Bay. Here the ice was flat, but the snow became deeper, so that the loads sank in. It also did not support the dogs, so they had a hard time pulling. At the peninsula at the entrance of Dickson Bay we found a hut where we took quarters for the night. Distance covered about 15 km.

26/1.13. We moved over towards the left side of Dickson Bay, where we expected to find two huts. However, we got dense fog, so we had to stop. Distance covered about 13 km.

27/1.13. Walked on in the fog at the west side of Dickson Bay. Later in the day it began to snow. Made camp at Giant Chair. Distance covered about 11 km.

28/1.13. Got no further because of snowstorm.

29/1.13. Still stuck as weather prevented moving on.

30/1.13. The storm abated, but fog instead. At noon we marched on. The snow had now become so deep that we had great difficulty in working our way on. Made camp for the night about in the middle of the alluvial soil (Norwegian: marslandet). Distance covered about 9 km.

31/1.13. Still fog, but continued on. The snow got deeper and deeper, so we felt compelled to leave a sledge behind with what clothes we could spare. A dog ran away from us. Made camp at Battyes glacier. Distance covered about 5 km.

1/2.13. Clear weather. It appeared that we were on the left side of the glacier. Went up the river valley. The snow here even deeper and climbing unusually difficult, even with only 1 sledge. Distance about 6 km.

2/2.13. Continued towards Mount Sir Thomas. Distance covered about 5 km.

3/2.13. Reached the first ledge today on the descent to West Fjord. Here Rognli froze both feet. When Rognli noticed this, one foot had already blistered, the other was massaged with snow as long as Rognli could stand the pain. We considered transporting Rognli to the hut in West Fjord, but did not feel it was responsible to continue with a sick man. A dog died the same day.

4/2.13. Rest and prepare for the return trip.

5/2.13. Rognli tried to walk, but it became apparent that it was impossible for him. He was therefore put in a sleeping bag, and this in turn [was put] in the tent, and in this wrapping [was placed] on the sledge. We had to leave all the equipment behind, except a tent, a sleeping bag and the most necessary provisions, until we reached the hut in Dickson Bay, where we left provisions. The home march began at 8 a.m. and lasted all day and night.

6/2.13. Walked until 1 a.m. and made camp at Table Taage (Tåkefjell).

7/2.13. Went on to the hut at Dickson Bay, which we reached late in the afternoon. Another dog died.

8/2.13. Sent Tessem to Advent Bay for a horse, as both we and the dogs were exhausted. However, Tessem returned and informed [us] that the ice had broken around Cape Thordsen and there was heavy swell.

9/2/13. Had to stay in the hut because of fog and wind.

10/2/13. Because we were expecting the possibility that the ice was breaking, we decided to try to pull Rognli to Advent Bay. The dogs were of no use that day. In the middle of the fjord we encountered the same screw ice with heavy new ice in between as we had on the first day of marching. We finally did not manage to pull the sledge on. Jenssen then asked Tessem and Pedersen to go to Advent Bay for a horse, and they agreed. The horse was expected that same night around 8 p.m.

11.2.13. As no help came, Rognli and Jenssen tried to leave, but had to return to the sleeping bag because of fog.

12.2.13. At about 10 a.m. Tessem and Pedersen arrived with 5 men and gave the following explanation as to why help had not come earlier. They had encountered broken ice at Rævnes with open water, and had not reached Advent Bay until the 11th in the forenoon. One dog was so sick that it had to be shot. At 11 a.m. they started from Advent Bay. The trail conditions proved so difficult that 5 men had enough to do pulling the sledge and sleeping bag. The dogs were so miserable that we turned them loose. Rognli himself agreed to go. At Hyperihatten [hut close to Hyperit Waterfall northernmost in De Geerdalen], the sledge and sleeping bag were left behind. One dog stayed behind; despite repeated attempts, we did not succeed in bringing him along; he followed a bit, but kept jumping back. Rognli stayed in Longyear City for the time being, where a horse picked him up later. At 10 p.m. all participants had arrived in Advent Bay.

The reasons why it took us so long to cover a distance of about 180 km can be seen in the following:

Bad weather and bad sledging track, as we had fog, storm and blowing snow almost all the time. The large amounts of snow were also a major obstacle for us. In addition, the dogs were miserable and only one sledge proved useful. We suffered a lot from the cold at night because the sleeping bags were bad. As we had no opportunity to dry our clothes, they froze stiff, so that they felt like a tank. Since we found only one hut and took a long time to get there, our supply of kerosene ran out. We therefore had no way to prepare enough warm drinks on the last day of the outward trip. We burned 3 pairs of skis.

In regretting the poor result of the expedition, we are aware that we did everything in our power to bring relief.

Advent Bay February 27, 1913

Einar Tessem, Ingvar Jenssen, Jakob Rognli, Einar Pedersen.

(Tessem et al. 1913)

The brutal conditions existing at that time are evident from the above report, which shows the suffering of the men, the frostbite injury and ailment of Rognli, the exhaustion of the dogs, and the death of at least five dogs (later reports placed this number at seven and at eight).

Ritscher's dog Bella had joined this second relief expedition as well, and was among the dogs who had sadly frozen or starved to death (Ritscher, 1916, 28).

And so ended the efforts of the workers of the Arctic Coal Company to search for the Schröder-Stranz expeditioners.

## The Human Cost, and the Company's Accounting of Financial Expense

As for the owner of the Arctic Coal Company, American businessman John Munro Longyear (1850–1922), he seemingly came to have nothing but suspicion and complaints about German expeditions in general, including the Schröder-Stranz expedition in particular, as put forth in P.J. Capelotti's paper "The Train Has Left the Station" (Capelotti, 2002).

Leadership of the Arctic Coal Company, which had hosted and organized the January–February relief expedition, and which had evidently housed Ritscher in Longyear City after his arrival at the end of December, later stated that the expedition owed them unpaid debts for supplies and medical expenses provided on credit through April 9th, presumably including during Ritscher's stay at Advent Bay (Dole, 1922, vol. 2: 196–199). In the book *America in Spitsbergen* by Nathan Haskell Dole, which chronicles John Longyear's Arctic Coal Company, the Schröder-Stranz expedition is introduced as "the German Scietific [sic] Expedition, which, under the direction of Lieutenant Hans Schröder-Stranz had, in the later summer of 1912 before, gone to Northeast Land, not with the intention of making a dash for the Pole, but to make a study of the ocean-currents in that little-known region" (Dole, 1922, vol. 2: 191). There seems to possibly be some confusion of names, as Lieutenant Herbert Schröder-Stranz was the leader of the expedition, and Hans Schröder-Stranz was his brother, who resided at the family home in Stranz. Hans, per Herbert's written declaration, was Herbert's deputy and represented him during the expedition (Allgemeiner Plan 1912, 11).

And so, when the Arctic Coal Company leadership proceeded to vigorously seek reimbursement, they corresponded with "Schröder-Stranz," their letters most likely being sent to Schröder-Stranz's estate, probably to his brother Hans. Their pursuit of reimbursement is presented in the following passage from Dole's book, in which he summarizes the account as well as quotes letters of correspondence (Dole, 1922, vol. 2: 196–198):

REPUDIATION OF THE DEBT. This party of Germans had drawn supplies from the Superintendent on credit, and by April 9 their debt amounted to 4300 kroner, and 600 more for medical expenses. Schröder-Stranz sent a wireless message to Captain Ritscher, a member of the refugee-party at Longyear City, guaranteeing 3500 kroner; but, later, when [Scott] Turner discovered that the debt contracted amounted to 6000 kroner, and demanded a settlement of the account, Schröder-Stranz repudiated all personal responsibility, asserting that the matter had been entrusted to the German Committee of the "Hilfe für deutsche Forscher im Polareis."

This Committee refused to pay. The Deutsche Arctische Forschung Society [sic] had in the meantime been collecting funds through their Captain Berg for the unlucky members of the expedition, but, as the committee had separated from the parent-society, it refused to honor Schröder-Stranz's demand. He, on the other hand, claimed that he had signed his name, not in his personal capacity but as merely the head of the exploring expedition which had come to grief. He wrote the Arctic Coal Company:

"I reckoned also that that was already collected and that more was to be expected after the committee, Hilfe für deutsche Forscher im Polareis separated from the 'Deutsche Forschung' took all the money and income, and now makes difficulties as to its obligations. I do not feel at all obliged to pay these 6000 kroner, and am not able to, either. More has surely come in by this collection than can be used, and the collection is, as the name shows, for that kind of expense."

A WORTHLESS GUARANTEE. Turner immediately sent a wireless to the Arctic Coal Company at Spitsbergen declaring that Schröder-Stranz's guarantee was worthless, and ordering the Superintendent to refuse further advances of supplies without good and sufficient security. The Superintendent made no reply to this order, but went on allowing the Germans to take what they wanted.

Turner also wrote a stern letter to Schröder-Stranz expressing his surprise that he should have repudiated his guarantee. "This Company," he said, "saw a telegram from you guaranteeing 3500 kroner to be used in giving aid to the unfortunate members of your Arctic expedition. Acting at once upon this personal guarantee, our winter mining-force at Advent Bay furnished medical attendance, food, supplies, men and dogs, and help of every kind. These expenses will later be enumerated in an itemized bill.

"Now that we have done this work and spent this time and money in your behalf, after accepting your guarantee in good faith and without question, you deem it your privilege to repudiate your word in the matter and leave American shareholders, who have no interest in you or your expedition, to pay the bills.

"We have to say that this is a commercial Company engaged in mining and shipping coal. We are business people, and in the habit of operating along strict business lines. Knowing something of your reputation, our winter superintendent accepted your telegraphic word as security. We therefore hold you personally responsible in so far as we have expended money in the relief of your party.

"We know nothing about the 'Hilfe für deutsche Forscher im Polareis' or any other of your committees, and we have had no dealings with them, and if you do not discharge your obligations to this Company, we will take the matter up in the newspapers and law-courts of your country, and get redress and satisfaction in that way.

"We do not think it will look very well in print to see a German gentleman repudiating his guarantee for money concerned in the relief of destitute and suffering German subjects, and the responsibility for the loss incurred being placed on American subjects thousands of miles away."

He ended by threatening to begin proceedings against Schröder-Stranz in diplomatic circles, and to place the matter in the hands of competent German attorneys.

The Company's Norwegian lawyers offered to undertake the case on condition the actual outlays and a commission of five or ten per cent according to difficulties were assured. They thought the expense would be considerable. Turner found out through German lawyers that Schröder-Stranz claimed to have no available funds, though he lived in "an elegant five-room flat furnished with very expensive furniture," and his father had an estate in Stranz in Westphalia worth a million and a half marks, though it was heavily mortgaged.

(Dole, 1922, vol. 2: 196–198)

The Arctic Coal Company ultimately was reimbursed for its expenses, with its itemized bill being paid by the Germans (Dole, 1922, vol. 2: 199). An absolute disdain, however, lingered, as expressed in the following passage from the book (Dole, 1922, vol. 2: 198–199):

A BULL-HEADED BLUNDER. Some sympathy might have been felt for the participants of the expedition had it not been so foolishly managed. All their calamities were due to the grossest and most inexcusable ignorance. Everything was done in a "bull-headed way." Even the story that they told was regarded as ridiculous and caused no end of laughter among those who knew its details. The Coal Company sent out a relief-expedition at the earnest solicitation of their friends, but it failed to find them; some of its members were severely frostbitten. The Director of the German Scientific Station told Mr. Longyear the next Summer that the Schröeder-Stranz [sic] party was better supplied with provisions and had a more complete equipment than the rescuers had. Their report of the deaths and hardships were in reality cases of pure "funk."

(Dole, 1922, vol. 2: 198–199)

Meanwhile, as of the Arctic Coal Company relief expedition's unsuccessful return in February 1913, the plight of the stranded Schröder-Stranz Expedition members on the ship *Herzog Ernst* continued, and the men there still awaited rescue.

## Alfred Ritscher's Telegram Reaches Germany, and Kurt Wegener Offers Aid

Captain Alfred Ritscher had informed the expedition office in Germany about the unfortunate course of the Schröder-Stranz expedition via the wireless telegraph stations in Advent Bay and Green Harbour (Ritscher, 1916, 28). Beginning on January 7, 1913, the messages went to the headlines of the German press: "Ein Funken-telegramm von der Spitzbergen-Expedition" (A Radio Telegram from the Spits-bergen Expedition), "Leiden der deutschen Spitzbergen-Expedition" (Sufferings of the German Spitsbergen Expedition), "Das Schicksal der Spitzbergen-Expedition" (The Fate of the Spitsbergen Expedition). From that time on, the radio messages went back and forth between Germany and Norway.

When the radio telegraph equipment of the German Geophysical Observatory in Ebeltofthamna happened to work again after many failures on January 23 (Wegener, 1914a, 24), the crew also learned that the Schröder-Stranz expedition had suffered hardships (Wegener, 1913, 1914b). Immediately, the station leader Kurt Wegener offered to organize a transfer of the distressed men to Cross Bay. However, his offer was rejected in the next radio message, because the following day a rescue expedition with four dog sledges would leave from the Arctic Coal Company American coal mine in Advent Bay (Fig. 5.1).

On January 27, three days after the Arctic Coal Company relief expedition had departed for the second time, Wegener received an official request from the Foreign Office in Norway to join the men of Ernest Mansfield's Northern Exploration Company on an expedition from the British marble quarry in Kings Bay, 25 km away. It would not be until mid-February when Wegener would be able to reply and thus also join the effort.

**Fig. 5.1**  Radio telegraph station of the German Geophysical Observatory in Ebeltofthamna (*Source* Wegener, 1914b, Table 15)

# Theodor Lerner Yearns to Help

After the call for help for the ill-fated Schröder-Stranz expedition had reached Germany by radio telegram on January 7, 1913, the journalist and polar explorer Theodor Lerner (1866–1931) immediately felt personally addressed (Lerner, 2005, 227–288). He had observed the launch of Andrée's balloon expedition in Spitsbergen in 1897 and was present at Wellman's airship trials in 1906 and 1908. In addition, he had led the German expedition to the Northern Arctic Ocean on the *Helgoland* in the summer of 1898. He was, so to speak, a connoisseur of Spitsbergen.

However, Lerner had earlier made himself a *persona non grata* in Berlin government circles and was therefore categorically excluded from the search by officials (Przigoda, 2012, 84). Thus, within the next three months, he would take it upon himself to organize a Frankfurt relief expedition with private funds (Villinger, 1929, 5–8).

Lerner was not viewed in a very favorable light by Fridtjof Nansen, who seemed to belittle and mock the "magazine-scribe", calling him "odd" and a wannabe "great estate-occupier" who claimed land and ocean as his own, and describing Lerner's activities on Bjørn-Øen (Bear Island) in 1899 as those of a "terrible tyrant" who had "declared himself ruler of this fog-island" (Nansen, 1920, 11, 16–18). Nansen stated that, as "the grand warrior," Theodor Lerner had menaced fellow Germans and Russians alike, and was indifferent to the Norwegians, whom he "threatened with... magazine-rifles if they dared come to the island," despite their having been hunting and maintaining houses there "from before [the time] little Theodorus I himself had seen the light of day." That fact, said Nansen, did not matter to this "absolute ruler," "We, Theodorus rex," who had proclaimed that the property now belonged to him— "This zealous prince [who] governed thus with a hard hand," this "fog-prince" who, it turned out, could not face an "Arctic sea's winter" and so ultimately "lost his domain" after his short "lustrous reign."

In Spitsbergen, Lerner was not looked upon very respectfully by the leadership of the Arctic Coal Company, either, who stated that, in October of 1907, John Longyear had received an "amusing letter" from Lerner that "would have delighted Mark Twain" (Dole, 1922, vol. 2: 390), and that previously "His most sensational exploit was in trying to corral the coal-deposits on Bear Island" (Dole, 1922, vol. 2: 392). Evidently, he had left a questionable impression on the Americans as well: "This Lerner was a regular comic-opera hero," was their later take on the individual (Dole, 1922, vol. 2: 391).

Now, however, this "comic-opera hero" was about to embark on his own expedition to join the several rescue efforts converging upon Spitsbergen.

## Notes on Original Material and Unpublished Sources

Einar Rotvold's and Olaf Henriksen's diaries are in the Library Archives at the Norwegian Polar Institute (NPI) in Tromsø.

1. E. Rotvold diary, 12 January 1913, (NPI), D00125
2. E. Rotvold diary, 13 January 1913, (NPI), D00125
3. E. Rotvold diary, 14 January 1913, (NPI), D00125
4. E. Rotvold diary, 17 January 1913, (NPI), D00125
5. E. Rotvold diary, 18, 19, 20, 22, and 25 January 1913, (NPI), D00125
6. E. Rotvold diary, 28 January 1913, (NPI), D00125
7. E. Rotvold diary, 29 January 1913, (NPI), D00125
8. E. Rotvold diary, 31 January 1913, (NPI), D00125
9. O. Henriksen diary, 2 January 1913, (NPI), D-305/D00305_0001
10. O. Henriksen diary, 16, 19, 20, and 21 July 1912, (NPI), D-305/D00305_0001
11. O. Henriksen diary, 24, 25, and 26 August 1912, (NPI), D-305/D00305_0001
12. O. Henriksen diary, 3 January 1913, (NPI), D-305/D00305_0001
13. O. Henriksen diary, 5 January 1913, (NPI), D-305/D00305_0001
14. O. Henriksen diary, 12 January 1913, (NPI), D-305/D00305_0001
15. O. Henriksen diary, 21 January 1913, (NPI), D-305/D00305_0001
16. O. Henriksen diary, 23 January 1913, (NPI), D-305/D00305_0001
17. O. Henriksen diary, 24 January 1913, (NPI), D-305/D00305_0001

## Unpublished Sources and Original Material

Allgemeiner Plan, (1912), Allgemeiner Plan der Deutschen Arktischen Expedition durch die Nordostpassage (Taimyr-Halbinsel) und durch den Stillen Ozean. Schröder-Stranz estate, Kasten 26, Nr. 13, Leibniz Institute for Regional Geography, Leipzig, Germany.

Henriksen, O., Diary, Olaf Henriksen's diary written while he was the Green Harbour Telegraph Station Manager, "Dagbok for Olaf Henriksen, 1911–13" ["Diary for Olaf Henriksen, 1911–13"], dated 14 August 1911 to 2 May 1913, D-305, D00305_0001. Library Archives, Norwegian Polar Institute, Tromsø, https://brage.npolar.no/npolar-xmlui/handle/11250/2600678. Accessed 24 July 2021.

Rotvold, E., Diary, Einar Rotvold's expedition diary written during the Schröder-Stranz expedition of 1913, "Dagbok for August Stenersen og Einar Rotvold, Tromsø: Treurenberg Bay fra 1. januar—7. juni 1913: Schröder-Stranz-ekspedisjonen 1912–13", "Einar Rotvold har skrevet dagboken", ["Diary for August Stenersen and Einar Rotvold, Tromsø: Treurenberg Bay from 1 January—7 June 1913: The Schröder-Stranz Expedition 1912–13", "Einar Rotvold has written the diary"], dated 12 January 1913 through 7 June 1913, D00125. Library Archives, Norwegian Polar Institute,

Tromsø, https://brage.npolar.no/npolar-xmlui/handle/11250/2426394. Accessed 24 July 2021.

Tessem, E., Jenssen, I., Rognli, J. and E. Pedersen, (1913). (Translated from Norwegian to German by Volkert Gazert and later translated to English by Cornelia Lüdecke.) Ritscher estate, Cornelia Lüdecke, München.

# References

Capelotti, P. J. (2002). The train has left the station: archaeological formation processes and the abandoned coal mining village of Pyramiden, Svalbard. In Bouzney, E. (Ed.), *International Scientific Cooperation in the Arctic: Proceedings of a Conference Devoted to the Centenary of the Swedish-Russian Arc-of-Meridian Expedition to Spitsbergen and 125 Years of V.A. Rusanov's Birth*. Moscow: Scientific World, pp. 116–127.

Dole, N. H. (1922). *America in Spitsbergen: The Romance of an Arctic Coal-Mine*. In Two Volumes. Boston: Marshall Jones Company. Volume II. Facsimile reprint by Scholar Select.

Lerner, T. (2005). Berger, F. (Ed.) *Polarfahrer: Im Banne der Arktis*. Zürich: Oesch/Kontrapunkt.

Nansen, F. (1920). *En ferd til Spitsbergen*. Kristiania: Jacob Dybwads Forlag. Digital archive, National Library of Norway, Oslo, Online archives. https://www.nb.no. Accessed September 28, 2023.

Przigoda, S. (2012). Bergbau auf der Bäreninsel? Deutsche Rohstoffinteressen und die Erkundung Svalbards (1871–1914). In: C. Lüdecke & K. Brunner (Eds.), Von A(ltenburg) bis Z(eppelin). Deutsche Forschung auf Spitzbergen bis 1914. 100 Jahre Expedition des Herzogs Ernst II. von Sachsen-Altenburg. *Schriftenreihe des Instituts für Geodäsie der Universität der Bundeswehr München*, Neubiberg, Heft 88, pp. 77–91.

Rave, C. (1913). *Tagebuch von der verunglückten Expedition Schröder-Stranz*. Schaffsteins Gründe Bändchen 49. Cöln: Schaffstein.

Ritscher, A. (1916). Wanderung in Spitzbergen im Winter 1912. *Zeitschrift der Gesellschaft für Erdkunde zu Berlin*, pp. 16–34.

Rüdiger, H. (1913). *Die Sorge Bay. Aus den Schicksalstagen der Schröder-Stranz-Expedition*. Berlin: Georg Reimer.

Tahan, M. R. (2021). *The return of the South Pole Sled Dogs: With Amundsen's and Mawson's Antarctic Expeditions*. Cham: Springer International Publishing.

Villinger, B. (1929). *Die Arktis ruft! Mit Hundeschlitten und Kamera durch Spitzbergen und Grönland*. Freiburg i. Br.: Herder.

Wegener, K. (1913). Die Hilfsexpedition von Cross- und Kings-Bay nach Wijde-Bay. *Petermanns Geographische Mitteilungen, 59*, 137–140.

Wegener, K. (1914a). Das Observatorium in der Crossbai 1912/13. In H. Hergesell (Ed.), Das Deutsche Observatorium in Spitzbergen, Beobachtungen und Ergebnisse I. *Schriften der Wissenschaftlichen Gesellschaft in Straßburg, 21*, 21–29.

Wegener, K. (1914b). Die Hilfsexpedition von cross- und Kings Bay, 21.II.–31.III.1913. In: A. Miethe (Ed.), *Die Expeditionen zur Rettung von Schröder-Stranz und seinen Begleitern geschildert von ihren Führern Hauptmann A. Staxrud und Dr. K. Wegener*. Berlin: Dietrich Reimer, pp. 69–101.

# Chapter 6
# The Scramble to Form Rescue Expeditions, and the Selection of Arve Staxrud as Leader (February 1913)

## The Aftermath of the Arctic Coal Company's Unsuccessful Second Attempt

The failed relief expedition of the Arctic Coal Company, led by Ingvar Jenssen, returned to Advent Bay on February 12, after having faced harsh conditions, having experienced many difficulties and injuries, and having lost several dogs who had died along the way.

Green Harbour telegraph station manager Olaf Henriksen recorded in his diary on the following day, February 13, that, as reported to him, the relief expedition had returned to Advent Bay with only 5 dogs alive, thus indicating that 8 had been lost (Olaf Henriksen Diary). He stated that Jenssen and his expedition members had been forced to abandon all their equipment on the ice, and that the sledges had all been lost as well—one abandoned, one broken on the ice, and one burned (for warmth).[1]

The Tromsø newspaper *Nord Norge* provided full coverage of the stranded Schröder-Stranz expedition as well as the attempted Arctic Coal Company relief expedition, based on Alfred Ritscher's telegrammed accounts from Advent Bay (*Nord Norge* 21 February 1913a, 1). The article (which was published on February 21) indicated the likeliness that Herbert Schröder-Stranz and his party of three men—Richard Schmidt, August Sandleben, and Max Mayr—had all perished, and expressed the strong possibility that the two biologists Erwin Detmers and Walter Moeser who had separated off along the east coast of Wijde Bay also may have risked their lives "thanks to their stubbornness," according to the article—the paper made it a point to say that the ice pilot August Stenersen had recommended a course along the west coast of Wijde Bay, which is what Ritscher and the rest had taken. Regarding the ailing cook Knut Stave and the brothers Julius and Jørgen Jensen, whom the article refers to as "the Tromsø-boys" (for this was their hometown), the newspaper described the Jensen brothers' staying behind with the ill Stave on the ship as "a companionship trait worthy of Norwegians." While a few of the dates vary, several added details are provided in the article, including a mention that Ritscher had instructed

M. R. Tahan and C. Lüdecke, *Stranded at the Top of the World*,
https://doi.org/10.1007/978-3-031-56288-4_6

Einar Rotvold and Stenersen to take additional provisions to Hermann Rüdiger and Christopher Rave at the hut after accompanying the weakened Wilhelm Eberhard back to the ship. The paper's impression given is that Rotvold and Stenersen at that time were "considerably closer to Advent Bay than Treurenburg [sic] Bay." As for the relief expedition sent out by the Arctic Coal Company under the leadership of "business manager Jensen [sic]," the newspaper article stated that the second attempt also failed and "7 of the dogs died."

Another article on a different page of this same newspaper (*Nord Norge* 21 February 1913b, 2) reported that Jakob Rognli, who had experienced frostbite in both his feet during the relief expedition, had "lost a toe" and therefore was incapacitated and "unable to work for a long time."

These newspaper articles and a couple of others, as shall be seen, would prove important enough for future expedition leader Arve Staxrud to save and later send as clippings to his trusted mentor Fridtjof Nansen. The stranded expedition and the failed expedition, it seems, were attracting much attention in Norway and much theorizing regarding strategy.

## Kurt Wegener Joins the Effort

After Kurt Wegener's January 23 offer to rescue the stranded Schröder-Stranz expedition had been rejected in favor of the Arctic Coal Company, he had subsequently received a message from the Foreign Office in Norway on January 27 requesting that he work with a proposed rescue expedition from the Northern Exploration Company, owned by British businessman Ernest Mansfield (1862–1924) (Wegener 1914, 71–77). Given the intermittent signals of the telegraph equipment at the German Geophysical Observatory in Ebeltofthamna (Cross Bay) at the time, Wegener was not able to answer until February 14, at which time he and his group were able to radio again from Ebeltofthamna and communicate their willingness to search for Schröder-Stranz and bring him to Kings Bay and Cross Bay. For this, however, they still needed detailed location information. A day later they received a response. The Norwegian relief expedition from the Arctic Coal Company had failed and had lost seven dogs and two sledges (as reported to him). Schröder-Stranz and three companions had intended to advance from the north coast of Northeast Land via Dove Bay to Wahlenberg Bay and via the Hinlopen Strait to the depot in Treurenberg Bay and finally overland to Cross Bay. The remaining members of the expedition would be in huts on the west side of Wijde Bay and on board the *Herzog Ernst*.

Wegener was only too happy to interrupt the monotonous station life with a search expedition. Due to the limited information available, Wegener and his two assistants Michaelis and Schwarz set out from Ebeltofthamna on February 21 and crossed Cross Bay to confer first with the English leaders of the Kings Bay venture, Millar and Mansfield's related brothers David and James Booth (Fig. 6.1).

The Kings Bay leaders gladly offered their help. After it was deemed that the men on the *Herzog Ernst* would probably be found to be safe, Wijde Bay was to be the

**Fig. 6.1** Departure of Kurt Wegener's search expedition at Ebeltofthamna. (Steinhagen, 2008, 70)

target of their search. They assumed that Schröder-Stranz and his companions had either already died or had saved themselves in the hunter's hut at the North Cape of Northeast Land. However, they could also already be on the *Herzog Ernst*. To search for the other men, the leaders first wanted to go across West Fjord to Wijde Bay and lay out depots until the time when they would receive more precise news from Germany. As Wegener recorded in his diary, the depot laying was very strenuous and had to be abandoned after a week of work due to exhaustion, frostbite, and lack of equipment:

> February 22. Kings Bay wants to take over depot laying and prepares everything. 5–7 p.m.: March to Lovén Islands (6 km) with two sledges. Kings Bay sledge turns out to be unusable.
>
> February 23. Dr. Wegener, Michaelis, Schwarz return to Cross Bay (30 km) to get another sledge and, since the equipment of the Kings Bay people also seems inadequate, as far as possible, better equipment. Millar, the leader in Kings Bay intends to lay out depot with five men and return to Lovén Islands the next day, but cannot find an ascent to Kings Glacier and therefore sleeps again at the house in Lovén Islands.
>
> February 24. Dr. Wegener, Michaelis, Schwarz leave Cross Bay for Lovén Islands (30 km) with sledges, furs and the like. Millar succeeds in climbing the Kings Glacier, setting out a depot 10 km from Lovén Islands.
>
> 25 February. Collectively bring all food and tents to 15 km southeast of Lovén Islands. I. tent camp.
>
> 26 February. Due to northeasterly storm, further march impossible. Michaelis foot-sick, goes back to Cross Bay. From Millar's tent three men report sick and go to Lovén Islands to recuperate. Dr. Wegener and Olafson go with them to get replacements. Millar and three men remain in camp (II).

27 February. Depots advanced to 20 km. Two of Millar's men definitely remain behind. II. camp. Petroleum ([from] Tromsø) permanently frozen.

28 February. After very cold and stormy night, Millar brings his tent up to ours; then weary march by all back to Kings Bay. Light seaman Schwarz declares that he does not want to risk his health and returns to Cross Bay

(Wegener, 1913, 138).

After their arrival at Kings Bay, they again discussed how they could best continue and came to the conclusion that the second sledge trip should also lay out a depot (Wegener 1914, 77–78). This time, in addition to Wegener (35 years old) and the Briton Millar (20 years old), only the Swede Olafson (45 years old) and the Norwegian Abrahamsen (26 years old) wanted to participate. Although Millar was physically weak and not accustomed to exertion, he wanted to join the expedition at his urgent request (Wegener, 1913, 138). These three men were to receive the fur equipment from Cross Bay. After Wegener had found on the previous depot trip that he, as the responsible leader, could not exert enough influence on his expedition members and therefore they "at decisive points … acted contrary to his wishes or ideas" (Wegener, 1913, 138), he preferred that the oldest and most experienced— Olafson—or the leader of the Kings Bay people—Millar—take over the leadership of the coming sledge trip. But the three men decided that Wegener should be the leader, perhaps because, as he was an academic and thus a recognized German authority, no one wanted to give him orders (Wegener 1914, 78).

Thus, February ended with the combined Wegener and Mansfield relief expedition only succeeding to lay down depots in preparation for the actual rescue excursion.

## The Berlin-Kristiania Connection: Miethe and Nansen Meet Per the Berlin Relief Committee

After Ritscher informed the Berlin Committee for the German Arctic Expedition about the disastrous outcome of the training expedition, another committee was formed, which placed an "Emergency Call to Save the Schröder-Stranz Spitsbergen Expedition" (Notruf zur Rettung der Schroeder-Stranz'schen Spitzbergen Expedition) in the newspapers and solicited financial "help for German explorers in the polar ice" (Rüdiger, 1913, 171; Notruf 1913; Hamburger Nachrichten, 1913b). Since the ship's crew consisted of Norwegians, Norway was also very interested in a relief expedition.

While Kurt Wegener carried out the first relief action for the expedition in distress, Professor Adolf Miethe (1862–1927) was commissioned by the relief committee to travel to Kristiania (Oslo) to discuss the equipment and the procedure of the official relief expedition with the German legation, the Norwegian Foreign Ministry, and Fridtjof Nansen (Miethe, 1914, XIII–XIV; Hamburger Nachrichten, 1913a). Financial resources would be provided by Germany via collections, while Norway would take care of the search team.

**Fig. 6.2** Captain Arve
Staxrud in uniform and with
medal, 1913. (*Photographer/
Byline*: Finne; *Source/
Owner*: Norwegian Polar
Institute)

Whereas Miethe requested Nansen's expertise and help to organize a relief expedition, Nansen in turn suggested Captain Arve Staxrud as expedition leader (Nøis, 1929, 6) (Fig. 6.2). This recommendation from Nansen would later be recounted by Daniel Nøis, who would accompany Staxrud on the rescue expedition as his next-in-command (Nøis, 1929, 6, 9; Hilmar Nøis Diary SVB).[2]

Staxrud was considered an expert on Spitsbergen as a result of his trailblazing surveying and mapping expeditions on the frozen archipelago in 1906, 1910, 1911, and 1912 (Committee to Norske Geografiske Selskap 18 March 1913; Hoel, 1933; Stenersen, 1988). He had worked on the very first surveying expedition with Captain Gunnar Isachsen in 1906, and also worked with lecturer/geologist Adolf Hoel, who would later state that Staxrud had helped pioneer Norwegian exploration on Svalbard. If Herbert Schröder-Stranz was of noble birth and born on an estate manor, Arve Staxrud came from more humble beginnings and a much more modest background. He was born on his family's farm, named Staxrud, in Brandbu, Norway, on September 6, 1881, and was one of ten children. Arve Staxrud took *Examen Artium* (acquired his academic certification) in 1898 and graduated from military school in 1901 at the head of his class, and then completed the exam at the Military College in 1905, after which he rose to the rank of Captain in 1911. The significant portion of his military service was with bicycle infantry units. During that time, in 1910, he briefly served with cyclist company units in divisions with the French Army, as a lieutenant. He apparently also studied in Germany, Switzerland, and Italy. Staxrud excelled in his studies and did very well in the military. But ever since his youth he also had a parallel passion for nature, outdoor activity, and the surveying of unfamiliar land, and thus was attracted to the Norges Geografiske Oppmåling (Geographical Survey of Norway), working in its topographical department in 1903–1905, and as permanent detailer in its geodetic department in 1906–1908, and doing triangulation work in several varying locations throughout the country in 1909. To prepare for his first surveying expedition on Spitsbergen in 1906, he went to Stockholm where he studied the photogrammetric survey method of Professor Gerard De Geer at *Stockholms högskola* (Stockholm University College) with Isachsen. It is reported that, as a result of his 1906 expedition, he was awarded the Knight Cross of the Order of Saint-Charles by Prince Albert I of Monaco. His second expedition in 1910 also resulted in detailed mapping of previously undocumented area. Serving as captain in the military in 1911, Staxrud also worked with Adolf Hoel to co-lead the state-sponsored scientific surveying expeditions to Spitsbergen beginning in 1911. These expeditions, later also reportedly co-sponsored by private contributions as well as the Nansen Fund, continued into 1912 and resulted in impressive mapping, topographical surveys, geological surveys, fossil findings recordings, and botanical studies. By the year 1913, Staxrud had also already gained experience leading a rescue expedition to Spitsbergen—this one in the autumn of 1911 to save several hunters/trappers who had traveled to the Arctic in 1909 intending to spend one winter, and who had subsequently been overwintering ever since and were stranded, with their ship not returning to retrieve them, and with no way for them to return home. Prior to his arriving at their wintering location in the eastern part of Spitsbergen, some had, sadly, already died, and others had been given transport on a sealing ship. Staxrud thus took action and made legal efforts to bring attention to this case and to the responsibilities that shipowners and shipping companies had to their overwintering hunters/trappers in the Arctic. He was later described by his surveying partner Hoel (in 1933) as having exceptional management capabilities as well as a quick sense of comprehension, as having an understanding of how to instill discipline and ideal relationships as well as camaraderie, and as having high productive energy, a positive good-humored outlook, efficiency, and good decision-making abilities, as well as being an excellent expedition leader. (Hoel, 1933; Stenersen, 1988; Committee to Norske Geografiske Selskap 18 March 1913).

And, so, this was the individual who had been recommended by Nansen to lead the rescue effort to find the Schröder-Stranz expeditioners.

Arve Staxrud met with Miethe and Nansen and Norway's Minister of Foreign Affairs in Kristiania on February 17, and wrote a letter that very same night to the Foreign Minister, Nils Claus Ihlen (1855–1925), regarding that evening's meeting, in which he summarized and reiterated, in no uncertain terms, his recommendations and approach for the rescue expedition that would be formed to save the Schröder-Stranz expedition (Staxrud 17 February 1913). In this letter, Staxrud spelled out, first and foremost, that he, as part of this effort, was "willing only to lead a Norwegian rescue-expedition (Norwegian vessel, Norwegian partakers)," and, second, that an inquiry must go out to the Norwegian minister in Berlin stipulating "that the Norwegian government—after conferring with experts—has planned a rescue-expedition."[3] To this end, he requested that the Norwegian minister in Berlin provide information, as soon as possible, as "to what extent Germany will support this expedition or send out [their] own." It would be based on this response that he would then be "able to hire a sealing ship" and travel northwards "as soon as possible," which is what he recommended, rather than find "an ordinary ice-sea vessel from Tromsø, which, due to its lesser strength for breaking through ice, will arrive later and maybe too late to bring" help to those who needed it in the dire elements. Staxrud indicated the five Norwegians working on the expedition who, in addition to the group of German scientists, also required aid, and urged that Norway do everything it could for them. He then delved into the topic of leadership for the rescue expedition, stating that the primary element for any relief expedition was to locate a competent expedition leader, and stressing the importance and the necessity of granting that leader full discretion for selecting the crew and expedition members and for acquiring the appropriate equipment for the expedition: "The first condition for, that a rescue-expedition shall reach [its] goal is, that one finds a suitable leader and that one leaves him to equip the entire expedition. He will then also have to take on the full responsibility for the expedition." Staxrud emphasized the disadvantage of dividing the responsibility, and mentioned a few examples of expeditions that had failed as a result of attempting to do so. "I have had to discuss this matter thus extensively in order to show the necessity that a leader must have freedom of choice of both his crew for [the] vessel as of the means, which the circumstances require [one] to apply in order to reach the goal." As to the route taken to and across Spitsbergen, he wrote, these would be determined by the conditions of the ice: "The committee's plan to push forward to the Northwest-coast as early as possible, I find particularly favorable," he wrote. "However, the expedition must not be attached to this route as the only [one]." He continued: "A sled-expedition from the mouth of the Ice Fjord will perhaps be able to get there faster. This will depend on the ice conditions north of the Ice Fjord." Similarly, the timing as to when to depart would depend on the ice and weather conditions, Staxrud advised—the end of March or beginning of April being the concluded target: "With regard to the time for [the] expedition's departure from Norway, I have in consultation with over-winterers as well as with Arctic Ocean skippers, come to the conclusion, that one must make an attempt late in March or early in April." In regard to the captain who would command the ship, Staxrud insisted that "Nobody can, under the conditions that apply, measure up to Arctic Ocean skippers." And as regards the ship itself, "As vessel one should preferably have a sealer ship." Summarizing what at that time was known regarding the Schröder-Stranz expedition members' locations at and around Treurenberg Bay, Staxrud also proposed preliminary plans and possible routes for the sledging journey across the ice, referencing the difficulty of the ice conditions and the dependency on local knowledge for some of the glacier areas, and apparently enclosing a map which must have accompanied the actual letter. He also outlined his requirements for crew, vessel, and equipment, mentioning a seal-hunting vessel coming from Sandefjord and some of the equipment being provided in Tromsø, and recommending the following: "On the vessel travel 2 land-parties of 4 men equipped for longer sled-journeys.... Each land-party brings 2 sleds and dogs and has to consist of experienced and locally well-acquainted people." Thus, the indication here is that a crew and expedition members who had experience on Spitsbergen and who maintained knowledge of the local area, such as those who overwinter on the archipelago, were essential and a necessary component for the expedition and should be recruited, as should sled dogs for the sledging journey. Also the power of a sealing ship would be needed. Staxrud ended the letter with the urgent conclusion and call-to-action: "If an expedition is to leave for Tromsø at the end of March, one already has to get started with the preparations for the expedition."

And so a leader with sole authority, a crew of local experts, sled dogs with sledges, and a sealing vessel were the necessary components preferred and reiterated by Staxrud in his letter to the Foreign Ministry.

Following the writing of this letter by Staxrud, another meeting took place involving Miethe and Nansen and others on February 21, according to the Tromsø newspaper (*Nord Norge* 21 February 1913b, 2), which stated that the final decision about the relief mission would be made the following day. The paper also reported that Nansen recommended that reindeer be used as part of the rescue effort, and that Miethe agreed to the recommendation. In addition, the article reported that recent storms in Spitsbergen had prevented the launching of any other relief expeditions, and that, once the weather had improved, a third attempt from Advent Bay to reach Wijde Bay would be made.

Indeed, it seems that the people in Advent Bay and Green Harbour were brain-storming about how to reach the stranded expeditioners once the current storms abated. Telegraph station manager Olaf Henriksen, in his diary entry of February 18, wrote that the plan was to take their last two dogs who had recently given birth to puppies, along with the remaining equipment at Green Harbour, to Advent Bay so as to enable and equip two men to travel northwards in order to search and investigate (Olaf Henriksen Diary).[4] And on the following day, Henriksen reported that a suggestion had been made to send a flying expedition to Wijde Bay.[5] Such was the fervor to reach the stranded expeditioners.

## The Selection of Arve Staxrud to Lead the Norwegian Rescue Expedition

Fridtjof Nansen's proposal of Arve Staxrud to lead the relief mission, and Staxrud's own recommendations given to the Foreign Minister and Committee regarding how to organize a proper rescue expedition, resulted in his being selected as leader of the official rescue effort.

On February 23, 1913, the negotiations between the Berlin Relief Committee and the Norwegians were concluded, and their results were announced at a press conference shortly before Miethe's departure for Berlin (Hamburger Nachrichten, 1913a). His trip to Norway had been of a private nature and the German government had nothing to do with the official German relief expedition and therefore did not provide any support. The relief committee he represented, which included many distinguished men such as Count Zeppelin (chairman), Professor Hergesell, and court marschal von Breitenbuch, was entirely private. This is what Miethe literally said about the Schröder-Stranz endeavor: "Unfortunately, the expedition did not do honor to the German name. It is now apparent that the expedition was extremely poorly prepared and that none of the leaders had any idea of what wintering on Spitsbergen entailed! They acted completely in the blind. That is why their situation is now so desperate, if they are still alive at all" (Hamburger Nachrichten, 1913a). The three (as far as they knew at that time) Norwegians on board the *Herzog Ernst* would probably be safe, they thought, while Schröder-Stranz and his companions would probably have perished. Originally, the relief committee would have liked to have had a German to join the relief expedition as one of the leaders, but the Norwegian Foreign Ministry favored a purely Norwegian expedition led by Arve Staxrud (Fig. 6.3).

Hans Gazert (1870–1961), the physician of the German South Polar expedition (1901–1903) led by Drygalski approximately ten years earlier, was very upset about this Norwegian relief effort. In his opinion, it was "a slap in the face of every German polar traveller" (Lüdecke, 1995, 45). He felt personally affected, since he had unsuccessfully applied to participate as a physician with Polar experience. Norwegians were preferred to Germans only on the basis of recommendations (Gazert 24 February 1913). He and his friend and geographer Ludwig Distel (born 1874), who were both

**Fig. 6.3** Captain Staxrud (*Source* Der Zeitspiegel 1913)

members of the Munich Academic Alpine Club and had made many first ascents in the Alps, would have had enough competent people to proceed purposefully, he thought. However, the rescue company became Norwegian.

Thus, in Germany there were those who took umbrage at the fact that the rescue expedition would be led by a Norwegian and would not include any Germans. But the Germans were not the only ones to complain about the selection of the personnel. Criticism came from some in Norway itself—from Northern Norway, to be precise, and specifically from Tromsø (*Nord Norge* 21 February 1913c, 2 Editorial). In an editorial published in the same Tromsø newspaper on February 21, regarding which Norwegians to select and which dates to depart for the rescue expedition, a question is presented whether those several individuals from Kristiania who spend a few summers in Spitsbergen—July and August—and who are perceived as the best experts, have the necessary experience as those who have sailed there during the heaviest ice conditions from April to September. Stated twice in the article is the contention that the present time of year was too early to dispatch a rescue expedition, and that the expedition should not leave prior to April. This article, too, would catch Staxrud's attention.

Indeed, according to Daniel Nøis, who would later write a first-hand account of the expedition, at this time a stream of evaluations, analyses, and critiques of both the personnel and their chosen methods and equipment came from several quarters, and even the venerable Nansen was not spared (Nøis, 1929, 6).

## Notes on Original Material and Unpublished Sources

Olaf Henriksen's diary is in the Library Archives at the Norwegian Polar Institute (NPI) in Tromsø. Hilmar Nøis's diary SVB is in the Historical Archive at the Svalbard Museum (SVB) in Svalbard. Arve Staxrud letters of correspondence, written from and to Arve Staxrud, are in the Library Archives at the Norwegian Polar Institute (NPI) in Tromsø.

1.  O. Henriksen diary, 13 February 1913, (NPI), D-305/D00305_0001
2.  H. Nøis diary SVB, 1913, journal page 48, SVB-AP2
3.  A. Staxrud to Norway's Minister of Foreign Affairs Nils Claus Ihlen, letter, 17 February 1913, Norwegian Polar Institute (NPI), 1913_02_17_Redningsexpeditionen_Schroder_St[r]anz_1913_Hoels_og_Staxruds_Spitsbergenexpedisjoner
4.  O. Henriksen diary, 18 February 1913, (NPI), D-305/D00305_0001
5.  O. Henriksen diary, 19 February 1913, (NPI), D-305/D00305_0001

## Unpublished Sources and Original Material

Committee Letter to Norske Geografiske Selskap, 18 March 1913, Letter from Committee to Norske Geografiske Selskap board, 1913_03_18_Brev_til_ Norske_Geografiske_Selskap, Library Archives, Norwegian Polar Institute, Tromsø. Received 20 July 2021.

Gazert, H., (24 February 1913). Letter to Drygalski. Gazert estate, Volkert Gazert, Partenkirchen.

Henriksen, O., Diary, Olaf Henriksen's diary written while he was the Green Harbour Telegraph Station Manager, "Dagbok for Olaf Henriksen, 1911–13" ["Diary for Olaf Henriksen, 1911–13"], dated 14 August 1911 to 2 May 1913, D-305, D00305_0001. Library Archives, Norwegian Polar Institute, Tromsø, https://brage.npolar.no/npo lar-xmlui/handle/11250/2600678. Accessed 24 July 2021.

Nøis, H., Diary SVB, handwritten journal of recollections including an account of the 1913 Staxrud expedition, ("Hilmar Nøis. Fangstmann. 1909–1923, dagbok med erindringer fra fangstlivet") ["Hilmar Nøis. Trapper. 1909–1923, diary with recollections from the trapping life"], not dated (latest year referenced is 1942 in journal pages and 1950 in inserted sheet), SVB-AP2-Hilmar-Nøis-dagbok-1909–1923-compressed. Historical Archive, Svalbard Museum, Svalbard, https://svalbardmuseum.no/no/samlingene/historisk-arkiv/arkivprosjekt-fangst-og-annen-overvintringsvirksomhet-fra-perioden-1910-til-1970/. Accessed 6 June 2021.

Notruf, (1913). Notruf zur Rettung der Schroeder-Stranz'schen Spitzbergen Expedition. Schröder-Stranz estate, Kasten 26, Nr. 76, Leibniz Institute for Regional Geography, Leipzig, Germany.

Staxrud, A., 17 February 1913, Letter from Captain Arve Staxrud to Norway's Minister of Foreign Affairs Nils Claus Ihlen regarding plans for rescue expedition after Schröder-Stranz, with notes, "Redningsekspeditionen for Schröder-Stranz 1913" ["The Rescue Expedition for Schröder-Stranz 1913"], copybook and notes, 1913_02_17_Redningsexpeditionen_Schroder_St[r]anz_1913_Hoels_ og_Staxruds_Spitsbergenexpedisjoner, Library Archives, Norwegian Polar Institute, Tromsø. Received 20 July 2021.

## References

Der Zeitspiegel Berlin im Frühjahr. (1913). *Newspaper clipping*. Ritscher Estate, Cornelia Lüdecke, München.

Hamburger Nachrichten. (1913a). Professor Miethe über die Hilfsexpedition nach Spitzbergen. *Hamburger Nachrichten*, February 25, 1913.

Hamburger Nachrichten. (1913b). Notruf für die Spitzbergen-Rettungs-Expedition. *Hamburger Nachrichten*, March 11, 1913.

Hoel, A. (1933). Arve Staxrud. *Norsk Geografisk Tidsskrift [Norwegian Journal of Geography]*, *4*(6), 345–346. https://doi.org/10.1080/00291953308622018

Lüdecke, C. (1995). Die deutsche Polarforschung seit der Jahrhundertwende und der Einfluß Erich von Drygalskis. PhD Thesis. *Berichte zur Polarforschung* 158. Bremerhaven: Alfred-Wegener-Institut für Polar- und Meeresforschung.

Miethe, A. (Ed.). (1914). *Die Expedition zur Rettung von Schröder-Stranz und seinen Begleitern— Geschildert von ihren Führern Hauptmann A. Staxrud und Dr. K. Wegener*. Berlin: Dietrich Reimer.

Nøis, D. (1929). Med kaptein Staxrud på leiting efter Schrøder-Stranz og hans folk [With Captain Staxrud on the search for Schröder-Stranz and his people]. In: *And-Ungen*, July 1929, pp. 4–12. Andenes: Andøyposten. Library Archives, Norwegian Polar Institute, Tromsø. Received July 20, 2021.

Nord Norge. (21 February 1913a). Spitsbergen-ekspeditionen. En oversigt. [The Spitsbergen expedition. An overview]. *Nord Norge* (Tromsø), 21 February 1913, p. 1. (Newspaper clipping attached to letter from Arve Staxrud to Fridtjof Nansen dated 17 March 1913, NB Ms.fol. 1924:6:A:2). National Library of Norway, Oslo, online archives, https://www.nb.no. Retrieved July 6, 2021.

Nord Norge. (21 February 1913b). Undsætnings-ekspeditionen. Nansen anbefaler rensdyr. [The Rescue expedition. Nansen recommends reindeer]. *Nord Norge* (Tromsø), 21 February 1913, p. 2. (Newspaper clipping attached to letter from Arve Staxrud to Fridtjof Nansen dated 17 March 1913, NB Ms.fol. 1924:6:A:2). National Library of Norway, Oslo, online archives, https://www.nb.no. Retrieved July 6, 2021.

Nord Norge. (21 February 1913c). Mye skrik og litet uld sa manden da han klipte grisen. Editorial by H. C. Johannesen. *Nord Norge* (Tromsø), 21 February 1913, p. 2. (Newspaper clipping attached to letter from Arve Staxrud to Fridtjof Nansen dated 17 March 1913, NB Ms.fol. 1924:6:A:2). National Library of Norway, Oslo, online archives, https://www.nb.no. Retrieved July 6, 2021.

Rüdiger, H. (1913). *Die Sorge Bay. Aus den Schicksalstagen der Schröder-Stranz-Expedition*. Berlin: Georg Reimer.

Steinhagen, H. (2008). *Max Robitzsch. Polarforscher und Meteorologe*. Jakobsdorf: Die Furt.

Stenersen, H. (Ed./Publisher). (1988). Tingelstad-brødrene som kartla Svalbard: Arve og Olav Staxrud har satt varige spor etter seg. In *Brandbu'stikka: Lokalhistorisk Kvartalskrift for Brandbu. 1. KV. 1988 Mars NR. 11 (March 1988)*. (Svalbard-forskerne Arve og Olav Staxrud) (pp. 1–26). Biography Archive/Library Archives, Norwegian Polar Institute, Tromsø. Received July 20, 2021.

Wegener, K. (1913). Die Hilfsexpedition von Cross- und Kings-Bay nach Wijde-Bay. *Petermanns Geographische Mitteilungen, 59*, 137–140.

Wegener, K. (1914). Die Hilfsexpedition von Cross- und Kings Bay, 21.II.–31.III.1913. In A. Miethe (Ed.), *Die Expeditionen zur Rettung von Schröder-Stranz und seinen Begleitern geschildert von ihren Führern Hauptmann A. Staxrud und Dr. K. Wegener* (pp. 69–101). Berlin: Dietrich Reimer.

# Chapter 7
# The Enlisting of the Amundsens, and the Plight of the Stranded Expeditioners

## Roald Amundsen's and Leon Amundsen's South Pole Expedition Dogs: Lussi, Storm, and Obersten

Upon being selected as the leader of the Rescue Expedition to Spitsbergen, Arve Staxrud immediately began gathering the people, animals, equipment, and vessel to perform the rescue expedition as he had outlined in his letter to the Foreign Minister on February 17, 1913 (Staxrud 17 February 1913). This included securing the seal-hunting vessel from Sandefjord he had mentioned, which would turn out to be the *Hertha*—the ship he hired for the rescue expedition. Staxrud formally requested sled dogs, supplies, provisions, and personnel.

In his report about the rescue expedition that was published in Germany in 1914 in Adolf Miethe's book, Staxrud would later write that, based on Alfred Ritscher's report, Staxrud was to bring help to the men left behind in the huts of Wijde Bay and Treurenberg Bay and then take further action to rescue Schröder-Stranz's missing sledge expedition (Staxrud, 1914, 4–31). Before the relief expedition could come to the rescue, however, it was first necessary to clarify how best to reach Spitsbergen as early as possible and then to get overland to Wijde Bay with food and equipment on sledges (as Staxrud had outlined in his letter to the Foreign Minister). For the crossing of the Arctic Ocean to Spitsbergen, the ice-going whaler *Hertha* was chartered in Sandefjord.

For the sledging journey across the Spitsbergen ice, sled dogs would be absolutely necessary.

An effort to enlist the aid of the Amundsens—Polar explorer Roald Amundsen (1872–1928) and his brother and business manager Leon Amundsen (1870–1934)—now ensued. Roald Amundsen had recently returned from becoming the first person to reach the geographic South Pole, which he had accomplished with the help and through the work and efforts of his 116 sled dogs during the Norwegian Antarctic Expedition of 1910–1912 (Tahan, 2019, 2021). The attainment of the South Pole in December 1911 was the culmination of those efforts. He had left Antarctica in

M. R. Tahan and C. Lüdecke, *Stranded at the Top of the World*, https://doi.org/10.1007/978-3-031-56288-4_7

January 1912 with 39 surviving sled dogs—21 of whom temporarily remained in Australia to then work on Douglas Mawson's Antarctic expedition, and 18 of whom traveled on with the *Fram* to Argentina (Tahan, 2019, 595–606; Tahan, 2021, 1, 54, 139, 142–143). Amundsen himself had arrived in Kristiania from Buenos Aires in July of 1912 and now, as of February 1913, was on his lecture tour currently giving talks and presenting his lantern slides in the United States of America (Tahan, 2021, 234, 295) (Fig. 7.1).

The three surviving sled dogs from the 18 who had traveled on the *Fram* from Antarctica to Australia to Buenos Aires had just returned to Norway exactly one week prior to Staxrud's letter to the Foreign Minister; they had sailed into the Kristiania Harbour on February 10, 1913—these three famous, heroic dogs were Obersten (meaning "The Colonel"), Lussi (also spelled Lucie and Lucy), and Storm (Tahan, 2019, 595, 603–609; Tahan, 2021, 279–280, 291–293).

Obersten had been brought from his native Greenland as part of the community of 100 Greenland dogs purchased by Amundsen in September 1909 and transported to Norway in June/July 1910 (Tahan, 2019, 18–19, 53–54). Lussi was born on the *Fram* en route to Antarctica, to her mother Lucy (also sometimes spelled Lussi), on December 28, 1910, and was the only female puppy—of all the female puppies born on the ship—who was allowed to live and who was not purposely thrown overboard

**Fig. 7.1**  Portrait of the explorer Roald Amundsen in 1913. Studio portraiture by Anders Beer Wilse (*Photographer* Anders Beer Wilse; *Source/Owner* National Library of Norway)

into the sea (Tahan, 2019, 182–183, 259–260, 604). Storm was born during the voyage south to Antarctica as well, to his mother Else, during a terrific storm on November 11, 1910 (Tahan, 2019, 158). Both the mothers Lucy and Else had worked on the actual trek to the South Pole (which had begun with 52 dogs) and had been intentionally killed by Amundsen and his men on their way to the Pole (Tahan, 2019, 387, 405–407).

Oversten was the only surviving dog of the 17 who had been to the South Pole and of the 11 who had returned from it (Tahan, 2019, 587). While Oversten had served on the South Pole trek, working as part of Oscar Wisting's sledge team, Storm had worked on the King Edward VII Land Eastern Expedition, helping to pull Jørgen Stubberud's sled, and Lussi had worked at the base camp Framheim, during Adolf Lindstrøm's sled excursions around the winter camp; all three had worked valiantly and accomplished sledging journeys during the expedition in Antarctica (Tahan, 2019, 560–561, 588) (Fig. 7.2).

Personality-wise, all three dogs were unique, loyal, and dynamic: Oversten was as independent as he was strong, having attempted an energetic long-distance swimming escape soon after arriving in Norway, and had assumed diligent leadership responsibilities in his team during the sledging trek to the South Pole; Lussi was

**Fig. 7.2** Persevering sled dogs at the South Pole, December 15 (14), 1911, during Amundsen's Antarctic expedition. Oscar Wisting's team is seen here, with Oversten most likely on the right (Tahan, 2019, 497, 507). Wisting stands near the sledge (*Photographer* unidentified; *Source/Owner* National Library of Norway)

**Fig. 7.3** The South Pole dog Obersten ("The Colonel"), ready for his photography session (*Photographer* Anders Beer Wilse; *Source/Owner*: Norsk Folkemuseum/National Library of Norway)

active, adventurous, curious, and lively, embarking on long-distance excursions with her mother in Antarctica, and engaging in vivacious frolicking with her fellow living creatures on the ship; and Storm seemed steadfast, reliable, and enthusiastic, successfully working on the Antarctic discovery journeys and later helping to bring a sense of the expedition to interested Norwegians (Tahan, 2019, 2021).

Physically, Obersten was a large and powerful dog with a reddish-brown coat; Lussi was a good-sized steel-grey dog with a pointed head and intelligent, animated eyes; and Storm was a strong dog with proven pulling power and stamina (Tahan, 2019, 2021) (Fig. 7.3).

Upon their arrival in Kristiania, the three returning South Pole Expedition sled dogs had immediately been taken to the veterinarian appointed by Leon Amundsen—Bernt Anker Nielsen—who housed them for a few days at his location along Rosenkrantzgaten (Rosenkrantz' gate) near Karl Johans gate, where they received medical care, cleaning, and bathing, and they had also been assigned their old mates from the *Fram* to look after them (Tahan, 2021, 279–280, 291–293). The Antarctic expedition members—their former crewmates from the *Fram*—appointed to take care of them were Oscar Wisting, Jørgen Stubberud, and Martin Rønne, with the occasional help of Thorvald Nilsen. The sled dogs were then brought to Roald Amundsen's home, "Uranienborg," located in Svartskog, along the edge of the Bundefjord (Bunnefjord) in the idyllic setting of Baalerud (Bålerud), where they arrived on February 13th, (Tahan, 2019, 609; Tahan, 2021, 291–293) (Fig. 7.4).

The three dogs had begun to attract attention from their very first appearance in Norway. After disembarking at Kristiania Harbour, they had been followed by a veritable parade of children and adults who had become fascinated with the three

**Fig. 7.4** Roald Amundsen's home "Uranienborg," located in Svartskog, near Oslo, Norway, along the Bundefjord, where Lussi, Storm, and Obersten came to live after their return from the South Pole expedition (*Photograph* by Mary R. Tahan)

impressive dogs and had accompanied them to the veterinarian's clinic (Tahan, 2021, 279–280, 291–293).

Leon kept his brother Roald apprised of the dogs' wellbeing through frequent and consistent correspondence. He reported on their previous departure from Buenos Aires and the hope that they would arrive alive in Norway, on February 6. He confirmed their safe arrival in Kristiania and updated Roald as to their assigned caretakers, on February 10. And he updated Roald as to their arrival at his home, on February 13. Leon specifically recommended mating Obersten and Lussi and breeding the three dogs—efforts which evidently indeed ensued. (There was also a goal to have a small workforce of sled dogs ready for Amundsen's next voyage, which at this time was planned for the North Pole). Leon also posed questions to his brother regarding his plans for the dogs, even prior to their arrival, suggesting to Roald that, in addition to breeding, the sled dogs be utilized for business, including promotions and income revenue, and suggesting possibly selling them at a good price (Tahan, 2021, 292–294). To all this, Roald—who had taken great pains to bring his last three surviving Antarctic expedition dogs home from Argentina—would reply that he wished to keep the three dogs together and with him at his home (Tahan, 2021, 295) (Fig. 7.5).

Indeed, the three sled dogs had survived much together and by now seemed inseparable (Tahan, 2021, 297). They had survived the turbulent voyage to and from Antarctica, as well as the grueling expedition on the Antarctic continent. They had

**Fig. 7.5** Leon Amundsen (top center) and his brother Roald Amundsen (to Leon's left), as well as their brother Gustav (below him, to Leon's right) seated on the outside steps at Roald's home in Svartskog (*Photographer* unidentified; *Source/Owner* National Library of Norway)

survived a deadly illness at the Buenos Aires Zoo that had sadly taken the lives of their compatriots—the remainder of the 18 expedition dogs who had traveled to Argentina had all died tragically. And they had survived the long voyage north from Argentina to Norway (Tahan, 2019, 2021).

And, now, it was at this time that the dogs were called upon to travel to the Arctic to save the lives of the humans who were iced-in on Spitsbergen and to rescue the stranded members of the Schröder-Stranz expedition (Tahan, 2021, 297).

Their involvement began on the same day that Staxrud wrote his letter to the Foreign Minister following his meeting with Nansen and Miethe and the committee. On this same day of February 17, Leon Amundsen was writing a letter to Roald Amundsen in which he announced that a "similar incident" to the Robert Falcon Scott tragedy in Antarctica was now occurring in the Arctic in that the Schröder-Stranz expedition was in North Spitsbergen with little provisions and equipment, and that one rescue expedition had already failed and another one would now be dispatched (Amundsen Letters of Correspondence). "But to get hold of dogs and experienced people in a hurry is not easy," he stated.[1] Leon informed Roald that he had "offered the German minister my [his] assistance" including "Prepared supplies and pemmican" which could be borrowed and presumably his help to locate "one or two dog-drivers like [Sverre] Hassel or [Olav] Bjaaland or if need be Helmer [Hanssen]"—[all of whom were Antarctic expeditioners who had been to the South

Pole with Amundsen (Tahan, 2019)]. Leon also recommended Kristian Prestrud—another Antarctic expeditioner who had been appointed to lead the King Edward VII Land expedition (on which Hjalmar Johansen had served)—as a good candidate for expedition leader, and said, if he were not chosen, then Adolf Hoel was another name that had evidently been suggested. "I doubt it will be a success to rescue the expedition but everything must of course be done," he wrote, concluding that "Dogs will probably have to be taken from Archangel by railway via [Saint] Petersburg."[1]

Thus, Leon's news of the stranded German expedition came with the fear that an attempted rescue may have little result. There was particular sensitivity at this time, given the recent news of the tragic death of Robert Falcon Scott and his men during their return from the South Pole—news which had been disseminated upon the return of Scott's British Antarctic expedition ship *Terra Nova* on February 10th—the very same day that the three sled dogs had returned to Norway (Tahan, 2021, 280–284). Amundsen had preceded Scott to the Pole through the help of the Norwegian explorer's sled dogs, and Amundsen had also been criticized for his secrecy in going to Antarctica and for his rivalry with Scott (Tahan, 2019, 626–629, 133–135; Tahan, 2021,1–3, 5–7, 231–234). And so Leon's news of the stranded Schröder-Stranz expedition came with the analysis that the situation called for dogs which were not readily available—at this time, obtaining dogs from Archangel (Arkhangelsk, Russia) seemed to be the most viable option. As for dog-sledge drivers and expedition leaders, Amundsen's men were top-of-mind for Leon.[1]

A few days later, on February 20, Leon again wrote to Roald with an update on the Spitsbergen situation, saying that conferences were being held every day at the Foreign Ministry involving the German minister, Nansen, Hoel, Staxrud, and another German individual who had traveled to Kristiania (most likely he was referring to Miethe) in order to enable a rescue expedition to depart as soon as possible (Amundsen Letters of Corresponrnce).[2] He also reported that a telegram sent from Tromsø stated that Helmer Hanssen (who was in Tromsø) was willing to lead the rescue expedition provided he receive "permission" from Roald Amundsen "but he does not want Hoel or Staxrud so that is not likely to come to anything."[2] Leon thus seemed to imply that, given Hanssen's desire that Hoel and Staxrud not partake in the rescue efforts, and given the unlikeliness of their not doing so, Hanssen probably would not lead the expedition. Roald replied to Leon with a note written at the bottom of the very same letter: "Pity it is, that one did not let Helmer handle Spitsbergen."[3]

It is important to note here that Helmer Hanssen (1870–1956) was one of Roald Amundsen's two expert dog-sledge drivers, who had accompanied Amundsen to the South Pole, and who had been a loyal ally, supporter, and defender of Amundsen (Tahan, 2019). During the Antarctic expedition, according to Hanssen himself, Hanssen had gone so far as to pause the progress of his lead sledge dog team as they approached the location of the South Pole in order to ensure that Amundsen would ski ahead to the front of the caravan and would be the first to touch foot at the Pole (Tahan, 2019, 494) (Fig. 7.6).

Two days after Leon's letter about Hanssen, on February 22nd, Staxrud had been officially selected, although the announcement was not made publicly yet, and Leon

**Fig. 7.6** The five men who reached the South Pole with the sled dogs (some of the dogs are seen here behind the men), photographed on the deck of the *Fram* after arriving in Hobart, Tasmania, Australia, following the Antarctic expedition. Pictured left to right are Sverre Hassel, Oscar Wisting, Roald Amundsen, Olav Bjaaland, and Helmer Hanssen (*Photographer* John Watt Beattie; *Source/ Owner* National Library of Norway)

wrote to Roald with this news, as well as with the announcement that "The Dogs"— presumably the three South Pole expedition dogs—were very much needed (the request to Leon most likely had been made by Staxrud with Nansen's direct input): "It will be <u>Staxrud</u> who will lead the expedition and he would have liked to have Helmer and Hundene ["The Dogs"]; had Gubben ["The Old Man"—Nansen] had time he would have taken lead of the business himself. We are going to lend them some food and equipment" (Amundsen Letters of Correspondence).[4] Leon also mentioned in the same letter that it was probable more of the German expeditioners may have been lost. And he alerted Roald to the fact that he had sent a telegram to him that very day regarding this Spitsbergen matter.

On the following day, February 23rd, presumably in response to Leon's telegram to him, Roald sent a telegram to Leon, from Syracuse, New York, with the following direct and simple reply: "Yes Helmer command the dogs."[5] Thus confirmation was given from Roald to Leon that Helmer Hanssen was authorized to take the dogs for the rescue mission—indicating, therefore, authorization both for the sled dogs to work on the rescue expedition and for Hanssen to serve as their dog teamer.

Leon Amundsen, presumably after hearing from Staxrud, quickly sent a telegram to Helmer Hanssen that same day of February 23, from Kristiania to Tromsø, stating that "Staxrud is asking You most kindly to send him telegram response directly."[6] Hanssen in turn promptly sent a telegram back to Leon that day expressing to him in no uncertain terms that under no circumstances was he, Hanssen, going to partake in a rescue expedition led by Staxrud, and furthermore, he went so far as to say, he strongly advised that the sled dogs not be loaned to the expedition, either.[7] He would follow-up this telegram with a letter, he indicated, which would lay out the reasoning for his decision.

Evidently, the respected dog-sledge driver and South Pole veteran had a view of Staxrud that varied from Nansen's in regard to Staxrud's ability as an expedition leader, and that affected Hanssen's own potential role of working with the sled dogs who would be instrumental in the rescue of the German expedition. Thus, a controversy began to brew among some of the Amundsen group over the selection of Staxrud as leader of the rescue expedition, as seen through the subsequent series of letters of correspondence between Leon, Roald, Hanssen, and Staxrud.

## The Helmer Hanssen Controversy: Criticism from a Respected South Pole Explorer

Helmer Hanssen wasted no time—and withheld few descriptive words—in writing a letter to Leon Amundsen that fully detailed his justification, in his mind, for refusing to work on the rescue expedition with Arve Staxrud, even if that refusal meant that he would not be able to aid in a rescue mission and would not be there to accompany the three sled dogs who were requested to work on the Staxrud expedition (Amundsen Letters of Correspondence). The two-page letter to Leon, dated February 24, unveils a deep dislike and severe criticism of Staxrud by Hanssen, who spells out in great detail the specific source of his profound aversion to Staxrud.[8] Hanssen states in his letter that he had received a telegram from Staxrud that very day—February 24—informing Hanssen that Leon had given Staxrud permission for Hanssen to participate in the Spitsbergen rescue expedition, and that the three sled dogs (Obersten, Lussi, and Storm) would be loaned to Staxrud with the provision that Hanssen himself would take responsibility for the three sled dogs. Hanssen, in the letter, expresses his utter bewilderment as to how Staxrud could be under this impression, and states his unwillingness to send a reply to Staxrud. Regarding the telegram that had been sent from Tromsø, which Hanssen's letter reveals originated from Zapffe—most likely Fritz G. Zapffe (1869–1956), (the Tromsø pharmacist, expedition agent, and friend of Amundsen [Tahan, 2019, 20, 496–497; Tahan, 2021, 232])—and which indicated Hanssen's willingness to go on the rescue expedition if Roald Amundsen did not object, Hanssen points out that the telegram also specifically said he would not do so if it meant working with Staxrud, whom he describes as completely incapable as a leader. The fact that Staxrud continued with these plans to include Hanssen proved his impertinence, claims Hanssen in the letter, opining that had the people in Kristiania been as familiar with Staxrud and his qualifications as

the people of Tromsø were, then they would have understood that he was not qualified to take part in this mission. Hanssen adds further, in what he states is his effort
to clarify his position as being justified in the eyes of Leon and of Fridtjof Nansen,
whom he respected, that the reason he had disassociated himself from Staxrud was
because Staxrud had publicly criticized and verbally demeaned Roald Amundsen, in
Hanssen's very presence, calling Amundsen's Polar work unserious, saying that his
concealing information from his crew (such as when he secretly went to Antarctica)
was his usual method of operation, and questioning if Amundsen had even truly
reached the South Pole. Staxrud, according to Hanssen, had summed up his disdain
of Amundsen by describing him as one who was under the influence of delusions of
grandeur. It was these utterances about Amundsen—the man whom Hanssen held
above everyone else, says Hanssen in the letter—that caused Hanssen to completely
detest Staxrud with all his being. Although most people, indicates Hanssen, would
wave away these utterances as ridiculous, he did not want there to be any possibility
of Amundsen hearing them or feeling slighted or insulted by them, and so he asks
Leon, in the letter, not to repeat to Roald what he had said. As for Nansen, however,
Hanssen indicates his desire that Nansen know and come to realize the true type of
*wretch* he was holding close to himself, as it appeared to Hanssen that Nansen had
taken Staxrud under his wing. And so, states Hanssen in his conclusion to the letter,
he wanted Leon to comprehend why, despite the fact that this was a rescue mission
of this sort, Hanssen would not assist in what he alludes to as a cleaning up of a mess
on behalf of Staxrud.

From a reading of this letter, it seems that Hanssen believed that Staxrud was
under the wrong impression—and had publicized this wrong impression—that Roald
Amundsen overestimated his own greatness and importance. According to Hanssen,
this fact, combined with his own absolute loyalty to his South Pole commander
Amundsen, and combined with his assessment and assertion that Staxrud lacked
qualifications, caused Hanssen to conclude that he could not possibly involve himself
in Staxrud's expedition, even if it meant refusing to join a rescue mission of this
magnitude and importance. Furthermore, while wanting to spare Amundsen's feelings, Hanssen seems to have desired to inform Nansen of his belief that Nansen was
unwittingly fostering what Hanssen deemed to be a fraudster.

Most certainly it appears that Hanssen's opinion of Staxrud as offensive and
incapable greatly differed from that of Fridtjof Nansen, who had proposed Staxrud
as leader of the expedition, and who had engaged in comradery previously with him
in Spitsbergen, as had been observed by Olaf Henriksen and recorded in his diary
the prior year (Olaf Henriksen Diary).[9] It also seems that Hanssen's perception of
Staxrud conflicted with that of Adolf Hoel, who worked with Staxrud and who would
later praise Staxrud for what he described as his superb ability to lead and his ample
qualities of instilling discipline and teamwork, making decisions, and employing his
wits and good humor (Hoel, 1933).

It also appears that, while Roald Amundsen preferred that his trusted South Pole
expedition member Helmer Hanssen lead the rescue endeavor, as he had stated in the
previous correspondence, Staxrud—as the officially selected leader—had requested

that Hanssen only assist him, especially with the dogs, and had received no response from Hanssen, who did not deign to reply.

In addition, it is evident that Hanssen's intense loathing and questioning of Staxrud had even spurred him to recommend that the three sled dogs not be sent to work on the rescue expedition with him, as was Hanssen's advice in his telegram to Leon.

Hanssen had shown loyalty to Roald Amundsen throughout the South Pole expedition and consistently after its conclusion, continuing up through this very moment (Tahan, 2019, 2021). The usually even-keeled, calm, and steadfast sledge driver now really seemed to have become infuriated by Arve Staxrud and was leaping to Amundsen's defense. Ironically, a rare disagreement had happened in Antarctica between Hanssen and Amundsen, resulting in Amundsen's temporarily refusing to speak with Hanssen, and the source of that disagreement had been none other than Storm's mother Else (Tahan, 2019, 573–574). Now Storm was yet again in the eye of a storm of controversy.

And Leon found himself in a very delicate situation.

## Hardship on the Ship: The Travails of the Men Onboard the *Herzog Ernst*

While all the machinations of controversy continued to swirl in Germany and Norway, and sincere efforts of rescue diligently began to form, sadly, the stranded men of the Schröder-Stranz expedition—residing on the *Herzog Ernst* ship frozen-in at Treurenberg Bay—were suffering.

Einar Rotvold made a steady and faithful attempt to document the events unfolding around him and the daily hardships experienced on the ship, in his expedition diary, describing the men's travails as both dangerous and heartbreaking to them, and detailing especially the poor and declining state of health in which they found the cook Knut Stave—a sad predicament for which Rotvold blamed the expedition organizers (Einar Rotvold Diary). Rotvold's documentation is punctuated by a running commentary on the depressing environment and the sad and long days in which the men found themselves.

Einar Rotvold, Knut Stave, Hermann Rüdiger, Christopher Rave, August Stenersen, and Julius and Jørgen Jensen were surviving as best they could during the month of February, cooking hot meals, gathering and chopping wood for fires, melting snow for water, hunting animals for food, and attempting to stay warm in the freezing temperatures. While the two German expeditioners—the artist Rave and the injured geographer Rüdiger—remained in their aft cabin, the Norwegian crewmen—Rotvold, Stenersen, the two Jensens, and the ailing Stave—worked and resided in the forward section of the ship.

February began with "terrible weather" that did not allow a person "to go up onto the deck, except for necessities, and [then it] was nearly enough to lose your breath," wrote Rotvold in his diary entry of February 1st.[10] The result, as described by Rotvold, was confinement below deck that created a circumstance of isolation and a craving for normalcy: "... here must be sad for us to sit closed down in a small cabin. Ohhh, for a life!".

He recorded their dinner meal on that day as consisting of "stockfish and rice-soup," and ended his diary entry with his usual sign-off, "God with us. All well." Perhaps his signature statement was also a combination of faithfulness, sadness, and optimistic hope that all would indeed turn out well.

On February 2nd, Rotvold reported that "Knut [Stave] says he is better in the chest, does not cough as much, but the hip and the foot are worse now."[11] With hopes that the weather would improve[12], and braving the "Severe cold" to go onshore where they "fetched driftwood and snow for water-thawing," the men endured the "Sad days."[13] A trip taken by the brothers Julius and Jørgen Jensen from the ship to the animal traps on February 6th resulted in their returning empty-handed.[14] On the next day, Rotvold noted in his diary the lack of variety in the men's diet due to the minimal provisions. He also observed that "Knut [Stave] is still keeping himself in the bunk, every now and then up a bit."[15]

The change in diet for the men occurred on February 10th. "At midday one got sight of 3 bears [a] short [distance from] outside the ship," wrote Rotvold in his diary. "One shot after them and got the mother and a yearling cub, the other cub was wounded, but managed to get away. One followed it nearly midway out on the Bay, but had to return, as it blew up with bad weather again. Weather permitting, we will go tomorrow and look for it [the wounded cub who got away]. We brought the 2 bears [the carcasses of the mother and first cub] to the ship, skinned them, and took the meat for provisions."[16]

The taking of the bears' lives and the subsequent nourishment provided therewith seemed to have breathed new life into the men. Knut Stave managed to go up on deck and spend some time there the following day, as did the injured Hermann Rüdiger, whose frostbitten foot previously had been tended to by the artist Christopher Rave. The temperature was "3° cold," and the men ate "fried bear-meat and oat-soup" as their meal. Rotvold wrote thankfully about the animals who had given them nourishment: "Those two bears were very welcome, because now we thus have meat for some time."[17]

Christopher Rave, too, raved about the change in menu which had resulted from the successful polar bear hunt (later ascribing the taking of the polar bear and her cub and the menu change to the day of February 11th) (Rave, 1913, 69). Every now and then, he would later recall, Rave took his dog Jule for a walk. "The tiring loneliness of the long winter night gradually makes itself more and more noticeable" and the longing for home grew, he would later write (Rave, 1913, 69).

The next day, February 12, proved as beneficial for Stave and Rüdiger, who both again spent some brief time on the ship's upper deck (Einar Rotvold Diary).[18] Meanwhile, Rotvold and some of the other men ventured out on the ice to search for the wounded bear cub but could not find the injured animal. The temperature reading at the middle of the day was "10.5° cold."

The addition of more bear meat also came with added hours of daylight, and yet it remained "sad and dreary" for the men as of February 13.[19] Knut Stave was seen to have taken a turn for the worse on the following day "but he has a good appetite," reported Rotvold.[20] On the 15th of February Rotvold again wrote that "Knut [Stave] has been better again today, he says. We cannot understand what is ailing him."[21] The men struggled to comprehend Stave's physical condition even as they fought off claustrophobic conditions. Rotvold wrote on the 16th: "Here is sad and dreary, no space to walk a bit, only sit closed off in a small cabin. One can just about not get to draw fresh air, when there is such weather. Ohhh, for a life! At midday 17.5° minus."[22]

Stave was faring poorly, becoming less and less able to breathe. Rotvold described him as weakening on February 17, and noted Christopher Rave's attempts to help the ill Stave. The diarist lay the fault at the door of the expedition leaders, who he stated had not required any physician-conducted physicals to confirm Stave's state of health prior to the ship's departure. Rotvold wrote:

> Knut is worse now. He has been aft to [see] Rave and received iron pills today. We have nothing to help him with, as Rave has all vegetables, citric acid, and medicines by him aft. He has been denied citric acid, he says. We do not know what is ailing him; he has poor appetite, bad moods[,] etc. If it is tuberculosis he has, which we fear for, it is not much good for Stenersen and me, to be together with him, down in a small cramped cabin, and we no disinfectant have [we]. He is rather reckless with his spit. It does not work to talk to him, he just protests and gets angry. We have now for the first time noticed signs of scurvy on him, blue-red spots on the shin. It is a pity for him, poor man, to be getting so sick here at such a place, and one no remedies has. The expedition should be held responsible for that there were no physical examinations done before we went out, as we assume that Knut was ill before we went out to sea."[23]

While Stave gradually deteriorated and Rüdiger just managed to maintain, the other men felt the pressures of the situation, especially with the worsening weather. Rotvold's February 18th diary entry contains the following passage: "One has been forced to stay below in the cabin all day. It looks like the weather is getting worse and worse as the days pass by. It is a sad existence to be here; wishing only that the time was over, so that we could get to the civilized part of the world again, but it passes slowly. At midday 19° minus."[24]

Rave continued attempting to help Stave, administering drops of various remedies (it is not clear regarding the citric acid). On February 19, according to Rotvold's diary, Stave went to the rear of the ship "to [see] Rave and received some drops, what kind we do not know. He stays mainly in the bunk. One has boiled milk, which he drinks, but we are unfortunately soon out of that kind of commodity."[25] It seems by now the ship's stores were being depleted of certain vital provisions as well.

On the following day, with Knut Stave's situation becoming increasingly serious, Christopher Rave made a house-call to see Stave: "Rave has been forward and given Knut morphine and almond drops," wrote Rotvold. "He is not in a good shape, terrible heaviness in the chest, and he has difficulties drawing breath."[26]

That day of February 20, which held heaviness for Stave and for the crewmen both literally and figuratively, was also the day of Rave's birthday gathering, as he would later record in his book, and he later recalled that the birthday party for him, therefore, offered a successful change of pace for everyone (Rave, 1913, 69). He had baked two cakes with improvised ingredients, he wrote, to which cocoa was then added and served, and a Sunday-like mood had spread (It seems that Rotvold did not mention this party with Rave in his February 20th diary entry, writing only that Rave came forward in the ship to attend to Stave [Einar Rotvold Diary][26]).

Rave observed that the cook, Knut Stave, however; became increasingly emaciated and began to suffer from severe respiratory problems (Rave, 1913, 69).

It seems that where there once had been hope that Knut Stave would improve, now there was despair that his situation continued to worsen (Einar Rotvold Diary). Rotvold expressed an especially desperate sentiment on February 21st, stating the following: "Heavy sad days. Knut is poorly, he is suffering from shortness of breath, has high fever this evening 41°. He is very thirsty, we have cooked oat-soup, which he drinks. We have nothing whatsoever to help him with; nursing him in the best manner and understanding. It is sad and wearing for us, that he should get so ill, God help him poor man."[27]

The fact that the statement was made in what Rotvold recorded as "Ca. 30° minus" temperature gives a further clue as to the dispirited condition in which all the men on the ship found themselves.

## The Quiet Passing: The Death of Knut Stave

As of February 22, Einar Rotvold and the crewmembers of the *Herzog Ernst* had many things about which to be worried, not the least of them being the Norwegian cook Knut Stave's ailing health situation, their missing German expedition members who had vanished, their long-departed captain Alfred Ritscher who had not yet returned, and the frozen ship's dwindling food supplies.

On this day, in addition to worrying about his ailing Norwegian crewmate Knut Stave, Rotvold was also concerned about the two German biologists Erwin Detmers and Walter Moeser, who still had not reappeared since their departure during the group's attempt to travel southward together—at which time the two biologists had veered off on their own, choosing a different course to travel (Einar Rotvold Diary). In addition, he wondered about the fate of Captain Alfred Ritscher, who had left them at West Fjord with a promise to bring aid. As part of his diary entry that day, Rotvold wrote:

> Knut is in same state, but he had only 36.4° [Celsius] this morning. He was sweating terribly last night. But still slept pretty well. He is weak and faint, doing his business below, which we immediately through overboard. We have received sublimate, which we use in spit tray and spit cup. Rave has been forward with a glass morphine ma[crossed out] drops, which he is to take, ½ tablespoon 3 times daily. He drinks oat-soup and eats a bit [dehydrated] vegetable-crackers. This evening the temperature was 37.5°. Hoping that he will be well again. We are now out of sugar.

> – As the captain [Alfred Ritscher] has not come hereto, back from Advent Bay, and the agreement was, that he was going to come late in January or first half of February, dependent on circumstances, we fear that an accident has befallen him, in one way or another, [ice pilot

August] Stenersen and I have decided to go to Advent Bay in first half of March, in order to investigate the situation, as the agreement with the captain [Ritscher] was thus. We are at the same time worried for the two Dr. Detmerss [sic] and Moeser, who left us at Steyle Hook the 31st September last autumn, and one nothing has heard of them since, and there was poor new-ice all of Wijde Bay inwards.[28]

Unfortunately, Detmers's and Moeser's choice to take a different route had evidently been a tragic one.

Equally unfortunately, it seemed that Rave could no longer help Stave with the available medication (Rave, 1913, 69).

The feeling of impending tragedy continued on the next day of February 23 (Einar Rotvold Diary). Knut Stave was dying. It was a poignant picture of self-control and grace which Rotvold painted of his crewmate in his diary. Letters were attempted to be written. Words were encouraged. Goodbyes were said. And prayers were read. Rotvold wrote:

There has been fire alarm from the cabin today as well, but it was soot only, which was burning between the paneling in the ceiling. We got it immediately extinguished. Knut has been poorly today. Rave has given him caffeine-powder and aspirin-powder. 9 o'clock evening he got so bad that he was about to suffocate. He said, that it is surely over for me now. We have asked him whether he wants, that we should write down something for him for the home, but he answered only, that we should greet them at home, and that his receivable money and effects should be shared between his mother and his sister. He bid farewell to all of us, was in full consciousness, prayed Our Father, and asked God to bless him. He can just about not speak due to shortness of breath.

Rave was here and gave him morphine-injection, so he is now a bit calmer, lying as if in a torpor. Hard it is at home to lose a mate, but even harder here in the wilderness. God help him, that he must get better again. One has read a hymn and prayer for him, as he wished. He says, he is not afraid of dying, but he does not want to die either.[29]

In confronting their life-and-death situation, *how* they faced it was a matter of significance, and the words Rotvold expressed about Stave are exceedingly affecting—*he is not afraid of dying, but he does not want to die...*

On the following day, a Monday, there was a light breeze from the south. The temperature was, approximately, a minus 30° (C) to 35° (C). There seemed to be a stillness. In the midst of the respectful observance of the crew and the poignant endurance of the cook, Knut Stave quietly passed away. His death was recorded by Einar Rotvold on February 24: "Knut was calm until 3:30 o'clock, when he wanted something to drink, 6:30 o'clock he talked to us, that he wanted to pass water, and that his feet were a bit cold. One helped him out, and tipped the secretions overboard. We lit the fire in the stove and were about to boil coffee, when it suddenly went so quiet in the berth. He had then the poor man expired. The time was then 7:15 morning 24th February. He died quietly and calmly. The Lord bless him."[30]

Stave's passing left Rotvold and his remaining crewmates devastated. The grief they felt is palpable in the pages of Rotvold's diary, which speaks volumes in the minimal amount of words entered on the following day, February 25, and the simple statement made: "Here is sad and dreary, even worse than before."[31]

The frozen Arctic terrain made it impossible for the expedition men to give their deceased companion a proper burial or lay his body to rest in a grave dug into the frigid ground at that time of the season.

For this reason, the men sewed Stave into a sleeping bag made of reindeer skin and transported his remains from the ship to the Swedish House, where his body would lie until that time when a grave could be excavated (Einar Rotvold Diary; Rave, 1913, 69).

Rotvold recorded the somber proceedings in his diary on the 26th (Einar Rotvold Diary): "One has driven Knut over to the House and laid him into the Sauna-room, until the ground will be thawed enough, that one is able to bury him. It is sad and dreary now after he has passed away from us. We brought 3 bags coal back to the ship. One has fetched snow for water-thawing. The rest of the day we stayed quiet onboard."[32]

In their silence, the men witnessed the first day of sun on February 27. The appearance made an impact. "Around midday we saw the sun again, for the first time this year, we saw it for ca 10 min, when [then] it went behind the mountain," wrote Rotvold. "It was a wonderful sight for us after this long Polar night we have had."[33]

Though seemingly still grieving for the loss of their crewmate, the crewmen resolved to take action to free themselves. On February 28, a day when they were able to view the presence of the sun for a full 1–1/2 h—"not quite clearly, as there was hazy air over the mountain," but visible nonetheless—on this day they renewed their vigor to escape from their frozen ship. "We have now started to get ourselves ready for the journey to Advent Bay," wrote Rotvold that day.[34] It was still "Ca. 35° minus" but the motivation was there to attempt the departure.

They did not know about the set of events that Captain Alfred Ritscher had put into motion, or about the multiple attempts to rescue them. For those of the stranded expeditioners who were able to travel, they would now take it upon themselves to go seek help, and to attempt to find rescue for themselves and for their shipmates.

## Notes on Original Material and Unpublished Sources

Roald Amundsen letters of correspondence, written from and to Roald Amundsen and Leon Amundsen, are in the Manuscripts Collection at the National Library of Norway (NB) in Oslo. Olaf Henriksen's and Einar Rotvold's diaries are in the Library Archives at the Norwegian Polar Institute (NPI) in Tromsø.

1. L. Amundsen to R. Amundsen, letter, 17 February 1913, NB Brevs. 812:1
2. L. Amundsen to R. Amundsen, letter, 20 February 1913, NB Brevs. 812:1
3. R. Amundsen handwritten note to L. Amundsen, at bottom of L. Amundsen to R. Amundsen letter dated 20 February 1913, NB Brevs. 812:1
4. L. Amundsen to R. Amundsen, letter, 22 February 1913, NB Brevs. 812:1
5. R. Amundsen to L. Amundsen, telegram, 23 February 1913, NB Brevs. 812:1
6. L. Amundsen to H. Hanssen, telegram, 23 February 1913, NB Brevs. 812:3:7

7. H. Hanssen to L. Amundsen, telegram, 23 February 1913, NB Brevs. 812:1
8. H. Hanssen to L. Amundsen, letter, 24 February 1913, NB Brevs. 812:1
9. O. Henriksen diary, 24 and 25 August 1912, (NPI), D-305/D00305_0001
10. E. Rotvold diary, 1 February 1913, (NPI), D00125
11. E. Rotvold diary, 2 February 1913, (NPI), D00125
12. E. Rotvold diary, 3 February 1913, (NPI), D00125
13. E. Rotvold diary, 4 February 1913, (NPI), D00125
14. E. Rotvold diary, 6 February 1913, (NPI), D00125
15. E. Rotvold diary, 7 February 1913, (NPI), D00125
16. E. Rotvold diary, 10 February 1913, (NPI), D00125
17. E. Rotvold diary, 11 February 1913, (NPI), D00125
18. E. Rotvold diary, 12 February 1913, (NPI), D00125
19. E. Rotvold diary, 13 February 1913, (NPI), D00125
20. E. Rotvold diary, 14 February 1913, (NPI), D00125
21. E. Rotvold diary, 15 February 1913, (NPI), D00125
22. E. Rotvold diary, 16 February 1913, (NPI), D00125
23. E. Rotvold diary, 17 February 1913, (NPI), D00125
24. E. Rotvold diary, 18 February 1913, (NPI), D00125
25. E. Rotvold diary, 19 February 1913, (NPI), D00125
26. E. Rotvold diary, 20 February 1913, (NPI), D00125
27. E. Rotvold diary, 21 February 1913, (NPI), D00125
28. E. Rotvold diary, 22 February 1913, (NPI), D00125
29. E. Rotvold diary, 23 February 1913, (NPI), D00125
30. E. Rotvold diary, 24 February 1913, (NPI), D00125
31. E. Rotvold diary, 25 February 1913, (NPI), D00125
32. E. Rotvold diary, 26 February 1913, (NPI), D00125
33. E. Rotvold diary, 27 February 1913, (NPI), D00125
34. E. Rotvold diary, 28 February 1913, (NPI), D00125.

## Unpublished Sources and Original Material

Amundsen Letters of Correspondence, Manuscripts Collection, National Library of Norway, Oslo.

Henriksen, O., Diary, Olaf Henriksen's diary written while he was the Green Harbour Telegraph Station Manager, "Dagbok for Olaf Henriksen, 1911–13" ["Diary for Olaf Henriksen, 1911–13"], dated 14 August 1911 to 2 May 1913, D-305, D00305_0001. Library Archives, Norwegian Polar Institute, Tromsø, https://brage.npolar.no/npo lar-xmlui/handle/11250/2600678. Accessed 24 July 2021.

Rotvold, E., Diary, Einar Rotvold's expedition diary written during the Schröder-Stranz expedition of 1913, "Dagbok for August Stenersen og Einar Rotvold, Tromsø:

Treurenberg Bay fra 1.januar—7.juni 1913: Schröder-Stranz-ekspedisjonen 1912–13", "Einar Rotvold har skrevet dagboken", ["Diary for August Stenersen and Einar Rotvold, Tromsø: Treurenberg Bay from 1 January—7 June 1913: The Schröder-Stranz Expedition 1912–13", "Einar Rotvold has written the diary"], dated 12 January 1913 through 7 June 1913, D00125. Library Archives, Norwegian Polar Institute, Tromsø, https://brage.npolar.no/npolar-xmlui/handle/11250/2426394. Accessed 24 July 2021.

Staxrud, A., 17 February 1913, Letter from Captain Arve Staxrud to Norway's Minister of Foreign Affairs, Nils Claus Ihlen, regarding plans for rescue expedition after Schröder-Stranz, with notes "Redningsekspeditionen for Schröder-Stranz 1913" ["The Rescue Expedition for Schröder-Stranz 1913"], copybook and notes, 1913_02_17_Redningsexpeditionen_Schroder_St[r]anz_1913_Hoels_og_Staxruds_Spitsbergenexpedisjoner, Library Archives, Norwegian Polar Institute, Tromsø. Received 20 July 2021.


# References

Der Zeitspiegel Berlin im Frühjahr. (1913). *Newspaper clipping*. Ritscher Estate, Cornelia Lüdecke, München.

Hoel, A. (1933). Arve Staxrud. *Norsk Geografisk Tidsskrift [Norwegian Journal of Geography]*, *4*(6), 345–346. https://doi.org/10.1080/00291953308622018

Rave, C. (1913). *Tagebuch von der verunglückten Expedition Schröder-Stranz*. Schaffsteins Gründe Bändchen 49. Cöln: Schaffstein.

Staxrud, A. (1914). Die Staxrudsche Hilfs-Expedition für Schröder-Stranz. In A. Miethe (Ed.), *Die Expedition zur Rettung von Schröder-Stranz und seinen Begleitern—geschildert von ihren Führern Hauptmann A. Staxrud und Dr. K. Wegener* (pp. 1–68). Berlin: Dietrich Reimer.

Tahan, M. R. (2019). *Roald Amundsen's sled dogs: The sledge dogs who helped discover the South Pole*. Cham: Springer International Publishing.

Tahan, M. R. (2021). *The return of the South Pole sled dogs: With Amundsen's and Mawson's Antarctic expeditions*. Cham: Springer International Publishing.

# Chapter 8
# The Arve Staxrud Expedition Coalesces: Local Norwegian Experts, Sámi Guides, Experienced Sled Dogs, Reliable Reindeer, and an Ice-Worthy Vessel

## The Amundsen Connection: The South Pole Expedition Dogs are Recruited, and the Staxrud Controversy Continues

Roald Amundsen had authorized Helmer Hanssen to take his three surviving South Pole Expedition sled dogs on a rescue mission to the Arctic to attempt to save the Herbert Schröder-Stranz expedition in Spitsbergen. Yet Helmer Hanssen had declined to go, citing his loyalty to Amundsen and his perception of Arve Staxrud as not being an appropriate leader for this mission. And the three South Pole expedition sled dogs, who were important to Amundsen, and whom he fervently wanted to keep with him, were being requested by Staxrud. Thus, they were unwittingly placed at the center of the controversy.

Controversial comments and comrades' criticisms notwithstanding, Leon Amundsen determined that it would be in the best interest of everyone if the sled dogs took part in the rescue expedition. He wrote to Roald Amundsen on February 25, in reply to Roald's February 23rd telegram authorizing Helmer Hanssen to go with the dogs on the Staxrud rescue mission (Amundsen Letters of Correspondence):

> Thanks for the telegram [of February 23], but Helmer [Hanssen] will not partake [in the rescue mission], probably for the reason that the expedition will be led by Staxrud with "Hertha" of Sandefjord whereas one in Tromsø would have managed it from up there. Helmer advises against sending the dogs and about this I am now expecting a letter from him. However, I do not think it will be acceptable to not place the dogs at disposal for a rescue-expedition like this – I am afraid the criticism would raise itself rather quickly both from here [Norway] and from Germany as one of course has [Robert Falcon] Scott's tragedy in recent memory. As soon as Helmer's letter arrives, it has to be decided – I will maybe ask Gubben ["The Old Man" – Nansen] for advice.[1]

Whichever analysis and consultation he conducted, Leon must have immediately come to a decision to indeed send the dogs on the rescue expedition with Staxrud, even without the participation of Hanssen. The sled dogs were a crucial part of the rescue effort to save human lives, and he could not refuse the request. It seems that he had already made up his mind. It was a fait accompli.

M. R. Tahan and C. Lüdecke, *Stranded at the Top of the World*, https://doi.org/10.1007/978-3-031-56288-4_8

On the same day of Leon's letter to Roald, February 25th, an article appeared on the front page of the prominent Kristiania (Oslo) newspaper *Aftenposten* announcing the formation of the Staxrud rescue expedition and also announcing that the three sled dogs who had just returned from Antarctica would take part in this Arctic rescue mission (Aftenposten, 1913, evening: 1) (Fig. 8.1). The newspaper article featured two eye-catching photographic illustrations to accompany the story: The first was of the expedition ship *Hertha*, and the second was of one of the sled dogs, with a caption underneath reading "One of the South Polar dogs." (This illustration, interestingly, is the same one that this very same newspaper would later use to portray Obersten and to announce his special appearances at dog show exhibitions in Norway, in its June 22, 1913 morning edition (Tahan, 2021, 319 & 325). The article stated that eight dogs would work in the rescue expedition and that "Staxrud has obtained Roald and Leon Amundsen's permission to lead [trek] with the 3 of the South Polar dogs, who are currently staying [residing] by the Bundefjord" (Aftenposten, 1913, evening: 1). The article went on to say, about the dogs, that "It is first and foremost the red-brown 'Obersten', 4 years old, the only living 4-footed creature that has been to the South Pole. Then there is the bitch [female] 'Lucie' [Lussi] and the dog 'Storm', who were born on board 'Fram'." (Obersten actually most likely was 5 or 6 years old at the time (Tahan, 2021, 437). The article also made note that "From Helmer Hansen [sic], Captain Staxrud has not yet, [as of] this morning, received any reply to his inquiry" (Aftenposten, 1913, evening: 1).

This article was picked up by several other newspaper publications across Norway, who ran similar versions of the article—including in Trondheim on February 26, with the additional sub-headline proclaiming that "Amundsen's South Pole dogs join" (Nidaros, 1913, 1; Nidaros: Trøndelagen, 1913). Other newspapers followed suit, heralding the three dogs' participation in the expedition in the sub-headline of the article and thus proclaiming that as part of the major news: "Three of Amundsen's Dogs Join" declared a few other papers (Trondhjems Adresseavis, 1913, 1; Trondhjems Folkeblad, 1913b, 3).

Thus, the *official* statement, made publicly, was that the three dogs would be serving on the rescue expedition.

Three days after the *Aftenposten* article was published, on February 28, Leon wrote a letter to Roald informing him of his decision to send the sled dogs (Amundsen Letters of Correspondence). He began by summarizing Helmer's position as refusing to work with Staxrud's rescue expedition and citing the reason for this as being that Staxrud "has expressed himself unfavorably about your [Roald's] Polar-journey in the presence of many, also [in] Helmer's [presence] at [a] public restaurant [;] this I have actually not been allowed to say but one has to show forbearance with such expression of jealousy."[2] Leon went on to say that Hanssen intimated "that the journey with Staxrud is placed under incompetent leadership," and that "In Helmer you [Roald] have a loyal friend," going on to quote Hanssen in his letter: "He [Hanssen] states thus about Staxrud's statements: 'After such statements about the Man [whom] I put highest of all, You will understand, that I despise Staxrud of all my soul.' All this was not to be referenced." After updating Roald about Hanssen's loyalty to Amundsen and his negative representation of Staxrud, Leon then informed his brother that he

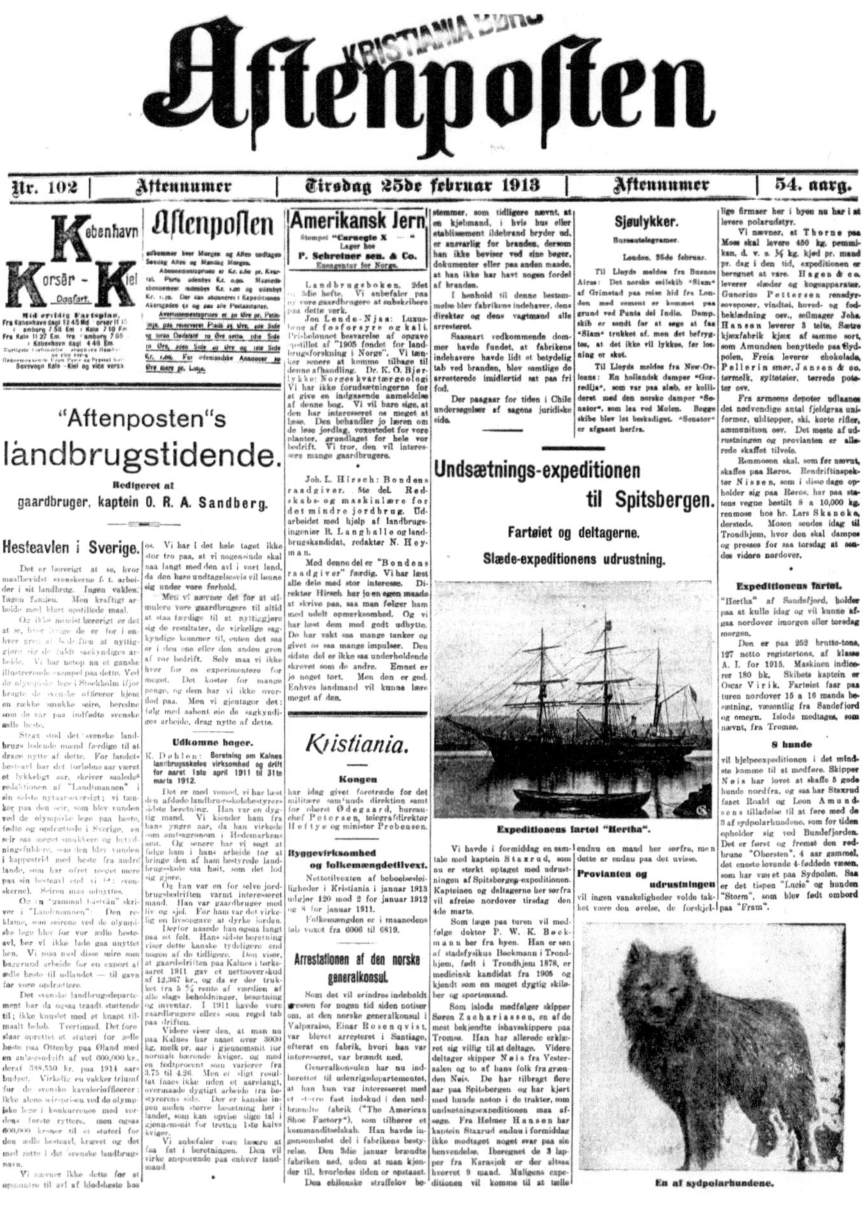

# Aftenposten

Nr. 102 | Aftenummer | Tirsdag 25de februar 1913 | Aftenummer | 54. aarg.

"Aftenposten"s landbrugstidende.

Redigeret af gaardbruger, kaptein O. R. A. Sandberg.

Hesteavlen i Sverige.

Kristiania.

## Undsætnings-expeditionen til Spitsbergen.

### Fartøiet og deltagerne.

### Slæde-expeditionens udrustning.

Expeditionens fartøi "Hertha".

Vi havde i formiddag en samtale med kaptein Staxrud, som nu er sterkt optaget med udrustningen af Spitsbergen-expeditionen. Kapteinen og deltagerne her sørfra vil afreise nordover tirsdag den 4de marts.

Som læge paa turen vil medfølge doktor P. W. K. Bøckmann her fra byen. Han er søn af stadsfysikus Bøckmann i Trondhjem, født i Trondhjem 1878, er medicinsk kandidat fra 1905 og kjendt som en meget dygtig skiløber og sportsmand.

Som islods medfølger skipper Søren Zachariassen, en af de mest bekjendte ishavsskippere paa Tromsø. Han har allerede erklæret sig villig til at deltage. Videre deltager skipper Nøis fra Vestersalen og to af hans folk fra grænden Nøis. De har tilbragt flere aar paa Spitsbergen og har kjørt med hunde netop i de trakter, som undsætningsexpeditionen maa afsøge. Fra Helmer Hansen har kaptein Staxrud endnu i formiddag ikke modtaget noget svar paa sin henvendelse. Iberegnet de 3 lapper fra Karasjok er der altsaa hvervet 9 mand. Muligens expeditionen vil komme til at tælle endnu en mand her sørfra, men dette er endnu paa det uvisse.

**Provianten og udrustningen**

vil ingen vanskeligheder volde takket være den øvelse, de forskjellige firmaer her i byen nu har i at levere polarudstyr.

Vi nævner, at Thorne paa Moss skal levere 450 kg. pemmikan, d. v. s. ½ kg. kjød pr. mand pr. dag i den tid, expeditionen er beregnet at vare. Hagen & co. leverer slæder og kogeapparater. Gunerius Pettersen renslyrsoveposer, vindtøi, hoved- og fodbeklædning osv., seilmager Joha. Hansen leverer 3 telte, Sætre kjæxfabrik kjæx af samme sort, som Amundsen benyttede paa Sydpolen, Freia leverer chokolade, Pellerin smør, Jansen & co. tørmelk, syltetøier, tørrede poteter osv.

Fra armeens depoter udlaanes det nødvendige antal fjeldgraa uniformer, uldtepper, ski, korte rifler, ammunition osv. Det meste af udrustningen og provianten er allerede skaffet tilveie.

Renmosen skal, som før nævnt, skaffes paa Røros. Rendriftinspektør Nissen, som i disse dage opholder sig paa Røros, har paa statens vegne bestilt 8 a 10,000 kg. renmose hos hr. Lars Skancke, dersteds. Mosen sendes idag til Trondhjem, hvor den skal dampes og presses for saa torsdag at sendes videre nordover.

**Expeditionens fartøi.**

"Hertha" af Sandefjord, holder paa at kulle idag og vil kunne afgaa nordover imorgen eller torsdag morgen.

Den er paa 252 brutto-tons, 127 netto registertons, af klasse A. I. for 1915. Maskinen indicerer 180 hk. Skibets kaptein er Oscar Virik. Fartøiet faar paa turen nordover 15 a 16 mands besætning, væsentlig fra Sandefjord og omegn. Islods medtages, som nævnt, fra Tromsø.

**8 hunde**

vil hjelpeexpeditionen i det mindste komme til at medføre. Skipper Nøis har lovet at skaffe 5 gode hunde nordfra, og saa har Staxrud faaet Roald og Leon Amundsens tilladelse til at føre med de 3 af sydpolarhundene, som for tiden opholder sig ved Bundefjorden. Det er først og fremst den rødbrune "Obersten", 4 aar gammel, det eneste levende 4-føddede væsen, som har været paa Sydpolen. Saa er det tispen "Lucie" og hunden "Storm", som blev født ombord paa "Fram".

En af sydpolarhundene.

**Fig. 8.1** Front-page article published in the *Aftenposten* February 25, 1913, newspaper announcing that the three sled dogs Lussi, Storm, and Obersten from Roald Amundsen's South Pole expedition would work on Staxrud's rescue expedition to Spitsbergen. (Norwegian Polar Institute) had decided he would loan the dogs to the rescue expedition, as this was the only thing and the right thing to do considering that the very lives of humans were in the balance, and that he was ordering Staxrud to look after the dogs very well.

Thus, Leon apprised his brother of his plan to send the South Pole expedition dogs, who had just come home from the southern region of the Earth, to the environs of the North Pole.

Roald Amundsen wrote a reply to Leon Amundsen on February 28th, writing on the back of Leon's letter to Roald dated February 13 which had announced the arrival of the three dogs Obersten, Lussi, and Storm at his home in Svartskog. In his reply Roald stated that he was glad that the three dogs were safe and had made it home, that he hoped they would now be able to safely complete the rescue mission in Spitsbergen, and that he wanted to keep all three dogs together at his home (Tahan, 2021, 295; Tahan, 2019, 609). Whereas during his Antarctic expedition Amundsen had slaughtered as many sled dogs as he had felt was necessary to expedite his journey to the South Pole, he now treasured his last three remaining dogs and wished to hold onto each one of them (Tahan, 2019, 626; Tahan, 2021, 207–208). It appears, by this time, he did not part with them lightly. Only a human need or a disaster of this proportion forced him to lend his dogs. Keeping the three dogs home and safe was of extreme importance to him.

But, as Leon moved forward with plans to send the sled dogs on the Staxrud rescue mission, and as Storm, Lussi, and Obersten became the focal points of national attention, viewed as potential rescuers and saviours of the expeditioners in the Arctic, their futures looked uncertain, due to the risky nature of the mission.

At some point between this date and the dogs' departure in early March, Leon decided to send only Lussi and Storm, not Obersten. Perhaps the last surviving South Pole sled dog was viewed as too valuable as both a resource for business and a veteran of the expedition to risk. Perhaps Amundsen expressed grave concern over his undertaking the dangerous rescue mission. Or perhaps there was another reason. As it transpired, Obersten remained behind at Svartskog and Leon proceeded to enter him in dog exhibitions and shows; only Lussi and Storm would venture to the Arctic to become rescuer dogs in search of the Schröder-Stranz expedition (Tahan, 2019, 609; Tahan, 2021, 297–298, 318–319).

And so, it was Lussi and Storm who would accompany Staxrud.

## Overwintering Trappers, Sámi Reindeer Walkers, Draft Reindeer, Sled Dogs, a Physician, an Engineer, and a Sealer

During the latter part of February, ever since his appointment as leader of the rescue mission, Arve Staxrud had been very busy pulling together all of the elements for the relief expedition to Spitsbergen. In addition to the three sled dogs whom he had sought to secure from Roald Amundsen, he had also reached out to Norwegian local experts whom he trusted would comprise an efficient and experienced crew with whom to trek through the Arctic—experts who fit the profile that he had recommended to the Foreign Ministry—and he had succeeded to enlist them. He also sought the supplies and equipment that would be needed. Government agencies, individuals, and businesses would be approached. And he hired the vessel that would take him through the northern ice. The news of this gathering of the expedition in terms of its members—both human and animal—and its equipment, as well as its provisions and ship, was released in the same newspaper article that had appeared in the Kristiania newspaper *Aftenposten* on February 25 announcing the participation of the three sled dogs (Aftenposten, 1913, evening: 1). In this article, the names of most of the expedition members, and the specifics of some of the supplies, as well as the description of the animals involved, were published (see Fig. 8.1). The information was based on an interview given that morning by none other than Staxrud himself, who, according to the article, spoke with the newspaper. At that time, and per the "conversation" the newspaper said it had with him, it was estimated that Staxrud and those expedition members located in Kristiania would depart for the North on March 4 and that the ship *Hertha* would depart from Sandefjord in southern Norway on February 26 or 27 (these dates would later change slightly). According to the newspaper, accompanying Staxrud out of Kristiania would be the physician "doctor P. W. K. Bøckmann", from the city, who, the article stated, was originally from Trondhjem (Trondheim), born there in 1878, the son of city physician Bøckmann in Trondhjem, and who, in addition to being a qualified medical doctor since the year 1905, was also recognized to be a sportsman and a skier of considerable skill. Dr. Bøckmann (also spelled Dr. Böckmann)[1] would therefore serve as the physician on the rescue expedition (Figs. 8.2 and 8.3).

The implication from the article also was that the three dogs who lived at Amundsen's home would go with Staxrud out of Kristiania as well.

The skipper Nøis (most likely Daniel Karolius Nøis, who would become second in command of the rescue expedition) would come from Vesteraalen (Vesterålen archipelago, in Nordland county, Northern Norway), along with two members of his family from the Nøis hamlet, in order to take part in the expedition, stated the newspaper article. (The participants with Daniel Karolius Nøis would turn out to be Johan Nilsen Nøis, Hilmar Andreas Nilsen Nøis, and Martin Pettersen Nøis—members of the Nøis family of trappers and hunters.) The Nøis family had many years of experience on Spitsbergen, the paper reported, and had much experience driving dog sledges within the geographical locations in which the rescue mission would now have to conduct its search mission (Fig. 8.4). The implication, thus, was

---

[1] The name of the expedition's physician, as announced in many major newspapers published in Norway at the time, is listed in the news articles as doctor P. W. K. Bøckmann (for example, *Aftenposten* 25 February 1913). Hilmar Nøis, in one of his diaries (NPI), mentions the physician's first name as Johan. Other Norwegian and German sources of the time, including Daniel Nøis, Christopher Rave, and Hermann Rüdiger, list his name as Dr. Bøckmann or Dr. Böckmann, with no first name given. Later reports and publications also listed no first name (see Norsk Polarhistorie.no/ Staxrud). The physician's name as it appeared in the newspapers across Norway at the time of the expedition, Dr. P. W. K. Bøckmann, is used throughout the text in this book.

**Expeditionens fartøi "Hertha".**

Vi havde tirsdag form. en samtale med kaptein S t a x r u d, som nu er sterkt optaget med udrustningen af Spitsbergen-expeditionen. Kapteinen og deltagerne her sørfra vil afreise nordover tirsdag den 4de marts.

Som læge paa turen vil medfølge doktor P. W. K. B ø c k-m a n n her fra byen. Han er søn af stadsfysikus Bøckmann i Trondhjem, født i Trondhjem 1878, er medicinsk kandidat fra 1905 og kjendt som en meget dygtig skiløber og sportsmand.

Som islods medfølger skipper Søren Z a c h a r i a s s e n, en af de mest bekjendte ishavsskippere paa Tromsø. Han har allerede erklæret sig villig til at deltage. Videre deltager skipper N ø i s fra Vesteraalen og to af hans folk fra grænden Nøis. De har tilbragt flere aar paa Spitsbergen og har kjørt med hunde netop i de trakter, som undsætningsexpeditionen maa afsøge. Fra Helmer H a n s e n har kaptein Staxrud endnu i formiddag ikke modtaget noget svar paa sin henvendelse. Iberegnet de 3 lapper fra Karasjok er der altsaa hvervet 9 mand. Muligens expeditionen vil komme til at tælle

endnu en mand her sørfra, men dette er endnu paa det uvisse.

**Provianten og udrustningen**

vil ingen vanskeligheder volde taket være den øvelse, de forskjel-

Trondhjem, hvor den skal dampes og presses for saa torsdag at sendes videre nordover.

**Expeditionens fartøi.**

"Hertha" af Sandefjord, holdt paa at kulle tirsdag og skulde af

**En af sydpolarhundene.**

Fig. 8.2 One of the many news articles announcing the names of the members of the Staxrud rescue expedition, including Dr. P. W. K. Bøckmann. This one (excerpt shown) was published in *Aftenposten: Ugens Nyt* on February 27, 1913. (National Library of Norway)

that they were particularly qualified to work in this environment, on this terrain, and under these conditions.

In addition, stated the paper, the well-known Arctic ice pilot Søren Zachariassen, from Tromsø, was ready to work on the expedition. And three members of the Sámi[2] community from Karasjok (Western Finnmark) also would take part in the rescue expedition. (Although not named in this article, these three individuals would be Per Hansen, Samuel Klemmetsen, and Johannes Kemi.) (Fig. 8.5).

As previously stated, it was mentioned in this article that Helmer Hanssen had not yet been heard from in terms of providing an answer as to whether or not he would take part. Thus, Staxrud had enlisted nine men, with the possibility of yet another joining.

---

[2] The term used by the newspaper article and widely used by others at that time was "Lapper" ("Lapps") but will not be used in this book, as the proper name is Sámi, and therefore, Sámi is the name used throughout this book text.

**Fig. 8.3** Physician Dr.
Bøckmann of Staxrud's
relief expedition. *Source* Der
Zeitspiegel (1913)

As it would turn out, later, the mining engineer Jakob Ellingsen (also spelled Jacob Ellingsen)[3] would also join the rescue expedition (Hilmar Nøis Diary NPI; Norwegian Polar Institute, 2003, 114). During the previous year, in 1912, as reported by Adolf Hoel, Jakob Ellingsen had worked on Hoel and Staxrud's topographical surveying expedition on Spitsbergen as one of two assistant geologists (Hoel, 1929, 20). Evidently he had a familiarity with the geology and the physical area.[4] Most likely, Ellingsen's previous experience on Spitsbergen with Staxrud and Hoel qualified him in Staxrud's eyes to be a part of the rescue expedition (Fig. 8.6).

As for the driving force to mobilize the rescue mission, in addition to the three South Pole expedition dogs from Amundsen's home near Kristiania, another five dogs would be brought by Nøis from Northern Norway, for a total of eight sledge

---

[3] Ellingsen's first name is spelled Jakob by Adolf Hoel in his book (1929) and another source, and it is spelled Jacob by Hilmar Nøis in one of his diaries (NPI) and by the Place Names publication (2003). The spelling Jakob is used in the text throughout this book.

[4] A recent genealogy project website reports Ellingsen's name and information as Jakob Andreas Martin Ellingsen, born in Hamarøy, Nordland, Norway, with dates of birth and death as 1886–1956, and with the profession of geologist, beginning his studies at the university in Oslo in 1908, finishing as a titled mining engineer in 1912, and working that same year with Adolf Hoel in Spitsbergen, Svalbard, on the Norwegian Spitsbergen Expedition (Hanseater.no).

**Fig. 8.4** Daniel Nøis (holding accordion), Johan Nilsen Nøis with one of the sledge-pulling dogs, Arve Staxrud (foreground) with one of Nøis's dogs, and Hilmar Nøis (standing), as seen later on Spitsbergen most likely after the expedition's sledge journeys. (Photographer/Byline: Arve Staxrud; Source/Owner: Norwegian Polar Institute)

dogs, according to the *Aftenposten* newspaper article (Aftenposten, 1913, evening: 1).

Staxrud would later record, in his report published in Germany in 1914, that, as the expedition was not able to procure enough trained sledge dogs in a hurry, Staxrud followed Nansen's suggestion to use reindeer to help pull the sledges (Staxrud, 1914, 6). In his published report, Staxrud mentions being only able to procure 20 trained sledge dogs quickly; this number is closer to the total number of dogs once the expedition was on Spitsbergen, having left Norway with 12 dogs and having added 6 dogs in Spitsbergen for a total of 18 dogs (Tahan, 2021, 298), as he specified in his telegram later sent from Spitsbergen on May 14th as quoted in the newspaper *Aftenposten* on May 15 (*Aftenposten* 15 May 1913, evening: 1–2) and—as shall be seen—he would write in a letter to Nansen on March 17th (Staxrud, 17 March 1913). Thus, the inability to have a sufficient number of sled dogs at this time necessitated the recruitment of reindeer as additional draft animals.

**Fig. 8.5** The three Sámi expeditioners Per Hansen, Johannes Kemi, and Samuel Klemmetsen as seen later in Green Harbour. (Photographer/Byline: Ernest Mansfield; Source/Owner: Norwegian Polar Institute)

## North Pole Provisions and a Bottlenose Catcher

Regarding provisions and equipment, the *Aftenposten* newspaper reported that a large supply of moss for the reindeer was included, as were food, supplies, and Polar equipment supplied by several named companies (Aftenposten, 1913, evening: 1).

According to the article, 450 kg of pemmican, enough to provide 1/2 kg of meat per person per day over the course of the expedition during the time it was estimated to last, was to be delivered by Thorne on Moss. Sledges, as well as food preparation and cooking apparatus, were to be provided by Hagen & co. Reindeer sleeping bags, windbreakers (anoraks), head clothing, and footwear would be supplied by Gunerius Pettersen. Three tents were to be provided by the sailmaker Johs. Hansen. The paper further proclaimed that biscuits, the same kind as those used by Roald Amundsen during his South Pole expedition, were to be provided by Sætre kjæxfabrik. Chocolate was to be supplied by (the famous) Freia; butter by Pellerin; and dry (powdered) milk, jams, and dried potatoes by Jansen & co. For clothing and weaponry, the army, through its depots, was loaning to the expedition the needed quantities of warm mountain uniforms, woolen blankets, skis, rifles, and ammunition. Most of these provisions, said the paper, had by now been supplied.

Still to be acquired, according to the news article, was the reindeer bog (peat moss) from Røros (in Trøndelag county). For that, the reindeer husbandry inspector Nissen, who was at that time remaining in Røros, had procured, on the state's behalf,

**Fig. 8.6** Featured here with the Nøis brothers Daniel Nøis and Johan Nilsen Nøis (center), and the three sledge-pulling dogs, appears to be most likely Jakob Ellingsen (on the far right), in a photo taken later on Spitsbergen probably toward the latter part of the expedition. (Photographer/Byline: Arve Staxrud; Source/Owner: Norwegian Polar Institute)

between 8000 and 10,000 kg of reindeer bog which he had ordered from Lars Skancke in Røros. According to the article, on that very day of the newspaper's publishing, the reindeer moss was being delivered to Trondheim for preparation—for steaming and pressing—so that it could be shipped to the northern part of Norway and then taken to the Arctic, ready for consumption.

Supplemental to these supplies that were noted in the *Aftenposten* newspaper article, additional major provisions were being obtained from Roald Amundsen—specifically, a portion of the provisions he had already gathered for his planned expedition to the North Pole—and these, too, were made available in Norway (Hamburger Nachrichten, 1913a).

As for the vessel that would transport all the humans and animals and their provisions northward for the rescue, the *Aftenposten* newspaper described the *Hertha* as having "252 gross tons, 127 net register tons, of class A. I. for 1915," with an engine that was capable of "180 hp" (Aftenposten, 1913, evening: 1) (Fig. 8.7). The vessel was to be captained by Oscar Virik and would also maintain a crew of 15–16 men who, like the ship, originated from the Sandefjord region. The ship's crew would be joined in Tromsø by ice pilot Zachariassen.

**Fig. 8.7** Ice-going whaler *Hertha. Source* Der Zeitspiegel (1913)

Daniel Karolius Nøis would later state, in his article published in 1929, that the *Hertha* recently up until the time of the expedition had been used primarily for whale catching—specifically the Northern Bottlenose Whale (Nøis, 1929, 6). Indeed, according to the Norwegian Polar Institute, the *Hertha* had a long and complex history and life itself, built in 1884 at Framnæs Mek. Verksted as the last sealer, and originally part of the Southern Norwegian sealing fleet, catching 1600 seals and 1 Bottlenose Whale in the Arctic waters in 1884 alone, then most recently catching 60 Bottlenose Whales in 1912, prior to being tapped to transport the Staxrud rescue expedition of 1913 (Norwegian Polar Institute/Hertha). Thus, the ship had gone from seal hunting to whale hunting, and now it would make its way through the ice to Spitsbergen to deliver expedition rescuers to the Arctic.

As Norway's newspapers continued to cover the coming together of the rescue expedition, including announcing the name of the physician from Kristiania and giving press coverage to the participation of the three sled dogs from Svartskog, for example, in Kristiania's *Social-Demokraten* on February 27 (Social-Demokraten, 1913, 2), and as newspaper articles continued to update the estimated dates of departure for the rescue mission, Professor Adolf Miethe arrived back in Berlin on February 27, where, as reported in one of the Norway papers, he published an article in which he specifically justified the rescue expedition's being composed entirely of Norwegians (Trondhjems Folkeblad, 1913a, 2). According to the paper, Miethe stated that locating German rescuers who possessed sufficient experience was not possible and thanked the Norwegians who were spearheading the effort, especially Fridtjof Nansen. Miethe also particularly criticized the planning approach of Schröder-Stranz in regard to his expedition. It was also mentioned in the news article that Nansen himself would assume the position of leader of the rescue expedition in the unlikely event that Staxrud was not able to carry through.

Meanwhile, also in Germany, an "Emergency Call for the Spitsbergen Rescue Expedition" was again published in the newspapers (Hamburger Nachrichten, 1913b). So far, the Senate and the citizenship (Bürgerschaft) of Hamburg had provided 3000 marks for the rescue operation; however, more funds would have to be raised through donations.

But for now, both in terms of leadership position and resources, it was all systems go for Staxrud.

## The Nøis Family of Trappers and Hunters: Daniel Nøis, Hilmar Nøis, Johan Nilsen Nøis, and Martin Pettersen Nøis

One of the immediate actions taken by Arve Staxrud after being selected as rescue expedition leader, according to Daniel Karolius Nøis (Daniel Nøis), was to send a telegram to Nøis inquiring as to whether Nøis could take part in the expedition, along with a couple of his people who were experienced in making overwintering trips with him on Spitsbergen, and if he, in addition, could find sled dogs for the expedition (Nøis, 1929, 6). The trapper/hunter and overwintering expert Daniel Nøis responded in the affirmative and informed Staxrud that he would bring with him Hilmar Andreas Nilsen Nøis (Hilmar Nøis) and Johan Nilsen Nøis (also called Johan Nilsen and Johan Nøis) as part of the expedition party. Not planned at that time, but occurring at a later date, a third companion joined—this was Martin Pettersen Nøis (also known as Martin Pettersen and Martin Nøis) (1875–1940)—(Svalbard Museum 2023a).

Hilmar Nøis was Daniel's nephew, and Johan Nilsen Nøis (1884–1958) was Daniel's brother (Nøis [slekt] lokalhistoriewiki.no; Svalbard Museum 2021a).

As requested, Daniel Nøis was able to gather five good dogs for pulling sleds, as well as his own two dogs who he already had with him (Nøis, 1929, 6). Thus, he sent a telegram to Staxrud apprising him of the fact that he now had seven sledge dogs in total and was currently running them together so as to enable them to become accustomed to working with one another. When Fridtjof Nansen read this telegram, according to Nøis, the famous explorer asked Staxrud if he had intentionally requested from Nøis that Nøis practice driving the dogs together beforehand. Staxrud replied to Nansen in the negative. As a result, Nansen quickly concluded that Staxrud had selected the correct person for this job.

Once again, based on Daniel Nøis's account, it seems that the importance of the sled dogs to the rescue expedition was being indicated here by Nansen—both in terms of their significance to the rescue efforts, and in terms of their relationship to the humans involved. How Nøis worked with the dogs and prepared them for the rescue mission was an indication of the character and capability of Nøis as a person.

Daniel Karolius Nøis (1880–1971) lived in Risøyhamn, a village on Andøya island, in the Andøy municipality, in Nordland county, Northern Norway (Andøya Avis, 1950, 1; Daniel Nøis Biography/Norwegian Polar Institute; Nøis [slekt] lokalhistoriewiki.no). He was born on his family's farm, on Andøya—the isle furthest north in the Vesterålen archipelago, just within the Arctic Circle. From his very early years, he had been building and sailing sailboats and trapping and hunting on Spitsbergen, extensively involved in overwintering expeditions to hunt and catch polar bears, polar foxes, and seals. He had also been leading sailing expeditions across the Arctic Ocean from Nordland, Norway, to Spitsbergen. He was a trapper, a skipper, a boat builder, and an overwintering expert, and he had accumulated this experience prior to being selected to serve as deputy commander on Staxrud's expedition to rescue the Schröder-Stranz expeditioners.

As previously discussed, Staxrud had stipulated in his letter to the Foreign Minister, on February 17th, that, in addition to sleds and sled dogs, the expedition would require people who had experience in and knowledge of the local area of Spitsbergen (Chap. 6). To this end, Daniel Nøis and his people were now recruited. The Nøis family members were local experts from Northern Norway experienced in overwintering on Spitsbergen.

Upon receiving the telegram from Staxrud requesting Daniel Nøis's participation in the rescue expedition, Daniel Nøis in turn sent a telegram to his nephew Hilmar Nøis, who at that time was working at a shipyard in Tromsø, and who immediately accepted the proposed assignment to accompany his uncle to Spitsbergen, as Hilmar Nøis would later record in his written personal recollections and memoirs (Hilmar Nøis Diary SVB; Hilmar Nøis Diary NPI).[3, 4] Hilmar Andreas Nilsen Nøis (1891–1975—Nøis [slekt] lokalhistoriewiki.no; Svalbard Museum 2021b) had experience in working on Spitsbergen as well. According to Hilmar, he replied to Daniel stating that he happily accepted the rescue job, as he had retained two dogs from his previous overwintering trip in 1912 (Hilmar Nøis Diary NPI).[5] And so, in the middle of February,[6] according to Hilmar Nøis, he agreed to join with his uncle Daniel Nøis in the rescue expedition for which Daniel Nøis would be the next-in-command[7] and for which the Nøis "overwinterers" would provide sledge dog-teams for the rescue mission.[8]

Hilmar Nøis also states in his recollections that Adolf Hoel was busy with geographical surveying and geological work on Spitsbergen and that therefore Hoel's partner Staxrud assumed leadership of the rescue expedition on behalf of Fridtjof Nansen (Hilmar Nøis Diary NPI).[9] He makes a point to state that Roald Amundsen had entrusted his two sled dogs Lusi [Lussi] and Storm to the expedition to work in the rescue and had given sledging equipment and Polar tents to the expedition (Hilmar Nøis Diary NPI).[10]

And so, as of February 25 or 26, according to a major Kristiania newspaper, having collected the men and seven dogs for the rescue expedition, Daniel Nøis set out with his companions—both human and canine—to make their journey (presumably from Risøyhamn) to Tromsø, where, after approximately more than eight days of travel, they would meet together with Staxrud (Aftenposten: Ugens Nyt, 1913, 1). Daniel Nøis himself would later report that, at the time, 20 draft reindeer also were acquired (from Northern Norway) for the sledging journey and that, as he was making his way to Tromsø, large amounts of reindeer moss and additional equipment were all being transported to Tromsø as well, where they would rendezvous with Staxrud (Nøis, 1929, 6). Hilmar Nøis specified that the 20 draft reindeer were acquired from Karasjok during February (Hilmar Nøis Diary SVB).[11] Thus, Northern Norway would be the meeting place for all the equipment, the ship, the men, the dogs, and the reindeer.

## Departure of the Ship, the Supplies, Staxrud, and the Sled Dogs

The sealing vessel *Hertha* set sail from Sandefjord, heading northward, on March 1, according to a Kristiania newspaper report (Aftenposten: Ugens Nyt, 1913, 1). The ship's estimated date of arrival in Tromsø, where it would take on Staxrud, the expeditioners, and the equipment, was March 9th. Presumably due to its rescue mission, it would not be levied any charges during its call at ports in Norway while en route to Spitsbergen, as recommended by the Ministry of Foreign Affairs and port authorities. According to Hilmar Nøis, it was planned to also pick up some supplies in Kristiania on its way to Tromsø (Hilmar Nøis Diary NPI).[12] Nøis specified that, what he described as the ship's full crew, 75 HP steam engine, three masts with reinforced build, and solid oak skin made it especially capable to break the northern ice and transport the rescue expedition members, whom it would meet in Tromsø (Hilmar Nøis Diary SVB).[13] From Tromsø, as reported by the newspaper, the ship and expeditioners would go on to Alten (today Alta, in Finnmark), where the vessel was scheduled to arrive on March 12th, and where it would pick up the three Sámi who had been recruited to work on the expedition (Hansen, Klemmetsen, and Kemi), as well as take 20 reindeer on board (these would be the draft animals in addition to the sledge dogs); the estimated date of departure from Alten (Alta, Finnmark) was March 14 or 15, and the estimated date of arrival at Spitsbergen was March 17 or 18 (Aftenposten: Ugens Nyt, 1913, 1). This was the planned schedule at the time, although, as shall be seen, this schedule would change due to weather delays.

On that same day that the ship sailed from Sandefjord, on March 1st, in Kristiania, Leon Amundsen wrote a letter to his brother Roald Amundsen describing how the three sled dogs at home spent the day pulling apparently delighted children in a sled across the ice (Tahan, 2019, 609; Tahan, 2021, 297). Leon Amundsen also, on that day, prepared not one but two letters to send to Helmer Hanssen in Tromsø (Amundsen Letters of Correspondence). In the first letter,[14] he replied to Hanssen's letter of February 24 in which Hanssen had refused to join the rescue expedition, citing his criticism of Staxrud as the reason for his reluctance, and furthermore stating that he felt Staxrud was presumptuous to believe that Hanssen would partake. In Leon's response to Hanssen, regarding Hanssen's reluctance, Leon assured him that he understood his reasoning. In fairness, however, he explained to Hanssen that "Staxrud wanted first to check whether my brother [Roald] agreed before he asked

You and it is probably thus that it came about that he sent telegram to America before he sent telegram to You." Thus, Leon seems to have been explaining why he thought Staxrud had written directly to Roald Amundsen regarding his request for Hanssen's assistance. Leon went on to assure Hanssen that "Your [Hanssen's] grounds for not joining I find under the circumstances completely satisfactory and I feel proud that my brother in You has such a faithful companion." The negative and demeaning "statement" that Hanssen had attributed to Staxrud regarding Roald Amundsen was described by Leon as "presumably only... expressions of petty and envy and not merely a little thoughtlessness" which did "not particularly hurt" Leon. Furthermore, he declared that he would have no problem in confronting Staxrud with this matter as well as informing Fridtjof Nansen about it. But withholding the sled dogs from Staxrud—that was another matter entirely. The dogs were too important to the rescue effort. They were too vital for the saving of lives. He would indeed send them. Here is what Leon told Hanssen:

> If I meet the person in question [Staxrud] I will nevertheless suggest to him Your reasons and at opportunity also mention the issue for Nansen. – Then comes the question of dogs. I do not dare do anything but lend them out as there are human lives at stake and one otherwise would expose oneself to severe criticism[,] not the least from the German side; had there not been such a necessity it is a matter of course that I would have refused to hand them over when having the information I have received from You.
>
> So thanks for Your letter – I am unhappy that it is not You who will be overseeing the dog-driving as then I know all would have passed in normal manner and in shortest time.[14]

And so, despite the misgivings regarding the dangers of the Arctic rescue mission, and the criticisms about its leader as brought forth by Amundsen's trusted companion, and despite knowing the value of the sled dogs who had returned from the South Pole expedition and their significance to Amundsen, Leon understood the utter importance of lending these dogs to the rescue expedition. And now he had stated this to Helmer Hanssen.

During this time, Staxrud was in Kristiania preparing for his trip. After writing this first letter, Leon wrote yet another on March 1st, in which he informed Hanssen that he had on this very day spoken to Staxrud and that he had "let him then know that the main reason that You [Hanssen] [had] not wanted to take part [in the rescue expedition] was the criticism he had expressed of the latest [Amundsen] Polar-expedition and he will then, regarding this, see You in Tromsø."[15] Thus Leon seems to have had a direct talk with Staxrud, regarding the Helmer Hanssen controversy, during the time of making the final arrangements that included the handing over of the sled dogs. According to Leon, he had broached the subject with Staxrud diplomatically: "I thought I should be allowed to say as much or was it perhaps against Your will?" he continued to Hanssen. "I understood You thus that I could do that[,] and it was done very carefully." Ever the consummate businessperson, Leon had stated what he felt must be said as he made Amundsen's equipment available to Staxrud and as he loaned Amundsen's cherished sled dogs to him.

In addition to borrowing the sled dogs from Amundsen, Staxrud was now also busy procuring the provisions made available by the Amundsens, for which, apparently, it seems the rescue expedition had to pay. Three days later, on March 4th, Staxrud wrote a letter to Leon Amundsen as well as a letter to Alex Nansen, who was the brother of Fridtjof Nansen and a noted attorney in Kristiania. In the letter to Alex Nansen (Staxrud, 4 March 1913b), Staxrud confirmed that, per his conferring with Leon Amundsen and their making available Roald Amundsen's provisions for use by the Arctic rescue expedition, Staxrud had obtained the requested provisions, which came to a total cost of 3260 kroner, for which he was now sending a check in payment via the Central Bank of Norway.[16] In the letter to Leon Amundsen (Staxrud 4 March 1913a), which was nearly identical to the one sent to Alex Nansen, Staxrud also informed Leon that he had on this very day paid the total monetary amount to Alex Nansen, and extended to Leon "the Expedition's heartfelt thanks" for his accommodation.[17] In terms of the provisions themselves, which Staxrud neatly itemized in both letters, these consisted of the following:

10 cases – 1,000 boxes of ½ kilo = 500 kilos of pemmican for humans.

3 cases – 60 boxes of 2 kilos = 120 kilos of meat pemmican for dogs.

9 cases – 180 boxes of 2 kilos = 360 kilos of fish pemmican for dogs.[17]

And so it seems that Staxrud conducted himself professionally as well, in writing, in regard to these transactions and the acquisition of supplies for the rescue mission.

In addition to the food provisions obtained from Roald Amundsen, according to Hilmar Nøis, Amundsen also lent sledges and equipment as well as Polar tents (Hilmar Nøis Diary NPI).[18] Additional equipment that was also collected and sent included 15 sledges—nine from Hagen & co and six from Fridtjof Nansen, according to a Kristiania newspaper (Aftenposten: Ugens Nyt, 1913, 1). This equipment, according to the paper, was sent out, destined for Tromsø, on March 4th or 5th.

Last—but not least—to leave for Tromsø, from Kristiania, were the leader Staxrud, the physician Dr. Bøckmann, and the two South Pole expedition sled dogs Lussi and Storm. They took the train to Trondheim, where they would transfer to a steamship and continue on to Tromsø the following evening (Aftenposten: Ugens Nyt, 1913, 1). The two men and two dogs departed from Kristiania on March 5, as later reported in the Kristiania newspaper *Aftenposten* (1913, evening: 4). A newspaper article reprint dated March 6 seems to give the impression they left on that day ("today") of the 6th (Aftenposten: Ugens Nyt, 1913, 1), but most likely it is in reference to the previous day of the 5th, as they arrived in Trondheim on the morning of March 6, as documented in a newspaper article hand-dated by Staxrud (Dagsposten, 1913).

By the time of Staxrud's departure from Kristiania, Leon had decided to loan to him both the younger dogs Lussi and Storm but not the older dog Obersten—it seems that the last surviving South Pole sled dog would not be risked on the rescue mission. "Captain Staxrud and doctor Bøckmann left this afternoon at 3:20 o'clock from here [Kristiania] to Trondhjem, where tomorrow they will organize the reloading [transshipment] of the expedition's equipment from the railway to the steamship, which departs from there tomorrow evening on express voyage [Hurtigruten] to Tromsø," reported the newspaper, subsequently adding: "And from here [Kristiania] Staxrud today brought [with him] Roald Amundsen's two dogs 'Storm' and 'Lucie' [Lussi]. 'Obersten' will not be joining [them] northward" (Aftenposten: Ugens Nyt, 1913, 1).

And so, armed with a physician/skier to help the injured expeditioners stranded on their ship and in their hut in Spitsbergen, and two powerful, trained, and famous sled dogs to work on the sledging journey across the Spitsbergen ice, Staxrud set forth on his rescue mission to save the remaining members of the Schröder-Stranz expedition. By March 5th, three important elements of his rescue mission were converging on the northern city of Tromsø: The sealing vessel *Hertha* with its ship's crew from Sandefjord, southern Norway; the Nøis family members of overwintering experts Daniel Nøis, Hilmar Nøis, and Johan Nilsen Nøis, along with seven trained sledge dogs, most from Vesterålen, Nordland; and Arve Staxrud himself, with Dr. P. W. K. Bøckmann, and the South Pole expedition sled dogs Lussi and Storm, from Kristiania and Svartskog. Still awaiting them in Northern Norway were Per Hansen, Samuel Klemmetsen, and Johannes Kemi, as well as 20 capable draft reindeer.

# Notes on Original Material and Unpublished Sources

Roald Amundsen letters of correspondence, written from and to Roald Amundsen and Leon Amundsen, are in the Manuscripts Collection at the National Library of Norway (NB) in Oslo. Hilmar Nøis's diary SVB is in the Historical Archive at the Svalbard Museum (SVB) in Svalbard. Hilmar Nøis's diary NPI is in the Library Archives at the Norwegian Polar Institute (NPI) in Tromsø. Arve Staxrud letters of correspondence, written from and to Arve Staxrud, are in the Manuscripts Collection at the National Library of Norway (NB) in Oslo.

1. L. Amundsen to R. Amundsen, letter, 25 February 1913, NB Brevs. 812:3:7.
2. L. Amundsen to R. Amundsen, letter, 28 February 1913, NB Brevs. 812:3:7.
3. H. Nøis diary SVB, 1913, journal page 48, SVB-AP2.
4. H. Nøis diary NPI, notebook 3, 1913 mid-February, pdf page 6, (NPI), D-307/D00307_3_0001.
5. H. Nøis diary NPI, notebook 3, 1913 mid-February, pdf page 6, (NPI), D-307/D00307_3_0001.
6. H. Nøis diary NPI, notebook 3, 1913 mid-February, pdf page 6, (NPI), D-307/D00307_3_0001.
7. H. Nøis diary SVB, 1913, journal page 48, SVB-AP2.
8. H. Nøis diary NPI, notebook 3, 1913 mid-February, pdf page 6, (NPI), D-307/D00307_3_0001.
9. H. Nøis diary NPI, notebook 3, 1913 mid-February, pdf page 6, (NPI), D-307/D00307_3_0001.
10. H. Nøis diary NPI, notebook 3, 1913, pdf page 7, (NPI), D-307/D00307_3_0001.
11. H. Nøis diary SVB, 1913, journal page 48, SVB-AP2.
12. H. Nøis diary NPI, notebook 3, 1913, pdf page 7, (NPI), D-307/D00307_3_0001.
13. H. Nøis diary SVB, 1913, journal page 49, SVB-AP2.

14.  L. Amundsen to H. Hanssen, letter a, 1 March 1913, NB Brevs. 812:3:7.
15.  L. Amundsen to H. Hanssen, letter b, 1 March 1913, NB Brevs. 812:3:7.
16.  A. Staxrud to A. Nansen, letter, 4 March 1913, NB Brevs. 812:1.
17.  A. Staxrud to L. Amundsen, letter, 4 March 1913, NB Brevs. 812:1.
18.  H. Nøis diary NPI, notebook 3, 1913, pdf page 7, (NPI), D-307/D00307_3_ 0001.

## Unpublished Sources and Original Material

Amundsen Letters of Correspondence, Manuscripts Collection, National Library of Norway, Oslo.

Nøis, D. Daniel Nøis Biography/Norwegian Polar Institute, "Nøis_Daniel_ Biografiarkiv_1", written by H. M. [possibly Haakon Aronsen, aka Haakon Magnus, per Gunnhild Holmen of the National Library of Norway, based on *Andøya Avis* 1964 article, according to communications from Odd Ivar Nøis Olsen of the Nøis Family and from Ivar Stokkeland and Petr Masat of the Norwegian Polar Institute, received 29 March 2022], Biography Archive/Library Archives, Norwegian Polar Institute, Tromsø. Received 20 July 2021.

Nøis, H., Diary NPI, handwritten journal of recollections recounting the events of the 1913 Staxrud expedition, Hilmar Nøis, "Minder og Fortelinger fra fangsttiden på Spitsbergen, Klade 3, 7 Mars 1970" [Hilmar Nøis, "Memories and Stories from the trapping time on Spitsbergen, Notebook 3, 7 March 1970"], dated 7 March 1970, D-307, ("Minder og fortelinger, Nøis, Hilmar, hefte 3"), D00307_3_0001. Library Archives, Norwegian Polar Institute, Tromsø, https://brage.npolar.no/npolar-xmlui/ handle/11250/274077. Accessed 20 July 2021.

Nøis, H., Diary SVB, handwritten journal of recollections including an account of the 1913 Staxrud expedition, ("Hilmar Nøis. Fangstmann. 1909–1923, dagbok med erindringer fra fangstlivet" ["Hilmar Nøis. Trapper. 1909–1923, diary with recollections from the trapping life"], not dated (latest year refer-enced is 1942 in journal pages and 1950 in inserted sheet), SVB-AP2-Hilmar-Nøis-dagbok-1909–1923-compressed. Historical Archive, Svalbard Museum, Svalbard, https://svalbardmuseum.no/no/samlingene/historisk-arkiv/arkivprosjekt-fangst-og-annen-overvintringsvirksomhet-fra-perioden-1910-til-1970/. Accessed 6 June 2021.

Staxrud, A., 4 March 1913a, Letter from Arve Staxrud to Leon Amundsen, NB Brevs. 812:1. National Library of Norway, Oslo, online archives, https://www.nb.no. Retrieved 6 July 2021.

Staxrud, A., 4 March 1913b, Letter from Arve Staxrud to Alex Nansen, NB Brevs. 812:1. National Library of Norway, Oslo, online archives, https://www.nb.no. Retrieved 6 July 2021.

Staxrud, A., 17 March 1913, Letter from Arve Staxrud to Fridtjof Nansen, with two newspaper article clippings enclosed featuring handwritten notation (*Nord Norge* [Tromsø], 21 February 1913, pages 1–2, and *Dagsposten*, Trondhjem [Trondheim], 6 March 1913), NB Ms.fol. 1924:6:A:2. National Library of Norway, Oslo, online archives, https://www.nb.no. Retrieved 6 July 2021.

# References

Aftenposten. (25 February 1913). Undsætnings-expeditionen til Spitsbergen. Fartøiet og deltagerne. Slæde-expeditionens udrustning. [The rescue expedition to Spitsbergen. The vessel and the participants. The sled expedition's equipment.]. *Aftenposten* (Kristiania [Oslo]), 25 February 1913, evening edition, p. 1. Library Archives, Norwegian Polar Institute, Tromsø. Received 22 July 2021. (Also in the National Library of Norway, Oslo, online archives, www.nb.no. Accessed 7 April 2022).

Aftenposten. (15 May 1913). Hjælpeexpeditionen paa Spitsbergen. Kaptein Staxruds beretning til "Aftenposten". [The rescue expedition on Spitsbergen. Captain Staxrud's account to "Aftenposten".]. *Aftenposten* (Kristiania [Oslo]), 15 May 1913, evening edition, pp. 1–2. National Library of Norway, Oslo, online archives, www.nb.no. Accessed 28 August 2012.

Aftenposten. (3 September 1913). Kaptein Staxrud hjemme igjen. Udtaler sig til "Aftenposten" om redningsexpeditionen. [Captain Staxrud back home again. Speaks to "Aftenposten" about the rescue expedition.]. *Aftenposten* (Kristiania [Oslo]), 3 September 1913, evening edition, p. 4. National Library of Norway, Oslo, online archives, https://www.nb.no. Retrieved 4 April 2022.

Aftenposten: Ugens Nyt. (27 February 1913). Undsætnings-expeditionen til Spitsbergen. *Ugens Nyt* published by *Aftenposten* (Kristiania [Oslo]), 27 February 1913, pp. 1–2. National Library of Norway, Oslo, online archives, https://www.nb.no. Retrieved 7 April 2022.

Aftenposten: Ugens Nyt. (6 March 1913). Hjelpeexpeditionen til Spitsbergen. [The rescue expedition to Spitsbergen.]. *Ugens Nyt* published by *Aftenposten* (Kristiania [Oslo]), 6 March 1913, p. 1. National Library of Norway, Oslo, online archives, https://www.nb.no. Retrieved 4 April 2022.

Andøya Avis. (6 October 1950). Fangstmann og båtbygger Daniel Nøis runder 70 år. [Trapper and boat builder Daniel Nøis turns 70 years old.]. Written by H. M. [possibly Haakon Aronsen, aka Haakon Magnus, per Gunnhild Holmen of the National Library of Norway, based on *Andøya Avis* 1964 article, according to communications from Odd Ivar Nøis Olsen of the Nøis Family and from Ivar Stokkeland and Petr Masat of the Norwegian Polar Institute, received 29 March 2022]. *Andøya Avis* (Andøya), 6 October 1950, p. 1. Library Archives, Norwegian Polar Institute, Tromsø. Received 29 March 2022.

Dagsposten. (6 March 1913). Hos Kaptein Staxrud. Han udtaler sig om Planerne—Man faar høre fra Undsætningsekspeditionen i Slutten af Marts. [At Captain Staxrud's. He speaks about the Plans—One will hear from the Rescue-expedition at the End of March.]. Written by Th. K. *Dagsposten* (Trondhjem [Trondheim]), 6 March 1913. (Newspaper clipping featuring handwritten notation by Staxrud, attached to letter from Arve Staxrud to Fridtjof Nansen dated 17 March 1913, NB Ms.fol. 1924:6:A:2). National Library of Norway, Oslo, online archives, https://www.nb.no. Retrieved 6 July 2021.

Der Zeitspiegel Berlin im Frühjahr (1913). *Newspaper clipping.* Ritscher Estate, Cornelia Lüdecke, München.

Hamburger Nachrichten. (1913a). Professor Miethe über die Hilfsexpedition nach Spitzbergen. *Hamburger Nachrichten*, 25 February 1913.

Hamburger Nachrichten. (1913b). Notruf für die Spitzbergen-Rettungs-Expedition. *Hamburger Nachrichten*, 11 March 1913.

*Hanseater.no*. Etterkommere av Hanseatene i Bergen: Geolog Jakob Andreas Martin Ellingsen. The Hanseatic Museum, and Slekt og Data Hordaland, "Hansa" Project. http://hanseater.no/tng/get person.php?personID=I14579&tree=hansa. Accessed 19 September 2023.

Hoel, A. (ed.). (1929). *Resultater av de Norske Statsunderstøttede Spitsbergenekspeditioner (Skrifter om Svalbard og Ishavet), Bind I, Nr. 1, Adolf Hoel: The Norwegian Svalbard Expeditions 1906–1926*. Utgitt på Den Norske Stats bekostning ved Spitsbergenkomiteen. Oslo: (I Kommisjon Hos) Jacob Dybwad. https://brage.npolar.no/npolar-xmlui/bitstream/handle/11250/173618/Skrifter001.pdf?sequence=1&isAllowed=y Accessed 19 September 2023.

Nidaros. (26 February 1913). Undsætningen til Spitsbergen. Amundsens Sydpols-hunder blir med. [The rescue to Spitsbergen. Amundsen's South Pole dogs join.]. *Nidaros* (Trondheim), 26 February 1913, p. 1. National Library of Norway, Oslo, online archives, https://www.nb.no Retrieved 7 April 2022.

Nidaros: Trøndelagen. (1 March 1913). Undsætningen til Spitsbergen. Amundsens Sydpols-hunder blir med. [The rescue to Spitsbergen. Amundsen's South Pole dogs join.]. *Nidaros: Trøndelagen* (Trondheim), 1 March 1913, p. 3. Digital Archives, National Library of Norway, Oslo, online archives, https://www.nb.no. Received 8 June 2021; retrieved 7 April 2022.

*Nøis (slekt)—lokalhistoriewiki.no*. Nøis Family. https://lokalhistoriewiki.no/wiki/Nøis_(slekt). Accessed 29 March 2022 and 20 June 2022.

Nøis, D. (1929). Med kaptein Staxrud på leiting etter Schrøder-Stranz og hans folk [With Captain Staxrud on the search for Schröder-Stranz and his people]. In: *And-Ungen*, July 1929, pp. 4–12. Andenes: Andøyposten. Library Archives, Norwegian Polar Institute, Tromsø. Received 20 July 2021.

Norsk Polarhistorie—Polarhistorie.no. Staxrud's Rescue Expedition. "Staxruds unnsetningsekspedisjon etter Schrøder-Stranz, 1913." ["Staxrud's rescue expedition after Schrøder-Stranz, 1913"]. (Norwegian Polar Institute, University of Tromsø, Troms County Municipality). http://www.polarhistorie.no/ekspedisjoner/Staxruds%20unnsetningseksp.html. Accessed 22 August 2012, and 19 September 2023.

Norwegian Polar Institute. (2003). *The Place Names of Svalbard*. Rapportserie nr. 122. G. S. Jaklin, technical editor. (O. Orheim, director; A. Urset, committee chair.) Peder Norbye grafisk. Tromsø: Norwegian Polar Institute. https://brage.npolar.no/npolar-xmlui/bitstream/handle/11250/173470/Rapport122.pdf?sequence=1&isAllowed=y. Accessed 19 September 2023.

*Norwegian Polar Institute*. Hertha. https://data.npolar.no/vessel/hertha. Accessed 24 July 2021.

Social-Demokraten. (27 February 1913). Hjælpeekspeditionen til de indefrosne tyskere. *Social-Demokraten* (Kristiania [Oslo]), 27 February 1913, p. 2. National Library of Norway, Oslo, online archives, www.nb.no. Retrieved 7 April 2022.

Staxrud, A. (1914). Die Staxrudsche Hilfs-Expedition für Schröder-Stranz. In A. Miethe (ed.), *Die Expedition zur Rettung von Schröder-Stranz und seinen Begleitern - geschildert von ihren Führern Hauptmann A. Staxrud und Dr. K. Wegener* (pp. 1–68). Berlin: Dietrich Reimer.

Svalbard Museum. (2023a). https://www.arkivportalen.no/contributor/no-SVAM_arkiv_000000000112. Accessed 3 March 2023.

Svalbard Museum. (2021a). https://www.arkivportalen.no/contributor/no-SVAM_arkiv_0000000 00361?ins=SVAM and https://bildearkiv.svalbardmuseum.no/fotoweb/archives/. Accessed 6 June 2021.

Svalbard Museum. (2021b). https://www.arkivportalen.no/entity/no-SVAM_arkiv_0000000 00022?ins=SVAM and https://www.arkivportalen.no/contributor/no-SVAM_arkiv_0000000 00032. Accessed 7 June 2021.

Tahan, M. R. (2019). *Roald Amundsen's Sled Dogs: The Sledge Dogs Who Helped Discover the South Pole*. Cham: Springer International Publishing.

Tahan, M. R. (2021). *The Return of the South Pole Sled Dogs: With Amundsen's and Mawson's Antarctic Expeditions*. Cham: Springer International Publishing.

Trondhjems Adresseavis. (26 February 1913). Undsætningsekspeditionen. Tre av Amundsens hunde blir med. [The Rescue expedition. Three of Amundsen's dogs join.]. *Trondhjems Adresseavis*

(Trondheim), 26 February 1913, p. 1. National Library of Norway, Oslo, online archives. https://www.nb.no. Retrieved 7 April 2022.

Trondhjems Folkeblad. (1 March 1913a). Undsætnings-ekspeditionen. Professor Miethe kritiserer Schrøder Stranz. [The Rescue Expedition. Professor Miethe criticizes Schrøder Stranz (sic).]. *Trondhjems Folkeblad* (Trondheim), 1 March 1913, p. 2. National Library of Norway, Oslo, online archives. https://www.nb.no Retrieved 4 April 2022.

Trondhjems Folkeblad. (1 March 1913b). Undsætningsekspeditionen. Tre av Amundsens hunde blir med. [The Rescue expedition. Three of Amundsen's dogs join.]. *Trondhjems Folkeblad* (Trondheim), 1 March 1913, p. 3. National Library of Norway, Oslo, online archives, https://www.nb.no. Retrieved 4 April 2022.

# Chapter 9
# The Journey Northward: Arve Staxrud and Dr. P. W. K. Bøckmann, Lussi and Storm, Daniel Nøis and Hilmar Nøis, and Per Hansen, Johannes Kemi, and Samuel Klemmetsen

## Arrival in Trondheim: The Staxrud Plan Is Unveiled

The rescue expedition leader Captain Arve Staxrud, the expedition physician Dr. P. W. K. Bøckmann, and the South Pole expedition sled dogs Lussi and Storm traveled by train from Kristiania en route north to Tromsø and arrived in Trondheim on the morning of March 6 (*Dagsposten* 6 March 1913).

On this occasion, Staxrud gave the only interview he would give to the press. He granted an interview to the newspaper *Dagsposten* in Trondheim on that same day of March 6—a busy day in which he would transfer his equipment and continue his travels northward. In the interview, he outlined his plans for the rescue mission and warned about the lateness of the season, contending that late March was the coldest time of year in Spitsbergen and proclaiming that, depending on the ice conditions, he hoped to reach the stranded expeditioners at the latest by mid-April.

The article that was published that day must have been significant to him, as he kept a copy of it for himself and would later send the article clipping to Fridtjof Nansen, 11 days after arriving in Tromsø (Staxrud 17 March 1913). On the clipping, above and surrounding the headline, Staxrud handwrote the following: "This is the only journalist, I have spoken with since the departure from Kristiania. 17/3 13 Arve Staxrud."[1]

The newspaper interview was presumably conducted at the Britannia Hotel, where Staxrud and Bøckmann had checked in upon arrival and where the newspaper reporter—Th. K.—evidently had sought him out for comments on his Spitsbergen plans (*Dagsposten* 6 March 1913)—this most likely being "the only journalist" to whom Staxrud had referred. Headlined "At Captain Staxrud's," the article seemingly afforded Staxrud the opportunity to reveal his plans in his own words. Featured in the report are his statements about the ship and crew, the route, the methodology, and the expedition members, including the animals, with a particular strategy revealed about the draft reindeer. Of special concern to Staxrud were the health and vulnerability of the stranded expeditioners who were injured with frostbite and their susceptibility to be stricken by scurvy. He was quoted in the newspaper as saying:

> The sealing ship "Hertha", which has a 16-man crew, will take us up as fast as possible and the landing will take place where the ice conditions are best, most likely the landing will take place in the Icefjorden [Ice Fjord] or in Cross Bay or by the Amsterdamøen [Amsterdam Island]. The vessel then goes out sealing and the land expedition[,] which consists of 9 men, of these [being] 3 Sámi[1] with approx.[imately] 20 reindeer[,] as well as 18 dogs[,] then go shortest route to the ice-locked vessel in Treurenburg [Treurenberg] Bay. The cabins in Wijde Bay, where some of the casualties have found refuge[,] will be investigated. Likewise will if possible an expedition be sent out from Treurenburg [sic] Bay over to the Nordøstlandet [Northeast Land] in order to search for Schrøder-Stranz [sic] and his three mates, if these have not returned to the vessel. One has at least certain hope of finding some of the unlucky expedition members, both Germans and Norwegians, alive but unfortunately it is feared, that some of these, particularly then those who are frostbitten, have died. An accidental injury like frostbite contributes to making people exposed to scurvy, because they have to stay put. The reindeer [we have] brought are therefore very useful for nourishment as fresh meat halts the scurvy.

> (*Dagsposten* 6 March 1913)

Thus the frostbitten and marooned expeditioners, both the German and Norwegian members, were top of mind for Staxrud, who feared for their fate and who had strategized the approach to Spitsbergen and the retrieval of the expeditioners. To this end, he had planned every aspect of the rescue mission and had selected the appropriate rescue team members—all experienced in their areas. He had strategized the use of 18 sledge dogs whom he would enlist to comprise the pulling force. Included in his strategy was the selection of reindeer, whom he intended to utilize for the multipurpose of motive power and meat protein—the 20 reindeer would pull sledges and then be used as meat to prevent scurvy among the frostbitten and injured men. In this last part of his strategy, he most likely was influenced by his mentor and supporter Fridtjof Nansen. Staxrud would later write, in his post-expedition report published in Germany in 1914, that the reindeer he had intended to use per Nansen's suggestion would also provide a meat reserve for emergencies (Staxrud, 1914, 6).

Interestingly, in the *Dagsposten* article published while he was still in Trondheim, Staxrud had mentioned that the sealing ship *Hertha* would proceed to hunt seals in the Spitsbergen waters after the rescue expeditioners had disembarked and during the time that the rescue expeditioners were to conduct their search on land. Evidently, the business of seal hunting would continue on the hired ship while the rescuers were working and searching on land. Treurenberg Bay, where the stranded expedition's ship lay locked in the ice, and Wijde Bay, where some of the expeditioners had taken refuge in huts, as well as Northeast Land, where Schröder-Stranz himself had disappeared, were the primary areas of target for the search-and-rescue team.

---

[1] Hilmar Nøis uses the name "Sámi" in one of his diaries (NPI), but uses the at-that-time contemporary term "Lapp" in his other diary (SVB); Daniel Nøis, in his article, also uses the older term "Lappene" or "lappen"; and Arve Staxrud uses the old term "Lapps" in his March 11th letter and in quoting A. E. Nordenskiöld; this term is no longer in use and will not be used in this book, as the proper name is Sámi, and therefore Sámi is the name used throughout this book text.

The *Dagsposten* news article went on to describe Staxrud's thoughts on the unpredictability of the weather and the ice conditions in the region (*Dagsposten* 6 March 1913). No one could gauge the conditions of the ice at any one moment, he was reported as saying, and furthermore, no one had ever attempted to reach Spitsbergen during a season such as the one before them. "It depended on a chance," the article continued. "If there was no ice up there, so that the landing could go ahead easily, one will by the end of March have reached the people; but [if] the conditions appeared less favorable, one would not be able to get there until the first half of April." It was expected, reported the paper, that Staxrud's expedition would send a telegram from Green Harbour at the end of the month of March, presumably updating the outside world as to the rescue efforts' progress.

The article itself continued with a realistic assessment of the challenges but a hopeful and optimistic prognostication of the outcome, centering on the extreme skill and competency of the expeditioners involved:

> There are probably many difficulties which will confront the expedition; but it is made up of such capable and experienced people, that the difficulties would have to be great, were they not to overcome them. Of the Arctic Ocean travelers who are going[,] some have overwintered several times, up to 4 years, in the regions, here in question. It should therefore for any other nation than the Norwegian be impossible to provide such experienced, completely capable people as here has been done. (*Dagsposten* 6 March 1913)

The point of this paragraph, then, is to convey the message that only Norwegians could provide this level of expertise in this area, and hence Staxrud's selection. Interestingly, these words echo Staxrud's own words written in his letter to the Foreign Minister on February 17 (Chap. 6).

As to the weather conditions, the article mentioned that the meteorological data of the time pointed to the fact that "the end of March is the coldest season" and that "the temperature has dropped right down to [minus] 47 degrees [Celsius]"—with conditions being variable and unstable (*Dagsposten* 6 March 1913).

As for the necessary supplies to work in the cold, the article reported that the clothing, equipment, weaponry, and ammunition were being loaned to the expedition by the Army stores. Transportation of the expedition's equipment and members to Tromsø was being provided as a donation, at no cost, by the national railway company Statsbanerne and the Trondheim-based shipping company Nordenfjeldske Dampskibsselskab, who partly owned and operated the Hurtigruten public coastal route steamship transport. In Tromsø, the expedition members and equipment would "be gathered" and from there they would set out for Spitsbergen, concluded the article.

Having successfully given this interview to the newspaper laying out his rescue plan to the public, and having spent the day reloading the expedition's equipment from the railway train to the Hurtigruten steamship, Staxrud, along with Bøckmann, Lussi, and Storm, left Trondheim in the evening, aboard the steamer destined for Tromsø.

## The Rendezvous in Tromsø: The Nøis Family Joins

Arve Staxrud, Dr. Bøckmann, Lussi, and Storm arrived in Tromsø on March 9, according to a Tromsø newspaper article (*Tromsø Stiftstidende* 11 March 1913b, 2). Based on the article, they apparently arrived to modulated fanfare and much public curiosity. The four had taken the Hurtigruten from Trondheim and were now finally at the rendezvous meeting point for most of the rescue expedition members. But they were also at the center of the place of controversy—the cradle of criticism from which negative comments had emanated from Norway regarding Staxrud, from both Tromsø and its treasured resident Helmer Hanssen. Interestingly, what drew attention to Staxrud at first, upon his arrival, was the fact that he was accompanied by the two South Pole expedition sled dogs Lussi and Storm. A newspaper article heralding his arrival began with this news: "On Sunday, captain Staxrud and doctor Bøckman [sic] came with the hurtigruten [sic] from Trondhjem [Trondheim]. They brought with them the two dogs 'Storm' and 'Lucie' [Lussi] belonging to Roald Amundsen, as well as a lot of equipment, ski-sledges, tents, provisions, etc. The two beautiful, stout dogs from the famous 'Fram-voyage' aroused general and enthusiastic interest." This was how the article began, and how it had announced Staxrud's arrival for his rescue mission to Spitsbergen. It was the dogs who were first and foremost garnering the attention.

Immediately after the positive statement about the arrival of the men, the sled dogs, and the equipment, the newspaper article seemed to attempt to bring the controversy about Staxrud to a conclusion—to smooth ruffled feathers, and to place the focus on the rescue efforts. "We do not intend—as we have stated before—to engage in any prior criticism or debate, especially since the expedition is a fact," continued the article. "Just a few things. One of Amundsen's companions stated, among other things: 'Staxrud has got the handsomest and most talented guys imaginable.'" Whether this quote was from Hanssen or another individual, and whether it referenced the two dogs or certain human members of the expedition, is unknown at this time. The article then launched into an assessment of said equipment—its qualities, how it was acquired, and where it was obtained. Noting that most of the equipment necessarily needed to be obtained in Kristiania and was hand-selected and supervised by the leader Staxrud, the article acknowledged its advantages: the special processing of the ski-sledges; the lightness and practicality of the tents, which consisted of exterior cloth lined with interior silk and a sensible sewn construction that allowed a small window for entrance; and the pemmican that was brought from Amundsen's own provisions. These items were part of the strategy to make sure that the expeditioners were appropriately equipped and that they maintained a light weight for traveling over an extended period of time. The draft reindeer acquired, too, were mentioned as "quite important" in that they would also serve as meat to place in depots along the way.

Then the article brought up the topic of Helmer Hanssen, and his being requested to work on the expedition. It seems that what had begun as controversy now turned to contrition—the newspaper article now actually issued an apology to Staxrud.

Explaining that previous sentiment had been based on a misunderstanding, the article went to great lengths to sort out the situation and recount the chronology of events. According to the paper, there had been "astonishment and criticism" when it had been perceived that Staxrud had sent a telegram directly to Roald Amundsen to ask him for his permission to have Hanssen work on the rescue expedition, rather than asking Hanssen himself first. The newspaper itself had been extremely critical of this *faux pas*, said the article. But they had been under the wrong impression and now wanted to correct the record. Staxrud, said the article, had actually telephoned Tromsø's city councilor first to request the name of any individual from Tromsø who might be willing to partake in the expedition, and the councilor had quickly recommended Helmer Hanssen. The expedition leader Staxrud had then first sent a telegram to Helmer Hanssen, and after that had called Leon Amundsen. It was Leon, as Roald Amundsen's brother and business manager, said the article, who had initiated sending a telegram to Roald Amundsen in America and had received, on the following morning, Roald's response "that both the two dogs could be included and that Mr. Helmer Hansen [sic] had his permission," said the article. Staxrud then had proceeded to send a second telegram to Hanssen and, as with the first telegram, received no response from Hanssen. It was Leon as business manager, and not Staxrud, who had sent the telegram to Roald Amundsen regarding Helmer Hanssen, the paper clarified. "There is thus absolutely no tactlessness committed by Captain Staxrud," proclaimed the paper, who extended their "sincere apology" to Staxrud for their negative "utterance" based on "erroneous" information.

And so it seems that the paper attempted to correct itself and apologize for prior criticism. Staxrud did not audaciously ask Amundsen for permission before asking Hanssen, as it had previously claimed, it now said. Indeed, Staxrud had telegraphed Hanssen first and had received no answer, and then had asked Leon, who took it upon himself to telegraph Roald for permission. (The article states that Roald had given permission for both the two sled dogs and for Hanssen, but, as seen in his telegram of February 23, he did not specify the number of dogs, as at that time there was mention of all three sled dogs going.) Staxrud had telegraphed Hanssen twice with no reply, the article further stated. Evidently, an offering of peace and resolution was being extended from the newspaper. Staxrud, however, is not quoted at all in the article, and there is no evidence that he gave them an interview as he had done with the Trondheim paper, which is in keeping with his letter to Nansen in which he later stated that the Trondheim reporter was the only journalist with whom he had spoken.

Indeed, at this time, Staxrud seems to have been selective with his press interaction, issuing a few interviews and statements—and later telegrams—as his main method of communicating, but being careful regarding the media and the messaging.

His important business at hand was to meet with the members of the expedition party who were scheduled to gather in Tromsø and to proceed to the North from there.

Staxrud had arrived in Tromsø immediately after the arrival of Daniel Nøis and his family members, who had been traveling for several days, from the Vesterålen region, so as to all meet together in Tromsø (Nøis, 1929, 6). The Nøis family members who rendezvoused with Staxrud were Daniel Nøis, Hilmar Nøis, and Johan Nilsen Nøis, as well as Martin Pettersen Nøis, who had joined later. Daniel Nøis had brought with him seven sledge dogs. Also rendezvousing with the Staxrud party was the engineer Jakob Ellingsen. But missing from the party was the ship that would transport them to the ice. *Hertha* had encountered severe storms along its way north, slowing down its travel up the coast, and causing it to require an inordinately long period of time to reach Tromsø.

The ship was drastically delayed, which in turn delayed the entire expedition party.

## The Nansen Factor: Revelations from the Leader

Staxrud used some of the delay time in Tromsø to write to his ardent supporter Fridtjof Nansen.

On March 11th, the day that the newspaper published the article about his arrival in Tromsø, Staxrud prepared a letter to Nansen in Kristiania informing the Polar explorer about the ship's delay, identifying to him the names of the three Sámi who would be working on the rescue expedition, and refuting criticism he had apparently received regarding his choice of reindeer as draft animals (Staxrud 11 March 1913). "'Hertha' has been somewhat delayed because of the strong storm, which the last eight days has raged around the coast," he began. "Captain Virik sends telegram today, that he hopes to be in Tromsø the 16th [of] this [month].—The harbor-master in Haugesund has on telegram request answered, that 'Hertha' is leaving Haugesund this morning—directly to Tromsø."[2] Updating Nansen that the gear and equipment which had been transported had all arrived successfully, Staxrud went on to inform Nansen that the names of the three Sámi who would be part of the rescue mission were Per Hansen, Johannes Kemi, and Samuel Klemmetsen. He then broached the subject of using reindeer (which had been recommended by Nansen as well) and his rationale for doing so. Citing Adolf Erik Nordenskiöld's praise of reindeer and the Northeast Passage explorer's first-hand account of his work with this animal as motive power, Staxrud defended his strategy to use reindeer for the rescue operation in Spitsbergen:

> In regard to the criticism, which has been put forward about the use of reindeer as draft-animals, it may perhaps be of interest to see, what A. E. Nordenskiöld writes on this from his overwintering 1872–73. I have not been able to get hold of his book, but in Ymer 1902 page 173 A. G. Nathorst writes [in Swedish]: "The reindeer partook from the journey's start, hitched to a pulka and was observed by Nordenskiöld with an entirely particular interest: 'I can safely recognize that it exceeded our expectations. The reindeer pulled, even though the Sámi [See footnote 1] explained, that it was not one of the best, more than 85 kg (a good reindeer pulls 128 kilogram), was calm and easily driven like an old workhorse, ate with greediness the moss which was brought along and provided, after this [moss] was all eaten, [the reindeer when] slaughtered [was] an excellent meat. With forty such draft-animals and the Parry Island [Parryøya] as starting point we could certainly have gotten pretty far north even under such unfavorable ice-conditions as those, which this year have been prevalent north of Northeast Land.' …"[3]

Having reassured Nansen of his resolve and rationale to use reindeer, Staxrud then turned to the subject of the other animals on whom he would rely for motive power: The sled dogs. "I have tried to get more dogs, but so far without success," he concluded to Nansen, thus ending his letter with this important matter still to be resolved.[4]

Evidently Staxrud, under the tutelage of Nansen, had fully embraced the use of reindeer for the multiple purposes he mentions. The reindeer would be utilized for some of the sledge-pulling, especially in light of the shortage of sled dogs, and they would be used for meat as well. It also appears that Staxrud was spending his delay time in Tromsø, while awaiting the ship, continuing to attempt to acquire the much-needed additional sled dogs to join the expedition party.

The tardiness of the ship and the telegram from its captain also made news in Tromsø. The newspaper which had published the article about the Staxrud expedition's arrival also printed a short statement, on the same page, announcing that *Hertha*'s arrival had been delayed due to storms and that it was now pushed to at least March 16th (*Tromsø Stiftstidende* 11 March 1913a, 2).

Still waiting, Staxrud took the opportunity to write to Nansen again on the following day of March 12, sending to him a telegram in which he alerted Nansen about the delay in the ship's anticipated arrival, updated Nansen that Staxrud was not responding to any negative comments or criticism from Tromsø, and thanked Nansen for his confidence (Staxrud 12 March 1913). "Hertha somewhat delayed due to storm," Staxrud wrote. "Master hoping be Tromsø 16. Some statements credited me Tromsø-posten 11 pure distortion. Answering on principle no Tromsø-paper. Sending a written account regarding various Tromsø-conditions[,] all in order our matters, just waiting for 'Hertha'. Heartfelt thanks for the trust that I have been shown."[5] It seems that Staxrud desired to communicate to Nansen what he termed as false reports from Tromsø, seemingly feeling that he had been maligned and reiterating his refusal to reply to Tromsø newspapers.

The 13th of March saw Staxrud writing to his partner Adolf Hoel with an urgent request (Staxrud 13 March 1913). "Send immediately Hammerfest the sextant you had lent out last Summer also a pocket-chronometer," he wrote in the telegram addressed to the university in Kristiania.[6] Evidently Staxrud was planning to conduct necessary surveying tasks along the way during the icy route to rescue the stranded expeditioners in Spitsbergen. He continued the telegram with the words "Ellingsen greetings"—thus extending greetings to Hoel from their colleague, the engineer Ellingsen, who had by this time joined the rescue expeditioners in Tromsø. And he concluded the telegram with a one-word request for Hoel to telegraph him back.

## Storm Delays and Stormy Relations

The day prior to the Tromsø newspaper article announcing the arrival of Staxrud, Bøckmann, Lussi, and Storm, on March 10th, Leon Amundsen in Kristiania was writing a letter to his brother and business client Roald Amundsen reassuring the

South Pole explorer yet again that his sled dogs should be safe in the hands of Arve Staxrud: "Obersten is home alone with Jørgen [Stubberud] and thrives excellently, the two others [Lussi and Storm] have traveled and Staxrud has been given strict orders to look properly after them," wrote Leon (Tahan, 2021, 298).

Two days after Staxrud's urgent telegram to Adolf Hoel, on March 15, while Staxrud still awaited his ship in Tromsø, Leon penned a letter to Helmer Hanssen, to which Hanssen would later reply on March 30 (Amundsen Letters of Correspondence). Based on the content of Hanssen's letter, it seems that the two men—Staxrud and Hanssen—had not spoken to one another or attempted any communication while Staxrud remained in the city.

Leon's letter to Hanssen alerted him to the fact that he had read in the papers that Hanssen was about to send a letter to him and he wanted to notify Hanssen that he would be absent for 14 days (Amundsen Letters of Correspondence). He also desired to update Hanssen that he had by now divulged to Roald Amundsen the status of the situation regarding Hanssen's feelings about Staxrud: "I have now referenced to my brother Your letters and one thing You can be certain of and that is that he will be feeling proud and happy when hearing about Your great devotion for him."[7]

To this Hanssen responded with a letter further reiterating that his words about Staxrud were true and intending to provide proof of his story's veracity (Amundsen Letters of Correspondence). In the letter he indicated that he was forwarding what he stated to be the statement made by Staxrud which he had described in his earlier letters (that is, the derogatory comments he had stated Staxrud had made about Roald Amundsen) and which now were in the form of a signed statement that he was sending to Leon.[8] Hanssen went on to say that, based on Leon's letter, he was under the impression that Staxrud had made no admission regarding making these statements. Furthermore, said Hanssen, Staxrud had not attempted to approach Hanssen while he was in Tromsø and had rather made great effort not to make contact with Hanssen, which pleased Hanssen greatly. In his letter Hanssen also seems to reference the Tromsø newspaper *Tromsø Stiftstidende* article that had announced Staxrud's arrival, saying that it had contained a carefully constructed criticism regarding Hanssen's not replying to the telegram sent by Staxrud (requesting him to join the rescue expedition), and that Hanssen had then felt it necessary to state publicly his reasoning for his refusal to reply, and to explain the source of his reluctance, which then led to further disagreement among other colleagues whom Hanssen seems to indicate were protesting on behalf of Staxrud. Hanssen in his letter insists to Leon that it is not out of personal mistreatment but on pure principle that he disregards Staxrud thusly, based on what Hanssen felt were words uttered by Staxrud which he, Hanssen, completely abhorred, and which had caused Hanssen to view Staxrud most negatively and as an undesirable person who had come to Tromsø. Hanssen ended the letter with an expression of his devotion to Amundsen.

During his extended delay in Tromsø, apparently maintaining no direct contact with Helmer Hanssen, and continuing to attempt to add more sled dogs to his expedition, Arve Staxrud also continued writing to Fridtjof Nansen, defending himself against Tromsø criticism and updating Nansen about the ship and the sled dogs (Staxrud 17 March 1913). On March 17th, he wrote the letter to Nansen in which he enclosed the two newspaper clippings he had saved to show to his supporter—the February 21st Tromsø editorial that he stated had been critical about the rescue expedition's members and timing, and the March 6th article featuring the interview he had granted in Trondheim (*Nord Norge* 21 February 1913, 2; *Dagsposten* 6 March 1913). (These two clippings also included the articles about the status of the stranded Schröder-Stranz expedition and the recommendation of Nansen to use reindeer in the rescue efforts [Chap. 6].) In the accompanying letter to Nansen, Staxrud stated: "I dare to send some comments regarding the criticism, which from Tromsø-quarters has appeared in regard to the rescue-expedition for Schröder-Stranz."[9] He also briefly referenced the previous rescue expedition and its legal proceedings. Staxrud went on to say that he had not yet heard anything regarding his expected ship *Hertha* since it had departed from Haugesund en route to Tromsø. The good news, however, was that he now had a total of 12 sledge dogs for the expedition: "I have obtained 3 dogs in addition to the 9, we had before," he wrote. But acquisition of dogs from Russia would not be possible, he informed Nansen: "I have investigated, whether it was possible to obtain driving-dogs from the Murmansk coast; but it was shown, that the assertions about this—appearing in Tromsø—were based on idle talk only.—In any case nobody is able to undertake to get the dogs." Promising to send accounts and reports to the Foreign Ministry right up until the time of departure from Hammerfest, Staxrud signed off.

From Staxrud's letter, this would indicate that the two South Pole expedition sledge dogs Lussi and Storm, who had been joined by the seven sledge dogs brought by Daniel Nøis, were now joined by another three dogs in Tromsø, for a total of 12 dogs from Norway (Tahan, 2021, 298).

Joining the party in Tromsø as well was ice pilot Søren Zachariassen (Hilmar Nøis Diary SVB; Hilmar Nøis Diary NPI), as Hilmar Nøis would recount later in his life, in a series of handwritten diary journals containing personal recollections (some of which were written in 1970)[10],[11]. Zachariassen was the Arctic Ocean captain known for having been the first to commercially ship coal from Spitsbergen and was thus credited with beginning the coal mining industry on the Arctic archipelago (Norsk Polarhistorie.no/Søren Zachariassen). He would serve as ice pilot for the voyage north.

The other bit of good news for Staxrud himself was that, on the following day of March 18, a letter to the Norske Geografiske Selskap (the Norwegian Geographical Society) board was sent by a committee tasked with review, in which the committee endorsed that Staxrud and Hoel continue their geological and oceanographical research of Spitsbergen (Committee to Norske Geografiske Selskap 18 March 1913). The committee's statement cited Hoel and Staxrud's studies of the region in 1911 and 1912, among other Norwegian surveying expeditions, as important scientifically and practically to Spitsbergen and Norway, and supported their planned continued scientific and mapping work programs and topographical and hydrographical survey work, pointing to the unfortunate Schröder-Stranz expedition as a prime example of the need for mapping of areas such as the north coast and east coast, both for emergency and planned overwintering purposes.[12] The committee's statement concluded with a recommendation of Hoel and Staxrud's continued state funding application.

## The Late Arrival of *Hertha* and the Quick Departure from Tromsø

The ship *Hertha*, which had been sailing north from Sandefjord in southern Norway, and which had been stopped along its way by stormy weather, finally arrived in Tromsø, where, according to Daniel Nøis, it was immediately loaded with the rescue expedition's equipment, and from where, on that very same evening, the men and dogs of the relief mission were ready to set sail northward (Nøis, 1929, 6).

The party now consisted of Staxrud, Bøckmann, Ellingsen, Daniel Nøis, Hilmar Nøis, Johan Nilsen Nøis, Martin Pettersen Nøis, Søren Zachariassen, the sled dogs Lussi and Storm, and ten other sledge dogs (Hilmar Nøis Diary SVB; Hilmar Nøis Diary NPI), for a total of 12 sledging dogs, although, according to Hilmar Nøis, the dogs totaled 17, which is most likely the count later after arrival on Spitsbergen.[13,14] Staxrud had written to Nansen on March 17th that he now had 12 dogs in Tromsø (Staxrud 17 March 1913); he had originally targeted having a total of 18 dogs, as he had stated in his Trondheim newspaper interview on March 6th; and so, there were still six more dogs who needed to be added to the party. (Staxrud would later telegraph from Spitsbergen on May 14, as quoted in the May 15th *Aftenposten* article, that he had added six dogs in Green Harbour, and had left Norway with 12 dogs [*Aftenposten* 15 May 1913, evening: 1–2; Tahan, 2021, 298]. Thus, at this time, in Tromsø, there were 12 dogs.) Of these dogs, Hilmar Nøis specifically remembered Lussi and Storm by name in Tromsø (Hilmar Nøis Diary SVB). He would later specify and describe them in one of his diaries as the two sled dogs, female and male, who belonged to Roald Amundsen, who had survived Amundsen's South Pole expedition and who were now part of the Staxrud rescue expedition that left for Spitsbergen.[15] Hilmar Nøis also specified in another diary (Hilmar Nøis Diary NPI) that Amundsen had entrusted his two sled dogs Lussi and Storm to the rescue expedition, to be part of the life-saving efforts.[16]

And so, having gathered and met all together in Tromsø, and having now loaded most of the equipment onto the welcomed ship, the men and sledge dogs boarded the *Hertha* and set sail for their journey north from Tromsø on March 19, as would later be recounted by Staxrud in his 1914 report (Staxrud, 1914, 10). Their departure was recorded in the newspapers (*Aftenposten* 3 September 1913, evening: 4). And, very importantly, it was being followed and tracked by the very interested parties in Spitsbergen, who awaited the arrival of the relief expedition. Green Harbour telegraph manager Olaf Henriksen was trying to remain updated as to Staxrud's whereabouts and activities (Olaf Henriksen Diary). He had recorded in his diary, seemingly mistakenly, that Staxrud had departed from Tromsø on March 3, but he also listed an entry for March 19 in which he recorded *Hertha*'s departure (presumably from Tromsø) that evening, although, for an unknown reason, there is a line drawn through this entry.[17,18] Evidently, and naturally, the rescue efforts were of great importance to those on Spitsbergen, and as of March 1, Henriksen had been reporting on Tessem's and Pedersen's dogs, who he recorded as having perished[19]—presumably these were dogs who had been on the second failed Arctic Coal Company rescue effort which had taken place with Einar Tessem and Einar Pedersen from January 24 to February 27. And so Staxrud's expedition could not come soon enough and was highly anticipated.

Interestingly, Henriksen was also acquainted with the Nøis family, whom he had witnessed speaking with Fridtjof Nansen in Green Harbour the previous year.[20] Now the Nøis family members were accompanying Staxrud on the rescue expedition to Spitsbergen. Staxrud had indeed gathered together the critical elements he had recommended to the Foreign Ministry—he had recruited the local overwintering experts, the trained and experienced sled dogs, and the sealing ship, and now he was on his way to pick up the northern reindeer experts and the draft reindeer.

## Encounter in Alten (Alta): The Sámi Reindeer-Herders from Finnmark

When the ship *Hertha* sailed from Tromsø on March 19, it made its way to Bosekop (Bossekop) in Alten (Alta) in Finnmark, where the last members of the rescue mission would board the ship. These members were the three Sámi—Per Hansen, Johannes Kemi, and Samuel Klemmetsen—who joined the expedition here, and the 20 draft reindeer who were brought on board (Staxrud, 1914, 10; Nøis, 1929, 6; Hilmar Nøis Diary SVB; Hilmar Nøis Diary NPI).[21,22]

According to Hilmar Nøis, the three Sámi expeditioners were reindeer drivers from Karasjok (Hilmar Nøis Diary SVB),[23] and thus they and the 20 reindeer had most likely traveled from Karasjok in Western Finnmark northwest to Bosekop, Alten. There at Bosekop the ship had picked them up.

Hilmar Nøis, who specified in his diary that Daniel Nøis was the deputy under expedition leader Staxrud, also specified that Per Hansen was the leader of the three Sámi expedition members who would work as the reindeer walkers and caretakers (Hilmar Nøis Diary NPI).[24] They would also be in charge of looking after the reindeer during the voyage across the Arctic waters to Spitsbergen (Hilmar Nøis Diary SVB).[25]

But prior to bringing the reindeer on board, the men had to build boxes to properly contain the animals—one box per reindeer—so as to house them in the hold of the ship (Hilmar Nøis Diary NPI).[26] These boxes were constructed, placed in the hold, and securely tied, so as to protect the reindeer (Hilmar Nøis Diary SVB).[27]

Also brought on board was a large amount of reindeer moss as food for the animals, as later reported by Staxrud (Staxrud, 1914, 12). According to Hilmar Nøis (Hilmar Nøis Diary NPI), this fodder amounted to 2000 kg reindeer moss and 500 kg hay; this was in addition to 100 kg senne-grass (or sennegrass [*Carex vesicaria*], used for insulating traditional Sámi winter footwear), which would be utilized for lining the *komager* (a soft boot made of reindeer skin without the pelt) and the *skaller* (a soft shoe made of reindeer skin with the pelt on the outside) and which would keep the men's feet warm during the journey.[28]

While in Bosekop, Hilmar Nøis would later record, there was a market event that was held, and Hilmar partook in a lively time which included convivial hospitality from the Sámi community, with some individuals performing joik (a traditional and cultural musical art of soulful singing, of great significance to the Sámi), and Hilmar seemed to have enjoyed his time there (Hilmar Nøis Diary SVB).[29]

After the ship had been loaded with the reindeer food, and the reindeer themselves had been brought on board and secured in stall-like boxes erected in the hold, the now 11 men boarded the ship, and *Hertha* proceeded a relatively short distance north from Alten to Hammerfest, where the expedition would take on the last of its equipment (Nøis, 1929, 6; Hilmar Nøis Diary NPI).[30]

The total number of sledge dogs remained at 12 at this point, as Staxrud had previously determined that there was no use in sailing on to Murmansk to search for more dogs there.

Here in Hammerfest, on March 24, Staxrud wrote another letter, presumably to Adolf Hoel, in which he praised Hansen, Kemi, and Klemmetsen and in which he also mentioned the newly-boarded draft reindeer (Staxrud 24 March 1913). "With the reindeer it is going excellently and the three Sámi are great guys," he wrote.[31] He went on to express that all was going very well on board. On the following day, March 25, Staxrud wrote to Fridtjof Nansen advising him that everything was ready for departure on that day to Spitsbergen (Staxrud 25 March 1913). In the telegram, he stated: "Leaving today. Reindeer thriving excellently on board[.] All has passed as wished. Sincerely[,] regards[,] Staxud"[32] (Figs. 9.1 and 9.2).

## Departure from Hammerfest with Reindeer and All Rescue Expeditioners

The expedition ship began to make its way northward from Hammerfest toward the open sea that would lead to Spitsbergen, but stopped to anchor at Rolsøyhamn during a storm, according to Daniel Nøis and Hilmar Nøis (Nøis, 1929, 6; Hilmar Nøis Diary NPI).[33] There the expeditioners waited for a Northwest gale to subside, so as to have better weather during their sailing journey across the Arctic Ocean. The main reason for this was the reindeer. The animals were now residing in their stables in the ship's hold, and the men therefore necessarily had to leave at least one of the hatches off so that the reindeer below could have fresh air to breath (Daniel states they kept the hatches off, and Hilmar states they kept one hatch off at all times). Moreover, the reindeer were tied to their stalls, and a smooth ride was needed so that the boxes would not be jostled and become loose.

The ship finally took to the open sea on March 27, when the weather had improved (Nøis, 1929, 6; Tahan, 2021). (According to Staxrud's official telegram, sent on May

**Fig. 9.1** Photograph of Staxrud's rescue expedition members aboard the *Hertha* prior to departure to Spitsbergen. Expedition commander Captain Arve Staxrud stands in the center, in front of the bridge, surrounded by his expeditioners, who include Hilmar Nøis, second-in-command Daniel Nøis, Johan Nilsen Nøis, Martin Pettersen Nøis, the mining engineer Jakob Ellingsen, the physician Dr. P. W. K. Bøckmann (not mentioned on the photo), the ice pilot Søren Zachariassen (top right on bridge), and the three Sámi reindeer walkers Per Hansen, Samuel Klemmetsen, and Johannes Kemi (at bottom in foreground). Note that a few of the 12 sledge dogs from Norway, who included the South Pole Expedition dogs Lussi and Storm and Nøis's draft dogs, can be seen in this photo. The 20 draft reindeer are below deck. Others pictured on the bridge and deck are the captain and crew of the ship. (Photographer: Unknown. Source/Owner: Svalbard Museum)

14 from Advent Bay after the return from the first trek across Spitsbergen, as quoted in the newspaper *Aftenposten* on May 15, the ship left from Hammerfest on March 27, where *Hertha* had taken on board the three Sámi, the 20 reindeer, and the physician Dr. Bøckmann, who had all joined there [*Aftenposten* 15 May 1913, evening: 1–2].) Thus, now *Hertha* had left with all its crew, expeditioners, and draft animals secure and ready to meet the ice.

**Undsætningsexpeditionen ombord paa "Hertha".**

1) Expeditionens chef, kaptein S t a x r u d.   2) Lægen, dr. B ø c k m a n n.   3) Islods Z a c h a r i a s s e n.
4) cand. min. E l l i n g s e n.   5) Hilmar N i e l s e n.   6) Bernh. F i n b a k k e n.   7) Helbert B r o r-
s t a d.   8) Johan N ø i s,   9) Martin N ø i s.   10) Daniel N ø i s.   I baggrunden ses de tre lapper, som
deltager i expeditionen.  De øvrige paa billedet tilhører "Hertha"s besætning.

**Fig. 9.2** The same photograph as above, as featured in the *Aftenposten* May 15, 1913, newspaper article, providing numbered identification within the image and name captioning below it, as follows: (1) Expedition leader Captain Arve Staxrud; (2) physician Dr. P. W. K. Bøckmann; (3) ice pilot Søren Zachariassen; (4) mining engineer Jakob Ellingsen; (5) Hilmar Andreas Nilsen Nøis; (6) Bernh. Finbakken; (7) Helbert Brorstad; (8) Johan Nilsen Nøis; (9) Martin Pettersen Nøis; and (10) Daniel Karolius Nøis. The three Sámi expeditioners Per Hansen, Samuel Klemmetsen, and Johannes Kemi are mentioned but not by name. *Hertha*'s crewmembers, too, are mentioned but not named. (National Library of Norway)

## *Hertha*'s Voyage North to the Ice of Spitsbergen

*Hertha* crossed the sea north from Norway to Spitsbergen, sometimes slowly, at low speed due to headwind, other times at better speed due to tailwind, using sail and engine, encountering drift ice along the way, and approached the ice edge near Spitsbergen on March 30 (*Aftenposten* 15 May 1913, evening: 1–2; Tahan, 2021; Nøis, 1929, 6; Hilmar Nøis Diary NPI; Hilmar Nøis Diary SVB). Meeting the drift ice southwest of Bjørnøya (Bear Island), the ship had to travel in a more westwardly direction so as to avoid and circumvent the ice. Experiencing good weather in general, it reached the ice that had a calming effect on the water on the night of March 31.

Some of its hard encounters with the ice floes caused those on board to have to hold on and steady themselves, as the ship's hull scraped against the mixture of new and old ice.[34,35]

On April 1, during a snow event, the ship experienced difficulty breaking through the ice, the floes of ice being so tightly packed together that it required great effort for the ship to be able to advance (Nøis, 1929, 6; *Aftenposten* 15 May 1913, evening: 1–2; Tahan, 2021). At this time, land was sighted: The crew spied the sea mark that Staxrud himself had installed at Kap Staratschin the previous summer (during a surveying mission), not anticipating at that time that his own expedition would become the first to sight his marker and be guided by it. It was ironic that Staxrud was guided by the marker he had erected to help others.

April 2nd saw a continuation of effort toward the Ice Fjord, at which point the ship once more became stuck and immovable (Nøis, 1929, 6). The ship in turn was seen by those on land, as reported by the Green Harbour telegraph manager (Olaf Henriksen Diary).[36]

There was such thick ice lying across the Ice Fjord, between Ice Fjord and Bell Sound, according to Hilmar Nøis (Hilmar Nøis Diary SVB; Hilmar Nøis Diary NPI), and simultaneously a powerful gale from the northwest, that the ship had to use all its resources—sails and engine—and spend much time to push its way through the drift ice.[37,38] The ship exerted its full strength to deliver the rescuers to the ice edge, as later recounted by Daniel Nøis (Nøis, 1929, 6).

Finally, on April 3, Green Harbour was reached (Staxrud, 1914, 12; Tahan, 2021; *Aftenposten* 15 May 1913, evening: 1–2; Nøis, 1929, 6). The ship was able to attain a landing at the solid ice edge just off Green Harbour. According to Staxrud's May 14th telegram from Advent Bay quoted in the May 15th *Aftenposten* paper, "the ship forced itself in to Green Harbour."

Staxrud would later record, in his official report, that when they reached the entrance to the Ice Fjord in Spitsbergen, it was still full of ice and they could only dock at the ice edge three kilometers from the coast at Green Harbour on April 3 (Staxrud, 1914, 12–13).

Hilmar Nøis would later recount that *Hertha* had finally forced itself into an open channel in the ice, at the firm edge of the ice in Green Harbour (Hilmar Nøis Diary NPI; Hilmar Nøis Diary SVB), although he would record the arrival date as April 2 rather than April 3.[39,40]

The rescue expedition could not have arrived soon enough for those awaiting it on Spitsbergen. The Green Harbour telegraph manager Olaf Henriksen (Olaf Henriksen Diary) recorded in his diary that day, April 3rd, that *Hertha* had arrived at the ice edge.[41]

Both Daniel Nøis and Hilmar Nøis reported that the reindeer were secured and appeared to do very well during the voyage to Spitsbergen (Nøis, 1929, 6; Hilmar Nøis Diary NPI), but Hilmar recorded that one reindeer sustained a leg fracture during the journey and thus had to be euthanized.[42] This may mean that there were possibly only 19 reindeer who landed at Spitsbergen or who remained shortly after landing.

Now, after all the coordinated organization, the extensive communications (and miscommunications), the long journey from Norway, and the eventful ship's voyage through the ice, the men, the dogs, and the reindeer had finally reached their destination in the Arctic and were ready to begin the actual rescue work.

## Notes on Original Material and Unpublished Sources

Roald Amundsen letters of correspondence, written from and to Roald Amundsen and Leon Amundsen, are in the Manuscripts Collection at the National Library of Norway (NB) in Oslo. Hilmar Nøis's diary SVB is in the Historical Archive at the Svalbard Museum (SVB) in Svalbard. Hilmar Nøis's diary NPI is in the Library Archives at the Norwegian Polar Institute (NPI) in Tromsø. Arve Staxrud letters of correspondence, written from and to Arve Staxrud, are in the Manuscripts Collection at the National Library of Norway (NB) in Oslo. Additional Arve Staxrud letters of correspondence, written from and to Arve Staxrud, are in the Library Archives at the Norwegian Polar Institute (NPI) in Tromsø. Olaf Henriksen's diary is in the Library Archives at the Norwegian Polar Institute (NPI) in Tromsø.

1. A. Staxrud to F. Nansen, letter, 17 March 1913, with enclosed two newspaper article clippings (featuring handwritten notation), NB Ms.fol. 1924:6:A:2.
2. A. Staxrud to F. Nansen, letter, 11 March 1913, NB Ms.fol. 1924:6:A:2.
3. A. Staxrud to F. Nansen, letter, 11 March 1913, NB Ms.fol. 1924:6:A:2.
4. A. Staxrud to F. Nansen, letter, 11 March 1913, NB Ms.fol. 1924:6:A:2.
5. A. Staxrud to F. Nansen, telegram, 12 March 1913, NB Ms.fol. 1924.
6. A. Staxrud to A. Hoel, telegram, 13 March 1913, Norwegian Polar Institute (NPI), 1913_mai_Telegrams_fra_Staxrud_Part3.
7. L. Amundsen to H. Hanssen, letter, 15 March 1913, NB Brevs. 812:3:7.
8. H. Hanssen to L. Amundsen, letter, 30 March 1913, NB Brevs. 812:1.
9. A. Staxrud to F. Nansen, letter, 17 March 1913, with enclosed two newspaper article clippings (featuring handwritten notation), NB Ms.fol. 1924:6:A:2.
10. H. Nøis diary SVB, 1913, journal page 49, SVB-AP2.
11. H. Nøis diary NPI, notebook 3, 1913 March, pdf page 7, (NPI), D-307/D00307_3_0001.
12. Committee to Norske Geografiske Selskap board, 18 March 1913, Letter, Norwegian Polar Institute (NPI), 1913_03_18_Brev_til_Norske_Geografiske_Selskap.
13. H. Nøis diary SVB, 1913, journal page 49, SVB-AP2.
14. H. Nøis diary NPI, notebook 3, 1913 March, pdf page 7, (NPI), D-307/D00307_3_0001.
15. H. Nøis diary SVB, 1913, journal page 49, SVB-AP2.
16. H. Nøis diary NPI, notebook 3, 1913 March, pdf page 7, (NPI), D-307/D00307_3_0001.
17. O. Henriksen diary, 3 March 1913, (NPI), D-305/D00305_0001.
18. O. Henriksen diary, 19 March 1913, (NPI), D-305/D00305_0001.
19. O. Henriksen diary, 1 March 1913, (NPI), D-305/D00305_0001.

20. O. Henriksen diary, 16 July 1912, (NPI), D-305/D00305_0001.
21. H. Nøis diary SVB, 1913, journal page 49, SVB-AP2.
22. H. Nøis diary NPI, notebook 3, 1913 March, pdf page 7, (NPI), D-307/ D00307_3_0001.
23. H. Nøis diary SVB, 1913, journal page 49, SVB-AP2.
24. H. Nøis diary NPI, notebook 3, 1913 March, pdf page 7, (NPI), D-307/ D00307_3_0001.
25. H. Nøis diary SVB, 1913, journal page 49, SVB-AP2.
26. H. Nøis diary NPI, notebook 3, 1913 March, pdf pages 7–8, (NPI), D-307/ D00307_3_0001.
27. H. Nøis diary SVB, 1913, journal page 49, SVB-AP2.
28. H. Nøis diary NPI, notebook 3, 1913 March, pdf page 8, (NPI), D-307/D00307_3_0001.
29. H. Nøis diary SVB, 1913, journal page 49, SVB-AP2.
30. H. Nøis diary NPI, notebook 3, 1913 March, pdf page 8, (NPI), D-307/D00307_3_0001.
31. A. Staxrud (to A. Hoel), letter, 24 March 1913, Norwegian Polar Institute (NPI), 1913_mai_Telegrams_fra_Staxrud_Part2.
32. A. Staxrud to F. Nansen, telegram, 25 March 1913, NB Ms.fol. 1924.
33. H. Nøis diary NPI, notebook 3, 1913 March, pdf page 8, (NPI), D-307/D00307_3_0001.
34. H. Nøis diary NPI, notebook 3, 1913, pdf page 8, (NPI), D-307/D00307_3_0001.
35. H. Nøis diary SVB, 1913, journal page 50, SVB-AP2.
36. O. Henriksen diary, 2 April 1913, (NPI), D-305/D00305_0001.
37. H. Nøis diary SVB, 1913, journal page 50, SVB-AP2.
38. H. Nøis diary NPI, notebook 3, 1913, pdf page 8, (NPI), D-307/D00307_3_0001.
39. H. Nøis diary NPI, notebook 3, 1913, pdf pages 8–9, (NPI), D-307/D00307_3_0001.
40. H. Nøis diary SVB, 1913, journal page 50, SVB-AP2.
41. O. Henriksen diary, 3 April 1913, (NPI), D-305/D00305_0001.
42. H. Nøis diary NPI, notebook 3, 1913, pdf page 8, (NPI), D-307/D00307_3_0001.

## Unpublished Sources and Original Material

Amundsen Letters of Correspondence, Manuscripts Collection, National Library of Norway, Oslo.

Committee Letter to Norske Geografiske Selskap, 18 March 1913, Letter from Committee to Norske Geografiske Selskap board, 1913_03_18_Brev_til_

Norske_Geografiske_Selskap, Library Archives, Norwegian Polar Institute, Tromsø. Received 20 July 2021.

Henriksen, O., Diary, Olaf Henriksen's diary written while he was the Green Harbour Telegraph Station Manager, "Dagbok for Olaf Henriksen, 1911–13" ["Diary for Olaf Henriksen, 1911–13"], dated 14 August 1911 to 2 May 1913, D-305, D00305_0001. Library Archives, Norwegian Polar Institute, Tromsø, https://brage.npolar.no/npolar-xmlui/handle/11250/2600678 Accessed 24 July 2021.

Nøis, H., Diary NPI, handwritten journal of recollections recounting the events of the 1913 Staxrud expedition, Hilmar Nøis, "Minder og Fortelinger fra fangsttiden på Spitsbergen, Klade 3, 7 Mars 1970" [Hilmar Nøis, "Memories and Stories from the trapping time on Spitsbergen, Notebook 3, 7 March 1970"], dated 7 March 1970, D-307, ("Minder og fortelinger, Nøis, Hilmar, hefte 3"), D00307_3_0001. Library Archives, Norwegian Polar Institute, Tromsø, https://brage.npolar.no/npolar-xmlui/handle/11250/274077 Accessed 20 July 2021.

Nøis, H., Diary SVB, handwritten journal of recollections including an account of the 1913 Staxrud expedition, ("Hilmar Nøis. Fangstmann. 1909–1923, dagbok med erindringer fra fangstlivet") ["Hilmar Nøis, Trapper, 1909–1923, diary with recollections from the trapping life"], not dated (latest year referenced is 1942 in journal pages and 1950 in inserted sheet), SVB-AP2-Hilmar-Nøis-dagbok-1909–1923-compressed. Historical Archive, Svalbard Museum, Svalbard, https://svalbardmuseum.no/no/samlingene/historisk-arkiv/arkivprosjekt-fangst-og-annen-overvintringsvirksomhet-fra-perioden-1910-til-1970/ Accessed 6 June 2021.

Staxrud, A., 11 March 1913, Letter from Arve Staxrud to Fridtjof Nansen, NB Ms.fol. 1924:6:A:2. National Library of Norway, Oslo, online archives, https://www.nb.no. Retrieved 6 July 2021.

Staxrud, A., 12 March 1913, Telegram from Arve Staxrud to Fridtjof Nansen, NB Ms.fol. 1924. National Library of Norway, Oslo, online archives, https://www.nb.no. Retrieved 6 July 2021.

Staxrud, A., 13 March 1913, Telegram from Arve Staxrud to Adolf Hoel, 1913_mai_Telegrams_fra_Staxrud_Part3, Library Archives, Norwegian Polar Institute, Tromsø. Received 20 July 2021.

Staxrud, A., 17 March 1913, Letter from Arve Staxrud to Fridtjof Nansen, with two newspaper article clippings enclosed featuring handwritten notation (Nord Norge [Tromsø], 21 February 1913, pages 1–2, and Dagsposten, Trondhjem [Trondheim], 6 March 1913), NB Ms.fol. 1924:6:A:2. National Library of Norway, Oslo, online archives, https://www.nb.no. Retrieved 6 July 2021.

Staxrud, A., 24 March 1913, Letter from Arve Staxrud (to Adolf Hoel), 1913_mai_Telegrams_fra_Staxrud_Part2, Library Archives, Norwegian Polar Institute, Tromsø. Received 20 July 2021.

Staxrud, A., 25 March 1913, Telegram from Arve Staxrud to Fridtjof Nansen, NB Ms.fol. 1924. National Library of Norway, Oslo, online archives, https://www.nb.no. Retrieved 6 July 2021.

# References

Aftenposten. (15 May 1913). Hjælpeexpeditionen paa Spitsbergen. Kaptein Staxruds beretning til "Aftenposten". [The rescue expedition on Spitsbergen. Captain Staxrud's account to "Aftenposten".]. *Aftenposten* (Kristiania [Oslo]), 15 May 1913, evening edition, pp. 1–2. National Library of Norway, Oslo, online archives, www.nb.no. Accessed 28 August 2012.

Aftenposten. (3 September 1913). Kaptein Staxrud hjemme igjen. Udtaler sig til "Aftenposten" om redningsexpeditionen. [Captain Staxrud back home again. Speaks to "Aftenposten" about the rescue expedition.]. *Aftenposten* (Kristiania [Oslo]), 3 September 1913, evening edition, p. 4. National Library of Norway, Oslo, online archives, https://www.nb.no. Retrieved 4 April 2022.

Dagsposten. (6 March 1913). Hos Kaptein Staxrud. Han udtaler sig om Planerne. – Man faar høre fra Undsætningsekspeditionen i Slutten af Marts. [At Captain Staxrud's. He speaks about the Plans. – One will hear from the Rescue-expedition at the End of March.]. Written by Th. K. *Dagsposten*, Trondhjem (Trondheim), 6 March 1913. (Newspaper clipping featuring handwritten notation by Staxrud, attached to letter from Arve Staxrud to Fridtjof Nansen dated 17 March 1913, NB Ms.fol. 1924:6:A:2). National Library of Norway, Oslo, online archives, https://www.nb.no. Retrieved 6 July 2021.

Nøis, D. (1929). Med kaptein Staxrud på leiting etter Schrøder-Stranz og hans folk [With Captain Staxrud on the search for Schröder-Stranz and his people]. In: *And-Ungen*, July 1929, pp. 4–12. Andenes: Andøyposten. Library Archives, Norwegian Polar Institute, Tromsø. Received 20 July 2021.

Nord Norge. (21 February 1913). Mye skrik og litet uld sa manden da han klipte grisen. Editorial by H. C. Johannesen. *Nord Norge* (Tromsø), 21 February 1913, p. 2. (Newspaper clipping attached to letter from Arve Staxrud to Fridtjof Nansen dated 17 March 1913, NB Ms.fol. 1924:6:A:2). National Library of Norway, Oslo, online archives, https://www.nb.no. Retrieved 6 July 2021.

*Norsk Polarhistorie – Polarhistorie.no.* Søren Zachariassen. "Søren Zachariassen: 1837–1915, Ishavsskipper fra Tromsø som innledet kulldriften på Spitsbergen." (Norwegian Polar Institute, University of Tromsø, Troms County Municipality.) https://polarhistorie.no/personer/Zachariassen,%20Soren.html. Accessed 12 May 2022.

Staxrud, A. (1914). Die Staxrudsche Hilfs-Expedition für Schröder-Stranz. In A. Miethe (ed.), *Die Expedition zur Rettung von Schröder-Stranz und seinen Begleitern - geschildert von ihren Führern Hauptmann A. Staxrud und Dr. K. Wegener* (pp. 1–68). Berlin: Dietrich Reimer.

Tahan, M. R. (2021). *The Return of the South Pole Sled Dogs: With Amundsen's and Mawson's Antarctic Expeditions.* Cham: Springer International Publishing.

Tromsø Stiftstidende. (11 March 1913a). "Hertha". *Tromsø Stiftstidende* (Tromsø), 11 March 1913, p. 2. National Library of Norway, Oslo, online archives, https://www.nb.no. Retrieved 4 April 2022.

Tromsø Stiftstidende. (11 March 1913b). Staxruds redningsekspedition. [Staxrud's rescue expedition.]. *Tromsø Stiftstidende* (Tromsø), 11 March 1913, p. 2. National Library of Norway, Oslo, online archives, https://www.nb.no. Retrieved 4 April 2022.

# Chapter 10
# The Stranded Expeditioners' Efforts to Save Themselves, and the Detrimental Rescue Activities of the Kurt Wegener Expedition

## Rave and Stenersen's Scuffle Regarding a Rifle

While the Arve Staxrud relief expedition had been forming and departing from Norway during the month of March 1913, the dire situation of the stranded Schröder-Stranz expedition members had remained perilous on Spitsbergen. The six men and several dogs remaining on the frozen ship *Herzog Ernst* were coping with extreme weather, frostbite injuries, dwindling supplies, and acute loneliness.

The Norwegian sailor Einar Rotvold continued to chronicle the day-to-day events in his diary (Einar Rotvold Diary), describing Treurenberg Bay on March 1st as "a dead and desolate place, [with] no trace to find of any living animal," and earnestly writing "Just wish time would pass quickly."[1] On that day, he recorded, the temperature was approximately –35 °C and the sun allowed 11 h of daylight. Dealing with "Heavy, sad days" during which they yearned "only [to be] at home again with ours" and fearing that only "God knows when that will be,"[2] the four Norwegian men nonetheless began to prepare for an attempt to rescue themselves and the two Germans. On March 4th, Rotvold reported that a ski-sledge was being built for a trek south to Advent Bay, and that, in addition to himself and August Stenersen, the two brothers Julius and Jørgen Jensen had also decided to make the attempt to reach Advent Bay.[3]

The implication, then, is that the injured Hermann Rüdiger and the artist Christopher Rave, who was looking after him, would both remain behind while the Norwegians ventured out to find rescue.

But as provisions were being gathered for the journey, a seemingly rare instance of dissent occurred between the Germans and the Norwegians on March 6th, described by Rotvold as follows:

> Rave has been forward and taken the rifle, which we have had in our keeping since last autumn, from us. He said, that he had to have 2 rifles left here, thus we should have only one rifle for 4 men, all the way to Advent Bay. Stenersen protested and stopped him on the way up and wanted to have the rifle back, he held him by one of his feet, but Rave just yelled,

M. R. Tahan and C. Lüdecke, *Stranded at the Top of the World*,
https://doi.org/10.1007/978-3-031-56288-4_10

"Let me go, if you want to live." Stenersen let him go then. Rave poured quite a bit out of himself up on deck, but Stenersen from all of it understood only, that he had left enough on the ice last autumn, for more than 500 kroner. Rave said as well that if he wanted to say a word, then - - - more did Stenersen not understand. Sounded like Stenersen should have failed in some or other fashion.[4]

The heated discussion and near physical altercation over the rifle, between Rave and Stenersen, and apparent difficulty in communication, ended with Rave having his own way and the Norwegians continuing to cooperate with him, because the following three days were spent with the four Norwegians improving their footwear for the winter walk,[5] attempting to supplement their diet with bear or fox—which could not be found,[6] and preparing to "go to Mossel Bay, to fetch the 2 cameras for Rave"[7]—presumably the equipment that had been abandoned on the ice the previous fall.

On March 10th, a day that began with "overcast air and snowflurry,[8] Rotvold and most likely some of the others "helped Rave with getting 1 Kg. Plasmon-crackers, 1 Kg. Pemmican and various other provisions over to the house, as he is going to move over there with Dr. Rüdiger early in April." Rotvold also reported that "Rave took sugar from the depot, he gave us some of it. We brought back to the ship 3 bags of coal, which we gathered up from the ground." The Norwegians spent approximately four hours that day assisting the two Germans by transporting goods to prepare for Rave and Rüdiger's relocation to the Swedish House—where Knut Stave's body lay.

The following day, according to Rotvold, in preparation for their upcoming journey to Advent Bay, Rotvold, Stenersen, and Jørgen Jensen "took the bag of provisions" packed for the trip and "pulled it up on the mountain-ridge, from there to bring it down to Mossel Bay, when we are going to fetch the cameras."[9] They then returned to the ship. Julius Jensen, in the meanwhile, was able to find two Polar foxes caught in the fox traps and brought those to the ship for future consumption.

On that same day of March 11, according to Rave, the weather was glorious as the Norwegian crew set off with a homemade sledge and a sack of provisions toward Mossel Bay to collect, for 100 kronor, the belongings left there by the two Germans, while Rave and Rüdiger were about to move to the more comfortable Swedish House, which was only an hour away on the other shore of Treurenberg Bay (Fig. 10.1) (Rave, 1913, 69).

## The Norwegians' Round-Trip March to Mossel Bay, and the Germans' Move from the Ship

It was actually March 12 when the four Norwegians made the round trip to Mossel Bay to retrieve Rave's cameras (Einar Rotvold Diary). Rotvold recorded that they left at 9:30 am, bringing the provisions bag they had placed on the mountain down to Mossel Bay and traveling in sunshine but in snowfall.[10] Arriving in Mossel Bay, where they found "hard conditions," they located the cameras and retrieved them along with a bag of clothing which they also loaded up onto the sledge for the return

**Fig. 10.1** House of the Swedish Arc-of-Meridian expedition on the eastern shore of Treurenberg Bay. *Source* Rüdiger (1913, 159)

journey to Treurenberg Bay. Driving through "frost mist until we were halfway on the mountain," they encountered "severe cold" on their way back to the ship, which they boarded that very evening at 6:30 pm. "It was quite a hard march, out and back in 9 hours," concluded Rotvold that day, adding that they "saw a lot of open water outside Verlegen Hook and westward."

On the day after their strenuous trip, March 13, the four Norwegians re-evaluated their physical condition and the likelihood of their being able to travel a longer distance at this time. "We have decided to remain here aboard the ship, in Treurenberg Bay, for the time being; as we felt yesterday on the march to Mossel Bay that we are not strong enough to go out on the trip to Advent Bay," wrote Rotvold in his diary.[11] The decision was a demoralizing one. Rotvold concluded: "We now have in the name of God to wait here for a relief, in one way or another. It will be many, long, sad days to remain here for us. Low on provisions."

To supplement their supplies, on the next day of March 14, Rotvold and Jørgen sojourned from the ship to the animal traps, where they found a Polar fox; meanwhile, Stenersen and Julius cooked for the expeditioners.[12]

The two Germans, in the meantime, were completely ready to move out of the ship. The relocation was accomplished on the following day of March 15, with special care taken for the injured expeditioner, as recorded by Rotvold. "Rave and Dr. Rüdiger have moved to the house," he wrote. "One helped them thereto; Dr. Rüdiger we pulled on the sledge, as it was hard for him to walk the long way there. We have received ca. 2 kg. sugar to share, from Rave."[13]

Most likely, Jule, the companion dog to Rave and Rüdiger, also helped in the moving of the two men from the ship to the house. Rüdiger described the dark yellow Jule's breed or "origin" as "not known" and stated that this was "a characteristic she probably has in common with most German draft dogs [taken on the expedition]" (Rüdiger, 1913, 104). It was later reported to Hilmar Nøis, according to his diary (Hilmar Nøis SVB), that Rave was assisted by a large Saint Bernard dog who helped pull Rüdiger on the sledge.[14] It is uncertain which dog was being referenced.

With the expedition party now split into two, the four Norwegians continued their routine on the ship while the two Germans began life at the house. Over the next three days Rotvold recorded weathering a storm while there was "No living creature to see, neither in the air nor on the ice,"[15] and the futility of catching any animals in this "poor hunting ground."[16] The monotony was broken somewhat on March 19th when Rotvold and the crew gave a thorough cleaning to the inside of the ship, and Rave came on board "and fetched 7 boxes of butter and various other goods,"[17] but the situation there for Rotvold was still "sad and dreary to live."

It was now the Easter holidays for the expeditioners, and, on the following day of Maundy Thursday, March 20, according to Rotvold, the Norwegians remained at rest. "We are celebrating Easter in our poor manner," he wrote. "We have no almanac for 1913, but we are hoping, that we are celebrating Easter at the right time."[18]

Their observance of the holiday continued through Easter Sunday—the first day of Easter,[19] as did their contemplation of the situation in which they now found themselves.

## The Norwegians' Departure: Heading Toward Advent Bay

On the second day of Easter (Easter Monday), March 24, the four remaining Norwegians decided unanimously to depart from the ship toward Advent Bay in order to obtain help and planned to make that departure on the following day, if the weather allowed (Einar Rotvold Diary). Rotvold was eager to announce the news to the two Germans and recorded that day: "Julius and I went in the afternoon on a trip to the house, and told Rave and Dr. Rüdiger [about] this, and [asked] whether they wanted to send mail with us. They were happy to hear, that we were going to Advent Bay, and Rave will be here early tomorrow with the mail. We got tea and plasmon-crackers from them; said goodbye to Dr. Rüdiger and went on board again."[20]

The visit was recorded by Rave as well (Rave, 1913, 74), who wrote that on that day Julius and Rotvold appeared at the Swedish House to tell the Germans that everyone was now leaving for Advent Bay and therefore left them the second rifle for hunting. Rave went back to the ship on the following morning, on the one hand to give the Norwegians letters for home and on the other hand to take more things for himself and Rüdiger on a sledge pulled by their dogs Jule and the black Caesar.

Rave's reciprocal visit to the ship was recorded by Rotvold (Einar Rotvold Diary) on that next day of March 25, the day that the Norwegians departed from the ship. "We are bringing with us, 2 woolen blankets each, 2 boxes of crispbread, 1 Box of

Plasmon-crackers, 2.5 kg. butter, 1 Box of tea, 2 packets coffee, 6 small cheeses, 3 cured sausages, some peas, grains, and flour," wrote Rotvold that day.[21] "No sleeping bags, as we have sleeping bags in the Wijde Bay house," he continued. (Recall that Rotvold and Stenersen, in their desperation to bring Wilhelm Eberhard quickly back to the ship, had left their sleeping bags stored at the hut on December 24, 1912 [Chap. 4]). Rotvold reported setting out that afternoon of March 25: "At 1 pm, we left the ship, we had our kit on a ski-sledge. Rave took several photographs of us before we left. One dog 'Lotte' we have with us."

And so, armed with their hand-improved footwear, essential provisions, and built sledge, and accompanied by one expedition dog, the four Norwegians departed for Advent Bay.

Now the two German men were all alone for the next while (Rave, 1913, 69). According to Rave's book about the expedition, at this time he devoted himself with leisure to painting. In the evenings, he and Rüdiger played board games or chatted about home. Because the large stove for cooking had to be heated up somewhat awkwardly, Rave now always cooked enough food to last for several days. "The softened dried vegetables mixed with pemmican and some diced bacon, cooked well until soft and then mixed with grated plasmon-cakes," continued to taste good to them even day after day, except that "on Sundays, instead of cake, there is hard bread fried in butter and cocoa," wrote Rave (Rave, 1913, 75).

While Rave began his tenure of solely caring for, cooking for, and looking after the ailing Rüdiger, the Norwegians began their long trek south, making their first stop in Mossel Bay on that first night of March 25th (Einar Rotvold Diary). But it was a restless night. "We made tea, and ate a little, went to bed, but we got no sleep due to smoke, as there was no proper stove in the house," reported Rotvold in his diary, presumably referring to the Polhem hut.[22] "Everything damaged and ruined. We hope with God's help, that we will have good weather and reach Advent Bay safe and sound."


## Wegener's Wayward Walk to Wijde Bay


During these first few weeks of March when the Schröder-Stranz expeditioners were struggling with survival and with attempts to rescue themselves, the Kurt Wegener rescue expedition was continuing its efforts to reach Wijde Bay, where they believed some of the stranded members would be. Wegener had departed from Ebeltofthamna (Cross Bay) on February 21 and had been laying depots from February 22 through February 28, at which time they had retreated to Kings Bay to reassess their efforts (Chap. 6). As of the beginning of March, they were plotting their next move.

It must be said here that, in the Arctic, everyone in distress is very dependent upon quick help. It is not for nothing that the hunters' huts on Spitsbergen usually provide food and firewood to enable one to survive for a few days until the weather changes or until the person can go hunting for themselves. If a person has availed themselves of this general offer, it is taken for granted that that person will replace the consumed or used-up goods prior to leaving the hut. Moreover, people are happy to come to the rescue immediately, because they would expect something similar to take place for themselves were they to be stranded in such a region. In addition, such rescue operations also provide a good opportunity to break the well-worn daily routine in a mining settlement. Setting up food depots is an extremely important contribution from relief expeditions. In addition, there is the search for written information in hunters' huts to find out who had been there and who went where with whom and when.

Unfortunately, the relief expedition of Kurt Wegener did not always follow this unwritten necessary protocol at the Polar huts, nor did it employ replenishment of supplies in order to not leave other expeditioners vulnerable. Though their intention was commendable, their practice did not always meet the requirements.

On March 1, the Wegener expedition recommenced its plans for its walk from Kings Bay to Wijde Bay. According to Wegener himself, on this day, everyone rested and prepared for the sledge trip (Wegener, 1913, 138–139). Since Millar first wanted to learn more details about Staxrud's expedition before setting out, Wegener got underway on March 3 in the direction of Cross Bay hoping that Robitzsch might have received new information at the observatory in the meantime. But Wegener had to turn back on the way, because he had fallen into the water while jumping from ice floe to ice floe. After a total of 40 km he rejoined the others and urged them to leave even without news from Staxrud.

The equipment, however, was far from perfect. In addition, they had only a small female dog named Freya, but they called her Lady, and a young Spitz, neither of whom was good for pulling a sledge; the two would, however, serve as emergency provisions (Wegener, 1914, 96, 78–79). But aside from this, the Spitz had a psychological significance and meaning for Millar. With the low temperatures, the kerosene would be frozen again, and they would have to rely on driftwood along the coasts for cooking. Then they had only one sleeping bag, for one person, which they expanded by adding an inset piece for a second man. They also sewed a second sleeping bag from reindeer skins for the two Scandinavian workers, but it only reached their chests and was not very suitable for the cold.

On March 4, the group finally set off from Kings Bay in the direction of Wijde Bay (Wegener, 1914, 81–82). From March 5 to 6, a snowstorm occurred that kept them stuck for the next few days. They built a snow cave and spent their time eating, grooming their feet, and mending shoes, which was probably due to their inadequate equipment. In addition, their cooking apparatus was useless without liquid kerosene.

After marching 30 km from the Lovén Islands through unknown terrain, they did not reach the 800 m high inland ice until March 8 and set up a depot there (Wegener, 1913, 139). Again and again, they had to search for a suitable route until they finally reached Wood Bay on March 12. Here they found driftwood and could finally prepare a warm meal again. "The equipment was already completely torn after this first week," was Wegener's conclusion (Wegener, 1914, 83).

When provisions threatened to run out, a lengthy discussion revealed that the men still wanted to continue until the next evening (Wegener, 1913, 139–140). However, if they had not shot any reindeer by then, they would turn back. Although they found shelter in the evening in an old dilapidated Russian hut and were able to dry their wet sleeping bags and fur clothing by a large campfire, discouragement spread. Fresh snow made it difficult to continue the march the next day. Fortunately, they were able to kill three reindeer in the afternoon and finally eat their fill again, which revived Olafson and Abrahamsen, who had wanted to turn back the day before. Two days later, the snow was already up to a meter deep, so they had to cut the sledge load in half and walk the trail twice. In the meantime, the shoes were falling apart and the two sleeping bags were full of holes, which could be patched more badly than well with the reindeer skins. By March 16, everyone was suffering from seizures, perhaps from the food, while Wegener and Abrahamsen were additionally plagued by toothache and rheumatism, and cough, respectively, and caught a severe cold. Provisions became scarce, and due to the lack of driftwood, there was only frozen food to eat. Finally, on March 17, they reached their northernmost point at Graa Huk, where the sea was frozen to the horizon (Fig. 10.2a, b).

After they had long since used the empty provisions boxes, sacks, and other superfluous items to make a fire, they now left behind the useless cooking apparatus and the lantern in order to further reduce the weight of the sledges. Crossing the mouth of Wijde Bay toward Mossel Bay was completely impossible because of the pack ice. This prevented them from continuing their march to the *Herzog Ernst* in Treurenberg Bay, where they had hoped to have the luxury of hot meals and a warm place to sleep during the cold nights.

In the meantime, all of them had developed frostbitten fingertips and a lively debate ensued about how to continue the march. One of them wanted to turn back, while the other one was concerned about the lack of provisions. Wegener himself, as leader, did not want to be responsible for the failure of the expedition if they turned back. In addition, shoes and sleeping bags still had to be repaired first. In the end, there was a compromise in that the men would only search for traces of Alfred Ritscher's group on the east coast of Wijde Bay on the way back. On March 20, they reached an abandoned hut at Wijde Bay where they discovered a message from Rüdiger and Rave, who had spent seven weeks here and then left for the ship (Wegener, 1914, 85–91). Based on this message, Wegener assumed that Dr. Erwin Detmers, the biologist, had had an accident and needed a doctor. Other messages from December 22–24, 1912, came from the group that Ritscher had sent back to the ship. From them it appeared to him that Eberhard was worried that the two Norwegians (Rotvold and Stenersen) would leave him behind on the way. He was already very weakened and could not keep up with them. When they left on December 24, they had left their sleeping bags behind for weight reasons.

Upon reading the brief reports, "we all came to the conclusion that any hardship or danger that could be averted no longer existed at this time (mid-March) for the Schröder-Stranz expedition" (Wegener, 1914, 92). After a day of rest, they took the available sleeping bags with them (those that the Norwegians had left behind in their haste to bring Eberhard back to the ship) and left their own dilapidated sleeping bags in exchange.

The switch would have consequences.

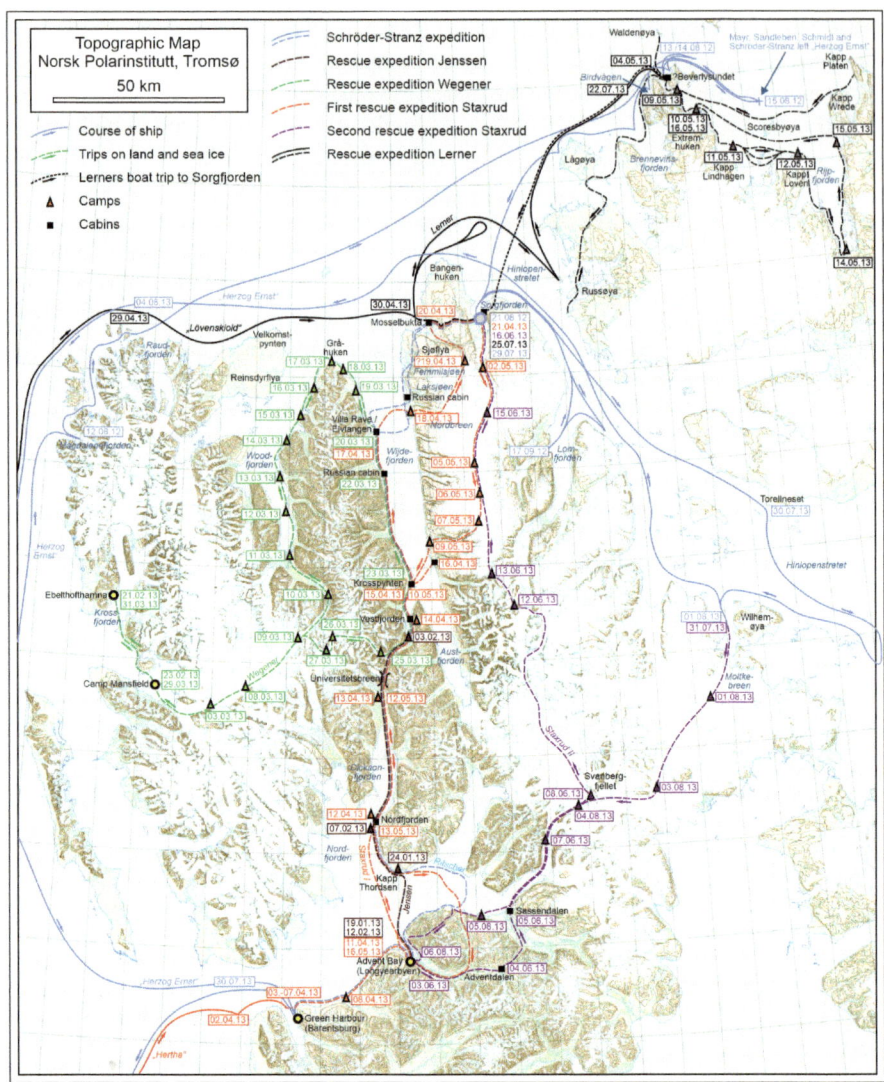

**Fig. 10.2  a** Travel map of the relief expeditions for Schröder-Stranz in northern Svalbard. *Source* Piepjohn (2021) [updated 2023]. **b** Member list of each relief expedition for Schröder-Stranz Svalbard. *Source* Piepjohn (2021) [updated 2023]

The trail continued through the Russian hut (March 22), where they conducted a futile hunt for reindeer, and the hut north of Cape Petermann (Cross Point hut) (March 23–24), where they found Ritscher's news from October (Wegener, 1914, 93–95). Here they had fresh reindeer meat and hides that probably came from Ritscher's group. Wegener's expeditioners assumed that some men would probably make their way from the ship to Advent Bay in the spring to telegraph home. Instead of hunting

**Fig. 10.2**  (continued)

Schröder-Stranz-Expedition (DAE)
1. *Biologist Dr. Erwin Detmers* †
2. *Engineer Wilhelm Eberhard* †
3. Jørgen Jensen
4. Julius Jensen
5. *Geographer Max Mayr* †
6. *Botanist Dr. Walter Moeser* †
7. Painter Christopher Rave
8. Captain Alfred Ritscher
9. Einar Rotvold
10. Dr. Hermann Rüdiger
11. *Lieutenant retd. August Sandleben* †
12. *Secretary Richard Schmidt* †
13. *Herbert Schröder-Stranz* †
14. *Knut Stave* †
15. Ice pilot August Stenersen

**Rescue expedition Jenssen**
16. Ingvar Jenssen
17. Einar Pedersen
18. Jakob Rognli
19. Einar Tessem

**Rescue expedition Wegener**
20. Abrahamsen
21. Millar
22. Olafson
23. Meteorologist Dr. Kurt Wegener

**First rescue expedition Staxrud**
24. Physician Dr. P. W. K. Bøckmann
25. Engineer Jacob Ellingsen
26. Per Hansen
27. Johannes Kemi (till Universitetsbreen)
28. Samuel Klemmetsen
29. Daniel Nøis
30. Hilmar Nøis (till Universitetsbreen)
31. Captain Arve Staxrud
15. Ice pilot August Stenersen (DAE)

**Rescue expedition Lerner**
32. Cameraman Sepp Allgeier
33. Rudolf Biehler
34. Captain Bottolfson
35. Gerhard Graetz
36. Journalist Theodor Lerner
37. Dr. Bernhard Villinger
38-44. 7 crew

**Second rescue expedition Staxrud**
25. Engineer Jacob Ellingsen
29. Daniel Nøis
30. Hilmar Nøis
45. Johan Nilsen Nøis
31. Captain Arve Staxrud

reindeer, which were numerous at the site, they used up the available supply of meat stored in the hut. They themselves did not want to go further to Advent Bay, but preferred to go home to report.

At some point here, during this time, most likely between March 24 and March 25, the Wegener expedition appears to have accidentally burned down the hunter's hut at West Fjord—a devastating disappointment for any party who would come along after them. Although it seems that this incident may not have been reported by Wegener, it would later be recorded by others, as shall be seen.

On March 25, the members of Wegener's expedition went on, and on the following day they fought their way up to a flat pass about 800 m high. Here their dog Freya (aka Lady) got away from them, whom they now would have liked to have as reserve provisions (Wegener, 1914, 96–97). Since the next depot was only 15 km away, there was no danger, "but it was necessary to leave the sledge and eat their way home through with the help of the depots" (Wegener, 1914, 98). After bad weather on March 27, they burned a spare ski and parts of the sledge during a break (Wegener, 1913, 140). The next day, the weather had improved, and they could see the surrounding area. The two Scandinavians proposed to cross the mountains on skis as quickly as possible, taking only the most necessary items with them. So they left the sledge with the equipment, and, after a few hours, they reached their last depot, the contents of which they ate almost completely. In the afternoon, fog came up, so they continued walking with the compass until 7 pm. At that point they could not continue because of the crevasses. Still, they had to move and walk back and forth to avoid freezing to death. Although he was strongly warned, Abrahamsen lay down for a while and froze the sole of his foot. When the weather cleared at 1 am, they continued on to the large depot at Diademberg. Here, too, they ate almost everything. Finally, after 70 km of uninterrupted marching, they reached the Lovén Islands at 5 pm. Another 6 km brought them to Kings Bay (March 30) (Fig. 10.3).

It took them eight hours to reach Cross Bay, 25 km away, partly by boat through the drift ice. It was March 31, when Wegener returned, exhausted. It was not until April 7 that Max Robitzsch managed to send a report about their rescue efforts to Green Harbour:

German Embassy Christiania. Went with Millar Olafson Abrahamsen over 3 crowns eastward, reached Wood Bay against will, exhausted forces and provisions in deep snow. Estuary Wijde Bay impassable because of pack ice, went back via Graa Huk, 26 days, lost sledge and all equipment in the mountains West Fjord. According to diaries Wijde Bay returned Rave, Rüdiger Detmers 23 November from hunter's hut 79, 40 to ship, request doctor and message to families. Foot Detmers not significantly improved by November. Eberhard Stinnessen [sic] Rotvodt [sic] returned from Cape Petermann hunter's hut 22 December to ship, passed hunter's hut 79, 40 December 24. All wintered Sorge Bay [Treurenberg Bay]. We found salt meat, reindeer, hides, good hunting in hunter's huts Wijde Bay. Distress not apparent. Therefore we returned. Request replacement sledging equipment and provisions Kings and Cross Bay. Wegener

(Wegener, 1913, 140).

While Wegener apparently was reporting what he understood from the diaries and letters left at the huts, there are obviously several misinterpretations and inconsistencies with the actual dates and course of events that had actually transpired, and some of his reporting conflicts with the previously mentioned sources.

Switching vital sleeping bags, burning down the hut inadvertently, losing their much-needed dog, and misconstruing communications—the relief expedition of Wegener had certainly left behind a series of conditions with which to contend.

**Fig. 10.3** After the return to Kings Bay. From left to right: Millar, Abrahamsen, Olafson, and Wegener. In the foreground, the second (small) sledge from Cross Bay. *Source* Wegener (1914, Table 19)

## A Rude Awakening: The Disappeared Sleeping Bags, the Mysterious Appearance, and the Tell-Tale Smoke

On March 26, six days after Wegener's rescue group had been at the Wijde Bay hut, Rotvold's group of weary, stranded expeditioners arrived there at the same hut, in anticipation of getting some much-needed rest, but receiving devastating news instead (Einar Rotvold Diary). They had traveled from Mossel Bay that day, leaving at 4:00 am and heading toward Wijde Bay, then briefly stopping "at Steyle Hook [sic] [where] we made a fire on the beach and made ourselves tea, and ate" before continuing on to make "the journey across Wijde Bay."[23] As Rotvold went on to describe in his diary:

> The first third, poor conditions, wet snow and an awful lot of pack ice, so it was heavy going for us to get through there. The rest of the way to 1ˢᵗ house we had level ice and good conditions. We came to the house at ca. 9 pm and found the door open; came into the house and found that there had been people [there], since we left it Christmas Eve 1912. We found a note in German, on the table, from which we learned that a German, D. Wegener [sic], with several [others] had been here; and left here again the 20th March 1913, in order to return to Kings Bay. They have taken our sleeping bags with them, and Capt. Sandleben's equipment, 2 journals, and some salted meat. We got into a pretty bad mood to find, that they had taken

the sleeping bags, which we had relied on were here, with them. Tired and exhausted as we are, we have to sleep 2 at a time, the other two have to sit up to tend to the fire in the stove, so as not to freeze, as it is a poor house. We are not in any good situation now, as we no sleeping bags have any longer. We cannot quite understand, what kind of expedition it has been, as we nothing hear about Captain Ritscher, and he was the only [one], who said to Stenersen and me, that he was going to come to [our] rescue from Advent Bay, as soon as it was in any way possible. Hoping with God's help, that we soon may find out about all these matters, concerning the expedition.[23]

The bleak humor in which the diarist and his colleagues now found themselves due to their discovery that their precious sleeping bags were gone is a reflection of the precarious circumstances in which the Wegener expedition's detrimental actions had now placed them. Rotvold's realization of his tenuous situation is matched only by his utter bewilderment as to who this Wegener expedition was.

Adapting to their new circumstances, Rotvold, Stenersen, Julius, and Jørgen "rested here in the house as best we could" on the following day of March 27, finding driftwood and chopping it for cooking and heat.[24] The ice pilot Stenersen, since the previous night, had been having trouble with his eyes and "has hardly been able to open them up," recorded Rotvold. Despite this, the men spied three wild reindeer that day and "got one of them," part of which they cooked for supper and part of which they planned to take along with them on the journey to the Cross Point hut the following morning "as we cannot know what it looks like over at the house, since we were last there." That night, Stenersen's eyes improved, and Rotvold expressed gratitude for having a sufficient amount of food.

After resting and eating again on the following day of March 28, Rotvold and his three companions left the Wijde Bay hut in the afternoon, hoping for and receiving cooperative weather.[25]

They arrived at Cross Point hut on the following morning of March 29, at approximately 3:00 am, and again virtually crossed paths with the remnants of the Wegener expedition group—but this time, there was a living soul to greet them: "The same expedition, that has been out in the Wijde Bay house has been here as well," reported Rotvold.[26] "They have left a dog here; it came out to meet us. We will bring it along to Advent Bay."

Most likely this dog is Freya, also named Lady, whom Wegener had lost three days prior, on March 26, up on the pass after leaving Cross Point hut, and whom he had regretted losing as reserve provisions. The dog who had gotten away must have double backed to the hut and was seeking shelter there when the Rotvold group had arrived. Fortunately Rotvold planned to take her with him.

But the men themselves were showing the signs of travel by now. "We are partly ailing and sore in the feet, more or less, as we are poorly equipped with footwear," wrote Rotvold on that same day.[27]

Encounters with the aftermath of the Wegener expedition group's visits continued on the following day of March 30, a Sunday which began with the Rotvold group's finding two Arctic foxes caught in previously-laid traps,[28] but which ended with sheer frustration for the men, as narrated by Rotvold:

About 4 o'clock pm, we noticed smoke rising from the house in the West Fjord. We thought then, that the expedition, which had been here before us, was still there. One went out there to the house in order to talk with them, and stay overnight there, and then continue on next morning; but we were shamefully disappointed. When we got there, the house was burnt down and it was only burning in the ruins which caused the smoke. They were wonderful people we had met! They had been rather reckless with the fire, when they left the house. We were forced to return to the Cross Point house, came thereto ca. 11 o'clock pm, ate a little, and then went to bed. It was 3 [metric] miles going for nothing.[29]

Rotvold and his mates had trekked to the house in the hopes of meeting people but were met with disappointment instead in the vision of a burned-down house. The sarcasm of the line regarding encountering such *lovely people* at the burnt-down hut is unmistakable and a rarity in Rotvold's diary—the disappointment in finding the ruins of the hut was deep on many levels. It is clear that Rotvold and his group believed that the same expedition who had been to the Wijde Bay hut had also been to the Cross Point hut and had been the one to carelessly set the West Fjord hut ablaze. It was apparently the Wegener expedition who had unwittingly burned the hut to the ground.

With nowhere else to go, Rotvold, Stenersen, Julius, and Jørgen went back that same night of March 30 to the Cross Point hut, after a seven-hour wasted trip, and there they would remain until April 2.[30] But the additional days at Cross Point were not without their further difficulties. The group had been prescient to bring the reindeer meat they had obtained from their hunt at Wijde Bay along with them to Cross Point, for indeed the previous expedition of Wegener had consumed the reindeer meat that had already been left there at the hut. And now, as of the following day of March 31, the men were beginning to run low on food, with no wild reindeer to be seen in the reindeer valley, and no tracks to be spotted in the extremely deep snow; "We will thus have to manage as best we can with the little we have here," recorded Rotvold in his diary that day, concluding with the words, "Cooking the last reindeer meat today for dinner. God with us. All well."[31]

# Notes on Original Material and Unpublished Sources

Einar Rotvold's diary is in the Library Archives at the Norwegian Polar Institute (NPI) in Tromsø. Hilmar Nøis's diary SVB is in the Historical Archive at the Svalbard Museum (SVB) in Svalbard.

1. E. Rotvold diary, 1 March 1913, (NPI), D00125.
2. E. Rotvold diary, 2 March 1913, (NPI), D00125.
3. E. Rotvold diary, 4 March 1913, (NPI), D00125.
4. E. Rotvold diary, 6 March 1913, (NPI), D00125.
5. E. Rotvold diary, 7 March 1913, (NPI), D00125.
6. E. Rotvold diary, 8 March 1913, (NPI), D00125.
7. E. Rotvold diary, 9 March 1913, (NPI), D00125.
8. E. Rotvold diary, 10 March 1913, (NPI), D00125.

9. E. Rotvold diary, 11 March 1913, (NPI), D00125.
10. E. Rotvold diary, 12 March 1913, (NPI), D00125.
11. E. Rotvold diary, 13 March 1913, (NPI), D00125.
12. E. Rotvold diary, 14 March 1913, (NPI), D00125.
13. E. Rotvold diary, 15 March 1913, (NPI), D00125.
14. H. Nøis diary SVB, April 1913, journal page 54, SVB-AP2.
15. E. Rotvold diary, 16 March 1913, (NPI), D00125.
16. E. Rotvold diary, 18 March 1913, (NPI), D00125.
17. E. Rotvold diary, 19 March 1913, (NPI), D00125.
18. E. Rotvold diary, 20 March 1913, (NPI), D00125.
19. E. Rotvold diary, 21 March 1913 and 23 March 1913, (NPI), D00125.
20. E. Rotvold diary, 24 March 1913, (NPI), D00125.
21. E. Rotvold diary, 25 March 1913, (NPI), D00125.
22. E. Rotvold diary, 25 March 1913, (NPI), D00125.
23. E. Rotvold diary, 26 March 1913, (NPI), D00125.
24. E. Rotvold diary, 27 March 1913, (NPI), D00125.
25. E. Rotvold diary, 28 March 1913, (NPI), D00125.
26. E. Rotvold diary, 29 March 1913, (NPI), D00125.
27. E. Rotvold diary, 29 March 1913, (NPI), D00125.
28. E. Rotvold diary, 30 March 1913, (NPI), D00125.
29. E. Rotvold diary, 30 March 1913, (NPI), D00125.
30. E. Rotvold diary, 2 April 1913, (NPI), D00125.
31. E. Rotvold diary, 31 March 1913, (NPI), D00125.

## Unpublished Sources and Original Material

Nøis, H., Diary SVB, handwritten journal of recollections including an account of the 1913 Staxrud expedition, ("Hilmar Nøis. Fangstmann. 1909–1923, dagbok med erindringer fra fangstlivet") ["Hilmar Nøis. Trapper. 1909–1923, diary with recollections from the trapping life"], not dated (latest year referenced is 1942 in journal pages and 1950 in inserted sheet), SVB-AP2-Hilmar-Nøis-dagbok-1909–1923-compressed. Historical Archive, Svalbard Museum, Svalbard, https://svalbardmuseum.no/no/samlingene/historisk-arkiv/arkivprosjekt-fangst-og-annen-overvintringsvirksomhet-fra-perioden-1910-til-1970/. Accessed 6 June 2021.

Rotvold, E., Diary, Einar Rotvold's expedition diary written during the Schröder-Stranz expedition of 1913, "Dagbok for August Stenersen og Einar Rotvold, Tromsø: Treurenberg Bay fra 1.januar—7.juni 1913: Schröder-Stranz-ekspedisjonen 1912–13", "Einar Rotvold har skrevet dagboken", ["Diary for August Stenersen and Einar Rotvold, Tromsø: Treurenberg Bay from 1 January—7 June 1913: The Schröder-Stranz Expedition 1912–13", "Einar Rotvold has written the diary"], dated 12 January 1913 through 7 June 1913, D00125. Library Archives, Norwegian Polar Institute, Tromsø, https://brage.npolar.no/npolar-xmlui/handle/11250/2426394. Accessed 24 July 2021.

# References

Piepjohn, K. (2021). *English version of the German map, first published in black and white* in K. Piepjohn (2012), p. 61, (updated 2023). From: Piepjohn, K. (2012). Weg-Zeit-Diagramm der Schröder-Stranz-Expedition und der norwegischen und deutschen Rettungsexpeditionen 1912/1913. In: Lüdecke, C. und K. Brunner (eds.), Von A(ltenburg) bis Z(eppelin). Deutsche Forschung auf Spitzbergen bis 1914. 100 Jahre Expedition des Herzogs Ernst II. von Sachsen-Altenburg. *Schriftenreihe des Instituts für Geodäsie der Universität der Bundeswehr München*, Neubiberg, Heft 88, pp. 59–68.

Rave, C. (1913). *Tagebuch von der verunglückten Expedition Schröder-Stranz*. Schaffsteins Gründe Bändchen 49. Cöln: Schaffstein.

Rüdiger, H. (1913). *Die Sorge Bay. Aus den Schicksalstagen der Schröder-Stranz-Expedition*. Berlin: Georg Reimer.

Wegener, K. (1913). Die Hilfsexpedition von Cross- und Kings-Bay nach Wijde-Bay. *Petermanns Geographische Mitteilungen, 59*, 137–140.

Wegener, K. (1914). Die Hilfsexpedition von Cross-und Kings Bay, 21.II.-31.III.1913. In A. Miethe (ed.), *Die Expeditionen zur Rettung von Schröder-Stranz und seinen Begleitern geschildert von ihren Führern Hauptmann A. Staxrud und Dr. K. Wegener* (pp. 69–101). Berlin: Dietrich Reimer, pp. 69–101.

## Chapter 11
# The Arrival of the Four Schröder-Stranz Expeditioners in Advent Bay, and the Welcoming of Staxrud's *Hertha* at Green Harbour

### Rotvold, Stenersen, and the Jensens Leave Cross Point and Reach Advent Bay

The month of April 1913 began with an abundance of snowfall and a lack of food for the four Schröder-Stranz expeditioners still taking refuge in the hut at Cross Point (Einar Rotvold Diary). Nonetheless, Einar Rotvold, August Stenersen, Julius Jensen, and Jørgen Jensen met the occasion. On that first day of April, the men went grouse hunting in the mountains and caught three grouse for supper, but the situation could not continue in this manner. "We are compelled to begin the trek again in the morning, weather permitting, as our provisions soon are running out," wrote Rotvold in his diary.[1] "It will be a hard march we will get, about 10 [metric] miles to the first house, without sleeping bags, God help us forward."

Most likely helping them on the march would be the two dogs Rotvold had previously reported about—Lotte, whom they had taken with them from the ship, and Freya (also called Lady), whom they had found at the Cross Point hut, left behind by Kurt Wegener's group (Chap. 10).

On the next day, April 2, the four Norwegians—and probably two dogs—left the Cross Point hut at 5:00 am and passed the burnt-down hut in the West Fjord, pausing there for tea before continuing on to "the bottom of the fjord... then up the mountain over to Dickson Bay," according to Rotvold.[2] Traveling in "good conditions... but heavy snow" they descended to Dickson Bay, arriving there at approximately 10:00 pm, and proceeded to begin the journey out of the fjord in "terribly bad conditions, deep snow and clogging [under the skis], so it was heavy going for us," continued Rotvold.

The following day of April 3 found them trekking through deep, slushy snow that must have slowed down their progress through Dickson Bay as their supplies continued to dwindle. "We now have nothing else to live on but tea and plasmon-flour soup [dehydrated vegetable powder or puree]," lamented Rotvold.[3] "Wish only that we were in Advent Bay. At 9 o'clock pm one arrived at Karl Eliassen's house, on the west side of the stream; but [there] wasn't much to find there, one wall and both windows and the stove were gone. We were then so tired and exhausted, that we could not go much further." But even rest was nearly impossible, especially without their sleeping bags: "One made tea, and laid down on the snow under the roof, with the few clothes we had, over us," stated Rotvold, "but we could not lie there long, before we had to get up again and start to walk ourselves warm, as it was too cold to lie down and sleep, even though it was mild weather." A little bit of Maggi broth revived them somewhat as they prepared for the morrow.

Little did they know on April 3 that the rescue expedition of Arve Staxrud had arrived at Green Harbour that very day, and that this would change what awaited them in Advent Bay.

At 2:00 am on the following morning of April 4, in the middle of an abundance of snowfall, the expeditioners resumed their journey and began their trek across to the east side of the stream so as to locate the hut identified by Rotvold as "Andersen's house."[4] They encountered, however, "poor" ice "and a lot of slush" that, according to Rotvold, caused the men to fall "through the ice several times, so that we had to make a fairly large detour out the fjord." Fortunately, they were able to cross the stream safely and by approximately 5:00 am had located the house. "We were happy, when we saw that it was in order, both stove and a good warm house," wrote Rotvold appreciatively in his diary that day. "Yes, we certainly needed it, because we had not tasted sleep, since we left the Cross Point house in Wijde Bay; a long and hard march it has been for us." Food, however, was another matter. "We found no foodstuffs here," he continued. "We thought we would find some rusks [hard biscuits], but no, we were bitterly disappointed." Nonetheless, the men had found warmth, safety, and a roof over their heads, under which they now prepared a meal of plasmon-soup and finally were able to catch some sleep. Afterward, during their day at the house, they used their gun to shoot two northern (Arctic) fulmars that they prepared for their supper. But they also encountered tracks of their colleague who had previously departed from this location. "It looks like our Capt. Ritscher has been here," reported Rotvold, "because we found his revolver on the floor, but he has not left any notes behind." The men spent the day drying their wet clothing, obtaining much-needed rest, and awaiting the weather to clear so that they could continue on the final leg of their trip south. It was still another approximately five (metric) miles to Advent Bay. But the men had already accomplished a very difficult part of the journey and were now refreshed, relatively speaking. "We will be going relaxed and easy thereto," concluded Rotvold in his diary on that same day.

And indeed the following day of April 5 found the expeditioners happily able to leave "the eastern Strøm-house" at about 8:00 am in conditions that were suitable for trekking, as reported by Rotvold.[5] Following several hours of travel, they were finally met by other human beings and helpful animals. "Ca. 3 o'clock pm after having passed Rævnæs, we saw people coming toward us," wrote Rotvold, specifying three named individuals "Mr. Mangham,[1] Mr. Louis, and Jensen, as well as a boy with [a]

---

[1] "Mr. Mangham" most likely refers to Bert Mangham, the winter superintendent for the Arctic Coal Company (Dole 1922, vol. 2: 17–19, 68, 200, 206, & 464), and "Jensen" may possibly refer to

horse, who came to welcome us. We were then told, that Capt. Ritscher was here, he had frostbitten hands and feet. We were well received in Advent Bay, got [a] sufficient [quantity] of food and necessary clothing, and a good room to sleep in." Their arrival included a touching reunion with their expedition captain, as described by Rotvold: "Capt. Ritscher was exceedingly pleased that we arrived hereto. The pleasure was mutual, for Stenersen and me, who had been together with him the longest and saddest time, and likewise were the last who parted with him on the Dickson Bay mountain. He has had a hard and excruciating march, but still shown himself as a strong-willed and energetic man, for which he ought to have all honor." Last but not least, the four Norwegians were also told about the numerous relief attempts. "There have been prepared several rescue-expeditions from here [Spitsbergen], which all however are unsuccessful, due to ice and weather-conditions," reported Rotvold. "The expedition from Norway arrived at Green Harbour the 3rd [of] April."

And so Rotvold, Stenersen, Julius, and Jørgen, who had managed to save themselves by enduring the long journey south from the ship and who had reached Advent Bay to report on the location of Hermann Rüdiger and Christopher Rave, now all knew of Staxrud and the Norwegian rescue expedition who had arrived to help them.

## *Hertha* Anchors at the Ice Edge in Green Harbour, and the Rescue Expeditioners Unload Supplies

The anchoring of Arve Staxrud's ship *Hertha* at the Green Harbour ice edge on April 3 was a momentous occasion—for many people on Spitsbergen, and for a multitude of reasons. First and foremost, the rescue expedition had arrived safely to begin its search for the still-stranded Schröder-Stranz expeditioners. But also importantly, the expedition ship had brought with it news from home for the many Arctic workers who had not seen or heard from their families and friends in quite a long time (Hilmar Nøis Diary SVB). Overwintering workers from the telegraph station and the various coal mining fields poured out onto the ice edge and to the docked ship to receive mail from the mainland and from their homes.[6] Many an overwinterer who had rushed to the *Hertha* stayed near the ship as they quickly tore into their mail, completely absorbed, reading the long-awaited letters from their loved ones (Nøis, 1929, 6). The rescue expeditioners who had arrived, meanwhile, wasted no time in unloading the ship (Figs. 11.1 and 11.2).

First to come off the vessel and onto the ice were the reindeer who had been confined in the hold since boarding the *Hertha* prior to March 24 and during the entire length of the voyage (Hilmar Nøis Diary NPI; Hilmar Nøis Diary SVB). These draft animals were now allowed to breathe fresh air and stretch their legs and then were immediately put to work pulling sledge-loads of equipment and supplies across the ice, as were the sledge dogs who up until now had been tied up securely

---

Ingvar Jenssen who led the first relief expedition—organized by the Arctic Coal Company—which had returned in February after failing to reach the stranded expeditioners (Chap. 5).

**Fig. 11.1** Unloading the expedition ship *Hertha* at Green Harbour. *Photographer/Byline* Arve Staxrud. *Source/Owner* Norwegian Polar Institute

on the ship.[7] Those sledge dogs included Lussi and Storm, from Roald Amundsen's Antarctic expedition, and the dogs brought by Daniel Nøis, as well as those who had joined in Tromsø (Chap. 9). Three of the reindeer were speedily employed by Per Hansen, Johannes Kemi, and Samuel Klemmetsen to go with one of these men to drive several items to the telegraph station.[8] And then all the men, dogs, and reindeer drove the complete contingent of supplies, provisions, and equipment from the ship across the ice to an old test-mining building on Cape Heer (the Schröder House or Schröder's House), which they utilized as their expedition headquarters and storage depot.[9] According to Hilmar Nøis (Hilmar Nøis Diary NPI), at this time there were 19 reindeer and 17 sledge dogs.[10] And according to Staxrud, in his May 14 telegram published in Norway's *Aftenposten* newspaper on May 15, 1913, there were 12 sledge dogs who had been transported from Norway and later another six who had been acquired in Green Harbour (Aftenposten, 1913, evening: 1–2; Tahan, 2021, 298), thus indicating a total of 18 sledge dogs. Staxrud also specified in his telegram, as reported by the newspaper, that the driving of equipment across the ice and to the shore was performed on challenging ice and in less-than-desirable conditions. Later in his published report (Staxrud, 1914, 13), Staxrud would say that they used 20 reindeer sledges and two dog sledges to transport all their equipment across the three kilometers of flat sea ice to shore (Fig. 11.3).

The reindeer, according to Daniel Nøis, were exceptionally helpful in the pulling of all the gear over the long distance and across the problematic ice, which had

**Fig. 11.2**  Photographed here at rest, the reindeer and dogs worked with the men to bring the supplies from the ship onto the shore. *Photographer* Arve Staxrud. *Source/Owner* Svalbard Museum

actually proven to be very uneven (Nøis, 1929, 6). And the sledge dogs did their part as well (Hilmar Nøis Diary NPI). In this way, the provisions and equipment, including the masses of reindeer moss, were all driven up to the Schröder House situated on the Heer-headland.[11] It took five full days to unload the ship (Hilmar Nøis Diary SVB),[12] and tremendous effort was exerted, during which time telegraph station manager Olaf Henriksen informed the Staxrud group of the good news that four members of the Schröder-Stranz expedition had made it safely to Advent Bay (Nøis, 1929, 6–7). According to Daniel Nøis, there was little more information to be gleaned. And according to Hilmar Nøis, the announcement came via a telegram sent from Advent Bay to Staxrud (Hilmar Nøis Diary NPI).[13]

Meanwhile, back in Advent Bay, on April 6—the day after reaching the mining settlement—Rotvold (Einar Rotvold Diary) and his three colleagues found themselves being visited by many people interested in knowing about the expeditioners and the German expedition.[14] On that day, Rotvold recorded that the four Norwegians had telegrammed their families with the happy news that they were safe. On the following day of April 7, Rotvold reported that a telegram sent to Staxrud regarding his coming to Advent Bay had been received by the expedition leader, but that Staxrud had not yet arrived.[15] He also reported that Henriksen had sent a telegram to Stenersen inviting the four Norwegians to Green Harbour. Thus, as of April 8, the four men were seriously considering making a trip to Green Harbour as soon as

**Fig. 11.3**   Transport of equipment ashore. *Source* Staxrud (1914, Tf. 1)

possible, as the telegraph equipment was not working as well in Advent Bay, and as the men were eager to hear from their families and to communicate with the outside world—especially with Tromsø: "We have read some newspapers, of which we see that there are many kinds of writings about the expedition and us," wrote Rotvold, noting that "we have, however, our own thoughts about it."[16] In addition to disclosing the whereabouts of the remaining German expeditioners to Staxrud and the rescue group, the Norwegians were also concerned about the state of the expedition and the remuneration that was owed to them, "because we hear many kinds of statements, which do not exactly put us in any good mood," continued Rotvold. "We are longing for responses from our families."

## Staxrud's Dog Sledge Trip from Green Harbour to Advent Bay, and His Crucial Meeting with the Surviving Schröder-Stranz Expeditioners

While the rescue expedition's men, reindeer, and dogs continued their tiring excursions pulling the loaded sledges of equipment across the Green Harbour ice during the several days following their arrival on April 3, Staxrud made ready to travel to

Advent Bay to meet with the four Schröder-Stranz expeditioners who had arrived there on April 5. His intention was to ascertain the specifics of the situation and to strategize the next steps for the rescue (Nøis, 1929, 7).

Since the local conditions were not likely to change anytime soon, Staxrud decided to set out from his docking location (Staxrud, 1914, 13–15). On April 8, together with Johan Nilsen Nøis and Dr. P. W. K. Bøckmann, Staxrud began his dog sledge drive to Advent Bay, using two sledges pulled by eight dogs each. (This indicates 16 sledge dogs going to Advent Bay at this time. Evidently, based on Staxrud's May 14, 1913, telegram published in the *Aftenposten* newspaper on May 15, the 12 dogs brought from Norway had by now been joined by the six dogs added in Green Harbour, as Staxrud implies taking 16 of the dogs in this statement in his 1914 report. Possibly two dogs remained in Green Harbour for the time being.) The physician Dr. Bøckmann had come along to look at the frostbitten foot of Captain Alfred Ritscher, who was currently recuperating in Advent Bay. Meanwhile, stated Staxrud, the Spitsbergen expert Daniel Nøis was to lead the reindeer caravan, which consisted of 14 reindeer sledges and six Sámi sledges (Pulken), each pulled by a reindeer. The sledges each weighed between 150 and 200 kg.

According to Daniel Nøis, he and the other men (Hilmar Nøis, Jakob Ellingsen, Hansen, Kemi, and Klemmetsen, as well as Martin Pettersen Nøis and Søren Zachariassen) remained behind in Green Harbour (with the reindeer and possibly a couple of the sled dogs) in order to pack the sledges with food and supplies for the trek north (Nøis, 1929, 7). The task of arranging the appropriate portions of provisions for all individual members and all three species of the rescue team—humans, dogs, and reindeer—was a rather complex one.

Staxrud, upon arriving in Advent Bay, conducted crucial meetings with Rotvold, Stenersen, and the Jensen brothers, as well as with Captain Ritscher, wherein they debriefed him on the status and location of the two remaining survivors Hermann Rüdiger and Christopher Rave and the last known locations of the seven men who had disappeared (Staxrud, 1914, 16; Einar Rotvold Diary; Nøis, 1929, 7; Hilmar Nøis Diary SVB; Hilmar Nøis Diary NPI).

First, Staxrud's group met with Ritscher and the four Norwegian sailors of the *Herzog Ernst* (Jørgen and Julius Jensen, Rotvold, and Stenersen) in Advent Bay on April 9, who told them that Rave and Rüdiger were staying in the house of the Swedish Arc-of-Meridian expedition (Staxrud, 1914, 16–18). Now the task was to bring help to Treurenberg Bay together with the doctor. Staxrud reported that, according to Ritscher, Schröder-Stranz had left the ship north of Scoresby Island. Dr. Erwin Detmers and Dr. Walter Moeser had last been seen on October 2 on the east coast of Wijde Bay, while Wilhelm Eberhard had disappeared on the Bangenhook Peninsula between Wijde Bay and Mossel Bay 10 km from Nordenskiöld's hut Polhem. Based on this information, Staxrud planned to search for the two biologists and Eberhard in the huts of Wijde Bay and Bangenhook Peninsula and to look for Schröder-Stranz on the coast of Northeast Land between Ryss Island in Murchison Bay and Cape Platen east of the Rijp Bay.

Rotvold recorded his meeting with Staxrud on April 9 in his diary (Einar Rotvold Diary), reporting that Staxrud, Dr. Bøckmann, and Johan Nilsen Nøis had arrived at

4:00 pm "with the dogs and 2 sleds" and that "The other party with the reindeer will be coming tomorrow to the Hotel-headland where they will meet up with Staxrud, to from there travel forth northwards."[17] Staxrud had brought both good news and bad news to Rotvold and his mates. He had delivered much-awaited personal correspondence, but had also imparted very concerning professional tidings. "It was a happy day for us, we received both letter and telegram from home," wrote Rotvold. "It is bad news we are hearing about the Schröder-Stranz Expedition; there is namely no funds, so there is little chance that we will get our hard-earned money. The captain is in a bad mood, which is no wonder."

According to Rotvold, a second meeting took place between the four Norwegian members of the Schröder-Stranz expedition and Captain Staxrud on the following day of April 10, during which it was decided that Stenersen would accompany Staxrud and the rescue expedition while the other three Norwegians would go on to Green Harbour to await Staxrud's return and then sail to Treurenberg Bay to retrieve the frozen-in ship: "We have been at Staxrud's today and talked with him, about our journey, from start to finish," wrote Rotvold, adding, "He thought we had done well. Stenersen will be going northwards with Staxrud, and stay onboard the ship."[18] Thus noting Staxrud's approval of Rotvold's group's actions, and announcing that the formerly stranded expeditioner Stenersen would now become one of the rescuers, Rotvold arrived at the conclusion that "It will probably be a long time before we get home again, maybe not until August..." Meanwhile, he also awaited the arrival of "The expedition's reindeer-party" who seemed to be delayed: "Bad weather and poor conditions are hindrances for that," concluded Rotvold, adding, "Hope they arrive tomorrow."

## Daniel Nøis and the Reindeer Caravan, from Green Harbour to Advent Bay

The reindeer party of Daniel Nøis, Hilmar Nøis, Ellingsen, Hansen, Kemi, Klemmetsen, and the reindeer did indeed form a caravan that started from Green Harbour on April 9—the day after Staxrud's departure—and marched toward Advent Bay, as reported by Daniel Nøis, encountering tremendous cold and snow along the way, having to camp in tents just past Coal Bay, and picking up additional reindeer moss on the following day of April 10, before arriving in Advent Bay on April 11—all with the fully packed sledges (Nøis, 1929, 7). Their journey from the Heer-headland with 19 reindeer was through frigid conditions and loose snow surface that prevented their skis and sledges from gliding smoothly over the icy terrain, according to Hilmar Nøis (Hilmar Nøis Diary NPI), who also, in his written account regarding this portion of the trip, mentions taking 17 sled dogs on the drive.[19] (Possibly two of the 18 dogs had temporarily remained in Green Harbour, as 16 had gone with Staxrud, per Staxrud's, 1914 report, and possibly one or two of those now accompanied Hilmar and the reindeer party to Advent Bay, as Hilmar references 17 dogs in both journal accounts.

There is also the possibility that one dog may have remained in Green Harbour, as shall be seen later, but this is not known for certain). The slow-going forced the party to spend that first night of April 9 out on the ice in −35 °C and in tents whose interiors quickly became rime-frozen. The men made use of reindeer-skin sleeping bags lined with flannel (Hilmar Nøis Diary SVB) and also were armed with lighter-weight down sleeping bags as well as silk inner tents in case those were needed.[20] When the party reached Advent Bay (on April 11), they camped on the ice at the Hotel-headland, from where the men could go to the Arctic Coal Company to speak with Ritscher and the four Norwegians from the *Herzog Ernst*.[21]

That evening, Daniel Nøis and his group visited the coal mining field and were informed of the news regarding the four Norwegian men who had rescued themselves from the frozen ship (Nøis, 1929, 7). Daniel Nøis made special mention of their account—of all the lost Schröder-Stranz expedition comrades, the difficult separation from Ritscher, the deceased crewmate Knut Stave, the life-and-death medical treatment of Rüdiger by Rave, and the disappearance of Eberhard, which had caused desperation and a near-death-experience for Rotvold and Stenersen.

Rotvold himself on this day of April 11 was reporting on the arrival of the reindeer party and on their camping at the Hotel-headland (Einar Rotvold Diary). "At 3 o'clock afternoon Staxrud and Stenersen drove with 2 teams of dogs out there," he wrote.[22] "We followed them to the camp. It is a whole caravan, that is trekking northward." After unloading some pemmican at the Hotel-headland, given that there would be pemmican available on the ship, the expedition was planning to begin its journey to Treurenberg Bay early on the following morning. Rotvold and the two Jensens were also planning to leave on that following day, heading toward Green Harbour, and taking with them Johan Nilsen Nøis and one reindeer (most likely one of the expedition's draft reindeer).

For Johan Nilsen Nøis had developed a fever en route to Advent Bay (Staxrud, 1914, 19), and by this time, he was severely ill with the mumps (Hilmar Nøis Diary NPI). He would not be able to proceed with the search party as planned. Stenersen was already going on the rescue mission to secure the frozen ship at Treurenberg Bay and prevent anyone from taking possession of it,[23] and he would also take Johan Nilsen Nøis's place on the expedition (Nøis, 1929, 7). Johan Nilsen Nøis would stay in Advent Bay and, evidently, per Rotvold, was scheduled to go with him to Green Harbour.

Meanwhile, on this day, in Green Harbour, Martin Pettersen Nøis and Søren Zachariassen remained with the cases of supplies that had been transported from the vessel to the Schröder House at Cape Heer (Nøis, 1929, 7; Hilmar Nøis Diary NPI). Their job now was to manage and guard the depot.[24] Simultaneously, the ship that had brought them there—the *Hertha*—awaited its chance to go out sealing in the Arctic waters where it had previously plied its whale-hunting trade (Hilmar Nøis Diary SVB). For the moment, however, the ship was imprisoned at Green Harbour, blocked by the ice.[25]

## Unexpected News of Great Expectations

In Advent Bay, the sledge dogs brought from Norway and the dogs engaged in Green Harbour were by now joined together and were prepared for the trek north (Nøis, 1929, 7).

Those dogs brought from Norway included Lussi and Storm, the Greenland sled dogs who had served on Roald Amundsen's Antarctic expedition and who were now part of Staxrud's rescue expedition (Hilmar Nøis Diary SVB; Hilmar Nøis Diary NPI). Regarding these two Polar dogs, Hilmar Nøis discloses a remarkable new detail in his diaries—he states that Lussi gave birth to puppies while on the rescue expedition trip; furthermore, per his documentation, some of these puppies would become an integral part of Spitsbergen and future rescue efforts.[26, 27]

This new fact regarding Lussi's giving birth to puppies while on the rescue expedition is affirmed by Staxrud in correspondence he would later send to Roald Amundsen (Staxrud 11 August 1913) regarding the addition of puppies on the expedition trip.[28]

While Hilmar Nøis and Staxrud do not state when Lussi gave birth to her puppies, two differing reports found later indicate different timeframes, ranging from as early as March (Morgenbladet, 1913, evening: 1) to April or May (Rüdiger, 1913, 187), as shall be further presented later in this narrative.

All the dogs of the Norwegian rescue expedition would later be described by one of the rescued Germans, Rüdiger, as varying in breed, from "Eskimo" to "Samoyed" to "mixed breeds," some of whom already had experience in Polar work, with those from Norway being smaller, faster, and more nimble than the German expedition dogs (Rüdiger, 1913, 187). In addition to the Greenland Dog breed of sled dog (Hilmar Nøis Diary SVB), the Gordon Setter hunting dog breed was also part of the Norwegian rescue contingent, according to Hilmar Nøis.[29]

Everyone in Advent Bay could now look forward to new action being initiated. A gathering of the minds had taken place. The rescue expedition had finally arrived. And certainly the animals involved in the expedition had stirred an interest in their own right—not least of whom were the reindeer (Einar Rotvold Diary). The moment was a meaningful one for Einar Rotvold. He likened it to a near mystical experience: "Here is a lot of life this evening; all people going outwards to the headland to see the reindeer. It almost looks like people heading to church out in the country."[30]

The Staxrud Expedition party—men, dogs, and reindeer—was ready to make the long journey north.

## Notes on Original Material and Unpublished Sources

Einar Rotvold's diary is in the Library Archives at the Norwegian Polar Institute (NPI) in Tromsø. Hilmar Nøis's diary SVB is in the Historical Archive at the Svalbard Museum (SVB) in Svalbard. Hilmar Nøis's diary NPI is in the Library Archives at the Norwegian Polar Institute (NPI) in Tromsø. Arve Staxrud letters of correspondence, written from and to Arve Staxrud, are in the Manuscripts Collection at the National Library of Norway (NB) in Oslo.

1. E. Rotvold diary, 1 April 1913, (NPI), D00125.
2. E. Rotvold diary, 2 April 1913, (NPI), D00125.
3. E. Rotvold diary, 3 April 1913, (NPI), D00125.
4. E. Rotvold diary, 4 April 1913, (NPI), D00125.
5. E. Rotvold diary, 5 April 1913, (NPI), D00125.
6. H. Nøis diary SVB, April 1913, journal page 50, SVB-AP2.
7. H. Nøis diary NPI, notebook 3, April 1913, pdf page 9, (NPI), D-307/D00307_3_0001.
8. H. Nøis diary SVB, April 1913, journal pages 50–51, SVB-AP2.
9. H. Nøis diary SVB, April 1913, journal page 51, SVB-AP2.
10. H. Nøis diary NPI, notebook 3, April 1913, pdf page 9, (NPI), D-307 / D00307_3_0001.
11. H. Nøis diary NPI, notebook 3, April 1913, pdf page 9, (NPI), D-307 / D00307_3_0001.
12. H. Nøis diary SVB, April 1913, journal page 51, SVB-AP2.
13. H. Nøis diary NPI, notebook 3, April 1913, pdf page 9, (NPI), D-307 / D00307_3_0001.
14. E. Rotvold diary, 6 April 1913, (NPI), D00125.
15. E. Rotvold diary, 7 April 1913, (NPI), D00125.
16. E. Rotvold diary, 8 April 1913, (NPI), D00125.
17. E. Rotvold diary, 9 April 1913, (NPI), D00125.
18. E. Rotvold diary, 10 April 1913, (NPI), D00125.
19. H. Nøis diary NPI, notebook 3, April 1913, pdf page 10, (NPI), D-307 / D00307_3_0001.
20. H. Nøis diary SVB, April 1913, journal page 52, SVB-AP2.
21. H. Nøis diary NPI, notebook 3, April 1913, pdf page 11, (NPI), D-307 / D00307_3_0001.
22. E. Rotvold diary, 11 April 1913, (NPI), D00125.
23. H. Nøis diary NPI, notebook 3, April 1913, pdf page 11, (NPI), D-307 / D00307_3_0001.
24. H. Nøis diary NPI, notebook 3, April 1913, pdf pages 9–10, (NPI), D-307 / D00307_3_0001.
25. H. Nøis diary SVB, April 1913, journal page 51, SVB-AP2.
26. H. Nøis diary SVB, 1913, journal page 65, SVB-AP2.
27. H. Nøis diary NPI, notebook 3, 1913, pdf page 39, (NPI), D-307 / D00307_3_0001.
28. A. Staxrud to R. Amundsen, telegram, 11 August 1913, NB Brevs. 812:1.
29. H. Nøis diary SVB, April 1913, journal page 55, SVB-AP2.
30. E. Rotvold diary, 11 April 1913, (NPI), D00125.

## Unpublished Sources and Original Material

Nøis, H., Diary NPI, handwritten journal of recollections recounting the events of the 1913 Staxrud expedition, Hilmar Nøis, "Minder og Fortelinger fra fangsttiden på Spitsbergen, Klade 3, 7 Mars 1970" [Hilmar Nøis, "Memories and Stories from the trapping time on Spitsbergen, Notebook 3, 7 March 1970"], dated 7 March 1970, D-307, ("Minder og fortelinger, Nøis, Hilmar, hefte 3"), D00307_3_0001. Library Archives, Norwegian Polar Institute, Tromsø, https://brage.npolar.no/npolar-xmlui/handle/11250/274077. Accessed 20 July 2021.

Nøis, H., Diary SVB, handwritten journal of recollections including an account of the 1913 Staxrud expedition, ("Hilmar Nøis. Fangstmann. 1909–1923, dagbok med erindringer fra fangstlivet") ["Hilmar Nøis. Trapper. 1909–1923, diary with recollections from the trapping life"], not dated (latest year referenced is 1942 in journal pages and 1950 in inserted sheet), SVB-AP2-Hilmar-Nøis-dagbok-1909–1923-compressed. Historical Archive, Svalbard Museum, Svalbard, https://svalbardmuseum.no/no/samlingene/historisk-arkiv/arkivprosjekt-fangst-og-annen-overvintringsvirksomhet-fra-perioden-1910-til-1970/. Accessed 6 June 2021.

Rotvold, E., Diary, Einar Rotvold's expedition diary written during the Schröder-Stranz expedition of 1913, "Dagbok for August Stenersen og Einar Rotvold, Tromsø: Treurenberg Bay fra 1.januar—7.juni 1913: Schröder-Stranz-ekspedisjonen 1912–13", "Einar Rotvold har skrevet dagboken", ["Diary for August Stenersen and Einar Rotvold, Tromsø: Treurenberg Bay from 1 January—7 June 1913: The Schröder-Stranz Expedition 1912–13", "Einar Rotvold has written the diary"], dated 12 January 1913 through 7 June 1913, D00125. Library Archives, Norwegian Polar Institute, Tromsø, https://brage.npolar.no/npolar-xmlui/handle/11250/2426394. Accessed 24 July 2021.

Staxrud, A., 11 August 1913, Telegram from Arve Staxrud to Roald Amundsen, NB Brevs. 812:1. National Library of Norway, Oslo, online archives, https://www.nb.no. Retrieved 6 July 2021.

## References

Aftenposten. (15 May 1913). Hjælpeexpeditionen paa Spitsbergen. Kaptein Staxruds beretning til "Aftenposten". [The rescue expedition on Spitsbergen. Captain Staxrud's account to "Aftenposten".]. *Aftenposten* (Kristiania [Oslo]), 15 May 1913, evening edition, pp. 1–2. National Library of Norway, Oslo, online archives, www.nb.no. Accessed 28 August 2012.
Dole, N. H. (1922). *America in Spitsbergen: the romance of an Arctic coal-mine*. In Two Volumes. Boston: Marshall Jones Company. Volume II. Facsimile reprint by Scholar Select.
Morgenbladet. (15 September 1913). "Oberstens" Barn. [Children of "Obersten".]. *Morgenbladet* (Kristiania [Oslo]), 15 September 1913, evening edition, p. 1. Digital Archive, National Library

of Norway, Oslo, received 30 March 2022; National Library of Norway, Oslo, online archives, https://www.nb.no. Retrieved 2 April 2022.

Nøis, D. (1929). Med kaptein Staxrud på leiting efter Schrøder-Stranz og hans folk [With Captain Staxrud on the search for Schröder-Stranz and his people]. In: *And-Ungen*, July 1929, pp. 4–12. Andenes: Andøyposten. Library Archives, Norwegian Polar Institute, Tromsø. Received 20 July 2021.

Rüdiger, H. (1913). *Die Sorge Bay. Aus den Schicksalstagen der Schröder-Stranz-Expedition.* Berlin: Georg Reimer.

Staxrud, A. (1914). Die Staxrudsche Hilfs-Expedition für Schröder-Stranz. In A. Miethe (ed.), *Die Expedition zur Rettung von Schröder-Stranz und seinen Begleitern—geschildert von ihren Führern Hauptmann A. Staxrud und Dr. K. Wegener* (pp. 1–68). Berlin: Dietrich Reimer.

Tahan, M. R. (2021). *The return of the South Pole sled dogs: With Amundsen's and Mawson's Antarctic expeditions.* Cham: Springer International Publishing.

# Part III
# Interspecies Cooperation and Perseverance: The Arve Staxrud Rescue Expedition

# Chapter 12
# The Staxrud Rescue Expedition: Embarking on the Sledging Trek North to Treurenberg Bay

## The Departure from Advent Bay

On April 12, 1913, the Staxrud Rescue Expedition was ready to make the arduous sledging journey north to reach the two stranded expeditioners at Treurenberg Bay and to search for the seven missing expeditioners near Mossel Bay and in Northeast Land; the rescue party comprised Arve Staxrud, Dr. P. W. K. Bøckmann, Daniel Nøis, Hilmar Nøis, Per Hansen, Johannes Kemi, Samuel Klemmetsen, the engineer Jakob Ellingsen, August Stenersen, three or two teams of sledge dogs (18 dogs were part of the expedition according to Staxrud in his *Aftenposten* telegram article, but the number is later given as 16 according to Staxrud's published report, 16 according to Daniel Nøis, and 17 according to Hilmar Nøis), and a caravan of draft reindeer (20 reindeer for 19 sledges according to Staxrud, and 19 or 18 reindeer according to Hilmar Nøis), as well as a calculated supply of food, medicine, and equipment (Staxrud, 1914, 19; Nøis, 1929, 7; Hilmar Nøis Diary SVB; Hilmar Nøis Diary NPI; Aftenposten, 1913, evening: 1–2).

Two dog sledges were being pulled by six dogs each, and one dog sledge was pulled by five dogs (thus indicating 17 dogs working as three sledge teams), according to Hilmar Nøis, who also specified that 18 reindeer pulled either a sledge or a pulk each—the expedition having mainly light-weight Nansen dog sledges that weighed approximately 28 kg, but also having approximately six or seven heavier, cumbersome reindeer-pulks that were pulled by some of the draft reindeer (Hilmar Nøis Diary SVB). Loaded onto the sledges were all the necessary provisions, which included dog pemmican, and all the equipment anticipated to be needed, as well as medicine for the expeditioners and for treating the injured they might find, plus two kayaks in case they encountered broken ice and open water during the return journey.[1] In good conditions of hard snow, where the reindeer-pulk could glide over the surface and not sink into the snow, one reindeer would be able to pull a total of 100 kg.

M. R. Tahan and C. Lüdecke, *Stranded at the Top of the World*,
https://doi.org/10.1007/978-3-031-56288-4_12

The number of dogs and number of dog sledge teams given by Staxrud in his published report (Staxrud, 1914, 19) differ from those given by Hilmar Nøis. In his report, Staxrud states that there were 16 dogs working with two sledges, thus indicating eight dogs per sledge team. He also states that 20 reindeer worked with 14 sledges and five Sámi pulks, thus indicating 19 sledges being pulled by a reindeer each, but he does not mention the role of the 20th reindeer. Recall that Hilmar Nøis had stated that one of the 20 reindeer had broken their leg during the voyage to Spitsbergen and had had to be euthanized (Chap. 9), that Daniel Nøis had stated that 19 reindeer came to Advent Bay (Chap. 11), and that Einar Rotvold had stated that he was to take one reindeer back with him to Green Harbour (Chap. 11).

The seeming discrepancy in the number of reindeer given in the various accounts, varying from 20 to 19 to 18, as well as in the number of dogs given in the accounts, varying from 18 to 17 to 16, is possibly due to the different timeframes, wherein the numbers probably depended on the date that the reindeer or dogs were counted. Thus, the differing numbers may have reflected the different number of dogs and reindeer at different times during the treks.

Of the nine men embarking on the sledging trip, the former stranded expeditioner Stenersen had replaced Johan Nilsen Nøis, who by this time was not even able to return to Green Harbour with Einar Rotvold and the Jensens, his case of mumps having become even more serious (Hilmar Nøis Diary NPI; Einar Rotvold Diary). Instead, he lay in the hospital at the Arctic Coal Company in Advent Bay.[2] Rotvold thus delayed his trip, but received on this day the one reindeer that he was to take with him to Green Harbour.[3] Most likely, this was one of the originally 20 draft reindeer brought by the rescue expedition (as Staxrud mentions only 19 reindeer sledges, and as the number of reindeer embarking toward Treurenberg Bay, as given by Hilmar Nøis in his diaries, varies from 19 to 18). The day of April 12 truly was a busy and momentous day for everyone involved, and Rotvold's diary reflects this—it is one of the only two diary entries during this time period that do not end with his typical signature statement. His sailor's refrain, "God with us. All well."—the usual hopeful sign-off that reflected his voice—does not appear at the end of this entry. The day's activities must have been frenetic indeed.

But Rotvold's unwritten words did not portend delay, for, on this day, the Staxrud expedition set out successfully from Advent Bay—they were a cooperative, concerted interspecies coalition of humans, canines, and reindeer, embarking on a life-saving mission.

## Across the Ice Fjord and to the Bottom of Dickson Bay; the Hard Ascent at Dicksonfjord; the Descent into West Fjord; and the Sending Back of Some of the Party

Thus, the Staxrud rescue expedition began their crossing of the Ice Fjord on April 12 (Staxrud, 1914, 19–20), covering 40 km in each of the first two days (Fig. 12.1).

**Fig. 12.1** Departure of the reindeer sledges from Advent Bay across the Ice Fjord. *Source* Staxrud (1914 Tf. 2)

As Staxrud would later report, near Lygtan they set up a depot of provisions for eight men for four to five days. They also deposited ammunition and two surplus kayaks there, because they could use the two kayaks which, according to Ritscher, were in Treurenberg Bay, to investigate the north coast. If the ice in Ice Fjord broke up when they returned, they could use the deposited kayaks to reach Advent Bay (The travel route is given in Fig. 10.2a, b in Chap. 10).

Another depot with 100 days' rations of provisions was established south of the watershed between Dickson Bay and West Fjord (Wijde Bay). Crossing the watershed at an altitude of 500 m was very difficult, as they could only manage the steep slopes with half a load and therefore had to make the trip twice (Fig. 12.2).

According to Daniel Nøis, during the very first portion of the trek across the Ice Fjord, on that first day of April 12, three of the sledge dogs who had been recruited in Green Harbour escaped, leaving the remaining dogs to pull the sledges (Nøis, 1929, 7). Once the party had arrived at the halfway point across the Ice Fjord, the men had stopped for coffee and food. After marching for nine hours, they stayed in a cabin at Dickson Bay and ate a meal of pemmican. Afterward, they proceeded to the bottom of Dickson Bay and established their camp for the night, pitching their tents and settling in after more than 13 h of driving. It was here that the three dogs who had escaped at the beginning of the journey finally returned to the men and were repatriated—they had been hovering around the party during the trek, giving Daniel

**Fig. 12.2** Rappelling of the dog sledges on the steep slope. *Source* Staxrud (1914: Tf. 5)

Nøis the impression that these three were smart to know that it was preferable to run along at the periphery than to partake in the pulling along with the dog teams. Now that the party had paused their driving, they could re-harness the three loose dogs. The reindeer, stated Nøis, pulled well, and the men quickly recognized that the reindeer had a better time with the lighter-weight Nansen sledges than the heavier reindeer sledges (pulks).

Hilmar Nøis's accounts of those first two days (Hilmar Nøis Diary SVB; Hilmar Nøis Diary NPI) specify the location in which they stayed at Dicksonfjord to have been in Cape Wick—also spelled Cape Vik—[4]and the trekking to have been extremely exhausting for the animals, who had a difficult task to pull the heavy sledges fully loaded with supplies.[5] According to Hilmar Nøis, the reindeer in particular worked the most intensely, as they required large amounts of nutrition to maintain their strength (Fig. 12.3), while the dogs were able to work well within the challenging circumstances, as they were suited to these conditions—from the Greenland Dog breed (such as Roald Amundsen's Storm and Lussi) to the Gordon Setter hunting dog breed (presumably such as Nøis's dog).[6] The men tried to give all the draft animals the necessary rest they needed. And they packed their sledges in a way so as to have to unpack only those that were required for a night's sleep or a day's respite. While camping at the bottom of Dickson Bay, Hansen, Kemi, and Klemmetsen demonstrated to Hilmar Nøis and the other men the best manner in which to dry the *komager* (traditional Sámi soft boot made from reindeer skin without the

**Fig. 12.3** Sámi expeditioners expertly worked with—and tended to—the extremely helpful and hard-working draft reindeer. *Photographer* Arve Staxrud; Source/Owner: Svalbard Museum

pelt) and the *skaller* (Sámi soft shoe made from reindeer skin, containing the pelt on the outside) without allowing the footwear to become stiff. As Hilmar Nøis reported, *komager*, lined with senne-grass, would be the best worn on skis, and *skallene* were worn in the tent. Taking care of this crucial winter footwear was important, and the helpful instruction received from the Sámi, as they imparted the best method, was welcomed by Hilmar Nøis and the rest of the group.

The difficult ascent up the mountain at Dicksonfjord, which occurred on the following day, according to Daniel Nøis and Hilmar Nøis, entailed a mighty effort, during which the men and the animals struggled with the loads and the incline, reaching the zenith of the University Glacier; the downhill drive itself was dicey, and the Sámi were instrumental in managing the descent, harnessing one reindeer behind the sledges so as to prevent a downhill collision (Nøis, 1929, 7; Hilmar Nøis Diary NPI; Hilmar Nøis Diary SVB). The severity of the struggle resulted in the decision to send Hilmar Nøis and Johannes Kemi, along with seven of the weakest reindeer, back to Green Harbour, in order to have more food available for the remaining humans and animals. This decision was made after the high threshold had been crossed and upon the descent toward the West Fjord in Wijde Bay,[7] at which time Hilmar Nøis and Kemi turned back south with the seven most exhausted reindeer, each reindeer harnessed to a pulk (the heavier sledges) loaded with equipment to return to Green Harbour.[8] This left seven men, with 17 dogs in three teams, and with 12 draft reindeer, to proceed north, according to Hilmar Nøis.[9], [10] While Daniel Nøis's and Hilmar Nøis's later accounts give seven as the number of exhausted reindeer that were sent back, Staxrud, in his published report (Staxrud, 1914, 20), states that eight reindeer were sent back with the two men, thus seeming to indicate that 11 reindeer remained on the trek north. Staxrud also specifies two dog teams rather than three.

That evening of April 14, the northward party reached the bottom of the West Fjord in Wijde Bay (Staxrud, 1914, 20–21; Aftenposten, 1913, evening: 1–2). They had begun the climb at nearly noon and completed their descent at 11:00 pm, arriving there exhausted from their journey (Nøis, 1929, 7–8). The men had hoped to spend the night in the hut at West Fjord, but found that it had apparently burned down recently, forcing them to spend the night in their tents again (Staxrud, 1914, 21). Unfortunately, Staxrud had now found the burned-down hut in which the Kurt Wegener group had stayed—the same hut that Rotvold and his group had disappointedly discovered on March 30. Staxrud would later report on this finding, in his telegram sent to the *Aftenposten* newspaper on May 14 (Aftenposten, 1913, evening: 1–2) (Fig. 12.4), indicating that Wegener's group probably had accidentally started the fire that had burned down the hut, and identifying the hut's owner. The newspaper quoted him as stating:

> At the arrival at Vestfjorden [West Fjord] we discovered, that the over-wintering hut there (hut no. 3 on the map), belonging to furrier Claus Andersen in Tromsø, was burnt down. The hut was erected in 1910 or 1911. Dr. Wegener's rescue expedition, in which 3 over-wintering Norwegians from dr. Mannsfield's [sic] station in Kings Bay took part, had been in the hut. This expedition started from Cross Bay [heading] for Treurenberg Bay the 22nd February and had left the aforementioned hut in Vestfjorden the 3rd April [sic]. The fire has probably happened by these people not properly extinguishing [the fire] in the stove when they left.

## Arrival at Cross Point Hut, and Dog Sledging to the East Side of Wijde Bay in Search of Detmers and Moeser

The party of Staxrud, Bøckmann, Daniel Nøis, Ellingsen, Hansen, Klemmetsen, and Stenersen now ventured further north in their pursuit of the "lost" expeditioners (Aftenposten, 1913, evening: 1–2; Hilmar Nøis Diary NPI). With the parting of Hilmar Nøis, Kemi, and seven of the reindeer, they had rearranged their sledges for their trip onward that would take them across the Wijde Bay ice.[11]

The rescuers next proceeded to the Cross Point hut, arriving there on the following day of April 15 (Aftenposten, 1913, evening: 1–2; Staxrud, 1914, 21–23). Here they found the hut to be in the best condition and were able to sleep in it that evening (Fig. 12.5).

At Cross Point hut, the Staxrud group found only news from Wegener's search expedition, as well as the writings from the four Norwegian sailors. In his report, Wegener summarized that Ritscher had been there with Eberhard, Stenersen, and Rotvold from October 8 to December 21, 1912, trying in vain to reach Advent Bay. Wegener's group had arrived there on March 23, 1913, and had found reindeer meat

**Fig. 12.4** Map of West Spitsbergen published in the *Aftenposten* newspaper article on May 15, 1913, quoting Staxrud's telegrammed report sent on May 14, and indicating the locations of the huts as numbered to correspond with Staxrud's telegram, as well as showing the rescue expedition's round-trip route from Advent Bay to Treurenberg Bay. According to the article, the numbered locations are as follows: Hut no. 1 = Wijde Bay (Second Valley) near Lake Walley [sic—Lake Valley]; Hut no. 2 = Cross Point hut; Hut no. 3 = West Fjord burned-down hut; Hut no. 4 = East Fjord (east of Cape Petermann) in Wijde Bay. (National Library of Norway)

Kart over Vestspitsbergen.

Den prikkede linje angiver undsætningsexpeditionens rute fra Advent Bay til Treurenberg Bay og tilbage.

and skins, which they had used themselves, since they had been in need. On March 25 (the day that Rotvold's group had set out from the ship on their journey south), Wegener's group would have left the hut again, taking Rotvold's skipper's book with them. The news from Stenersen, the two Jensen brothers, and Rotvold, per their report written on April 1, concerned their first short stay from March 29 to 30. They had returned to the hut that very same night of the 30th due to the hut on the West Fjord being burned down (Staxrud, 1914, 23). They could only salvage some objects lying around. The weather had deteriorated with snow and fog alternating, and their provisions had been running low. To make matters worse, they had no sleeping bag, as Wegener—who had disparagingly called it a "fur"—had taken it with him. On the day of April 2, the Norwegians had set off again. Staxrud concluded from the news that the east coast of Wijde Bay, where Detmers and Moeser had last been seen, had not yet been explored and thus determined that he would search there.

Sleeping inside the Cross Point hut that night was a welcomed pleasure for the Staxrud men after having camped outside of the West Fjord burned-down hut the

**Fig. 12.5** Sleeping in Cross Point hut. *Source* Staxrud (1914 Tf. 3)

previous night. It also afforded Daniel Nøis the opportunity to engage in a hunting expedition that evening, in a valley close by, during which he shot one wild reindeer that supplemented their food supply and allowed him to vary the diet (Nøis, 1929, 8).

The reindeer herders Hansen and Klemmetsen decided that it was best to give the draft reindeer a well-deserved rest for a day, and so they remained with them at the Cross Point hut on the following day. These working animals were untied and set free temporarily, with the reindeer walkers minding them.

On that same day of April 16, Staxrud, along with Daniel Nøis, Bøckmann, and Ellingsen, with three teams of dogs, sledged east of Cape Petermann, across Wijde Bay, in search of the two German doctors Erwin Detmers and Walter Moeser, but, unfortunately, they were not able to find them at the hut located there (Staxrud, 1914, 23; Nøis, 1929, 8) (Fig. 12.6).

When they found no traces of the two biologists in the area, Staxrud assumed that the two men had probably perished between this hut and Albert Dirkses Bay (today: Dirksbukta) to the north (Staxrud, 1914, 23). In his telegrammed report to the newspaper, regarding his trip to the east fjord in Wijde Bay, he stated: "One does not know any more about the fate of these two, but in all likeliness they have gone through the ice and drowned on the east side of Wijde Bay, south of Albert Dirkses Bay... whilst attempting to get southwards. They were last seen the 2nd October" (Aftenposten, 1913, evening: 1–2).

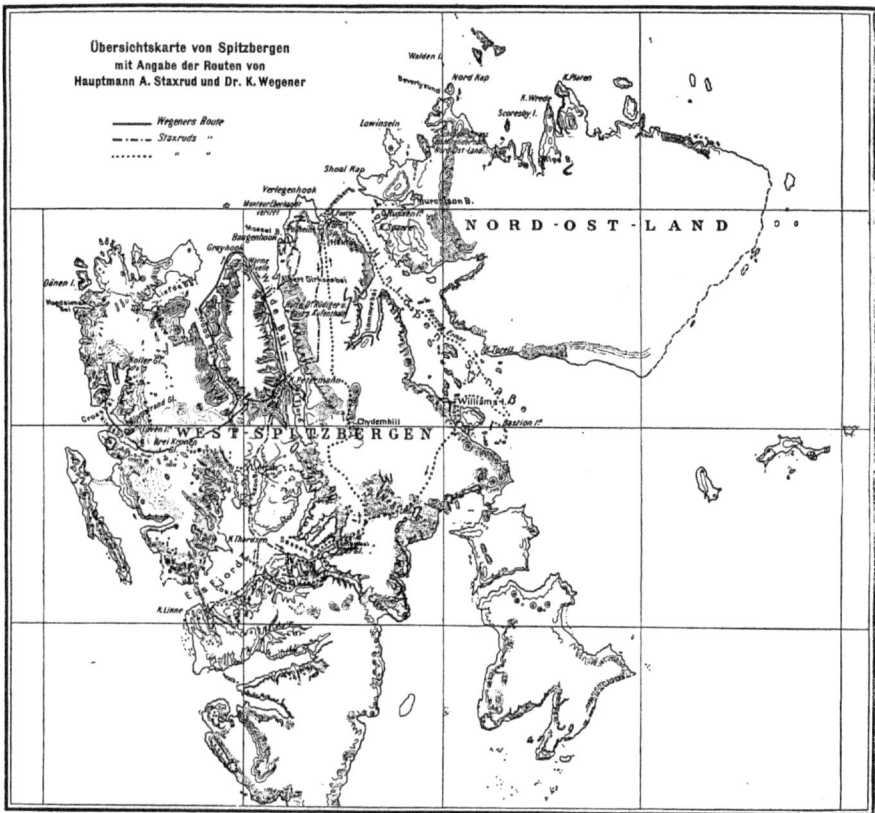

**Fig. 12.6** Route map of the search expeditions of Kurt Wegener and Arve Staxrud. *Source* Miethe (1914)

## Polar Bear Hunting, Wijde Bay Hut, and Traces of the Stranded

After Staxrud, the men, and the dogs returned to the Cross Point hut, the group established another depot there, with 100 days' rations for the journey back, before the entire party then continued on northward (Staxrud, 1914, 24–25; Nøis, 1929, 8). Along the way north, the expeditioners spotted a polar bear, and a hunt ensued, led by Staxrud, who, according to Daniel Nøis, was the designated person to shoot. As Nøis would later record, the dogs Suggen and Baysen were released and gave chase, while Staxrud chased the bear with several of the other dogs, causing the bear to quickly flee. The polar bear's near escape, however, was stopped by the unleashed dogs, which brought Staxrud in closer proximity to the bear prior to shooting. The bear was hit, but not fatally so, at which point all the dogs attacked the wounded bear en masse, resulting in Staxrud killing the bear with his revolver in order to euthanize the large animal. Both Staxrud and Nøis reported that the killing of this polar bear provided both the human expedition members and the dogs with urgently needed fresh meat. They took some of the meat with them as provisions for the onward journey.

Later on, as the party made its way across the ice, the Sámi reindeer herders found that one of the draft reindeer unfortunately could no longer follow along (Nøis, 1929, 8).

On the evening of April 17, the rescue party arrived at the Wijde Bay hut, where Christopher Rave and Hermann Rüdiger had stayed during October and November, and where the rescuers would now be able to spend the night (Staxrud, 1914, 25). At this hut in Second Valley, they found Wegener's message of March 20. In it he had listed where his group had placed its three depots. And he had indicated that the hunter's hut in the Wijde Bay had been abandoned. "I took some salt meat, Captain Sandleben's watch, 2 journals, 2 sleeping bags, and in exchange leave a sleeping bag for two men, which we have used up to now," he wrote (Wegener as quoted in Staxrud, 1914, 25). According to Wegener, Eberhard had returned to the ship with Stenersen and Rotvold, while Ritscher had departed for Advent Bay. "I intend, as our boots and clothes are torn, and provisions very scarce, to return to Kings Bay via West Fjord as soon as possible. Are all very exhausted" (Wegener as quoted in Staxrud, 1914, 26). With this final statement, Wegener ended his communication.

Staxrud also found that the four Norwegians had left a message at the hut as well, on March 28, showing the extent of the damage caused when ill-equipped expeditions help themselves to supplies at hunter's huts without providing replacements. The four men had set out from Treurenberg Bay on March 25 and made their way to the hut in Second Valley via Mossel Bay in order to obtain help for Rüdiger, arriving at Mossel Bay on March 26 (Stenersen et al. in Staxrud, 1914, 26). They did not know if Ritscher and the two biologists were still alive. In the hut, they would have found an incomprehensible letter. Apparently, they could not make any sense of Wegener's message, since he mentioned an injured Detmers who had to be rescued. The Norwegians knew that Detmers had already separated from Rave and Rüdiger at the very beginning of the departure from Treurenberg Bay. Then, the Norwegians complained that Wegener's group had taken their sleeping bags, on which they had relied. Fortunately, they had been able to kill a reindeer (on the following day of March 27) after their provisions had run out. Lastly, they mentioned that the cook Knut Stave had died (on February 24) and that Eberhard had perished on the way back to the ship near Mossel Bay on December 25. Rüdiger was "well" except for frostbite on his hands and feet.

Having digested this information, Staxrud and his party rested for the night at the Wijde Bay hut, which, according to Daniel Nøis, had been an approximately 40 km drive from Cross Point hut (Nøis, 1929, 8). The hut was uncomfortable enough, however, that the men elected to sleep in their tents. They were also able to digest some bear steak for dinner, which they shared with the dogs, who eagerly dined on the frozen meat.

## Unsafe Ice, Open Water, and the Exhausting Inland Route

On the following day of April 18, when Staxrud set out on the forward march to Bangenhook Peninsula, the route over the ice became too unsafe, so that the party had to go ashore at the small Russian hut on the eastern shore of Wijde Bay (Staxrud, 1914, 27–28). As Staxrud described it, "we encountered open water 6 km south of Albert Dirkses Bay," and, after reaching the Russian hut, traveled up to the inland ice, "where we came up to a height of 550 meters" (Aftenposten, 1913, evening: 1–2). The expeditioners continued across the inland ice toward Mossel Bay (Staxrud, 1914, 28). According to Daniel Nøis, the soft ice and open water forced them to undergo a treacherous two days of driving and climbing; negotiating the ice was so intense that they did not even consider hunting a female bear and her two cubs whom they spotted, and the hard slog as well as strenuous climb up the mountain were extremely exhausting for everyone, resulting in two reindeer becoming thoroughly depleted and having to be slaughtered (Nøis, 1929, 8). While the men had been able to find driftwood for fire and thus roast bear steak for a dinner, it must be presumed that food was short for the reindeer. After an 18-h rough tour filled with toil and exertion, the party made camp on the mountain. Daniel Nøis recounts taking a moment to enjoy the sight of the midnight sun appearing over the sea ice and to muse on the thought that they were now near 80° N.

According to Staxrud, from their vantage point on the mountain, the descent to Mossel Bay looked to be so steep that they first set up the camp to gather their strength for the next day's efforts (Staxrud, 1914, 28–29).

## The Steep Descent to Mossel Bay, the Old Moss, and the Arrival in Treurenberg Bay

On April 20, the expeditioners made the difficult descent to Mossel Bay (Aftenposten, 1913, evening: 1–2; Nøis, 1929, 8). The descent with the reindeer sledges worked surprisingly well; however, the snow was so hard in some steep places that the dogs had to be unhitched and two men would have the largest difficulties in getting the half-loaded sledge down the incline intact (Staxrud, 1914, 29).

In Mossel Bay, the party arrived at Adolf Erik Nordenskiöld's hut Polhem (established for his 1872–1873 expedition). It had been hoped by Staxrud and the men that the reindeer moss deposited there by Nordenskiöld in 1872/73 would still be edible for their reindeer (Staxrud, 1914, 28). Happily, the men found it. "Here we got new fodder for our draft reindeer, when it turned out, that a whole lot of the moss, that was put in depot here by Nordenskiöld in 1871 [sic] still was in good condition" (Aftenposten, 1913, evening: 1–2). The ravenous reindeer hungrily pounced on the 41-year-old moss (Staxrud, 1914, 29). The hard-working herbivores desperately needed the sustenance. While they eagerly munched on the moss, the extent of substantial nutrition they received from the four-decades-old supply was difficult to determine, indicated Daniel Nøis (Nøis, 1929, 8).

The reindeer were allowed to remain at the hut location in order to feed and recuperate, looked after by Hansen, Klemmetsen, and Ellingsen, while the rest of the men and the teams of dogs went on to Treurenberg Bay (Staxrud, 1914, 30–31; Nøis, 1929, 8; Aftenposten, 1913, evening: 1–2). According to Daniel Nøis, the sledge dogs were quick and agile and oftentimes enabled the men to simply stay seated on the sledges. And according to Staxrud, in his telegram that appeared in *Aftenposten*, the route taken was over land, and the distance covered was approximately 26 km.

Staxrud's sledging party reached Treurenberg Bay late that night of April 20, and, just past midnight, at 1:30 am on April 21, finally arrived at the Swedish house where the stranded Rave and the injured Rüdiger anxiously awaited rescue. The rescue expeditioners knocked on the door of the silent house, and waited for an answer.

## The Rescue, the Retrieval, and Rüdiger's Frostbitten Foot

Inside the Swedish house at Treurenberg Bay, Christopher Rave and Hermann Rüdiger were marking the days and the hours. They had remained here since Rotvold and his group had departed from the ship in late March.

According to Rave, after March had not been particularly cold, temperatures had dropped to as low as –40 °C in April (Rave, 1913, 76–77). It was not until the middle of the month that southerly winds brought Arctic spring. From a high vantage point, Rave could already see ice-free water off Treurenberg Bay. He and Rüdiger worried about their rescue and wanted to wait for help only until the end of July so as to still get to Norway by steamer. On the other hand, Rüdiger could not walk or kayak for long because of the frostbite on his feet and hands.

After midnight on the night of April 20, as the day turned into April 21, during the wee hours of that morning inside the hut, Jule—one of the two stalwart dogs—became restless and began to bark. There was another voice as well. "Mister Rave!" someone shouted from outside (Rave, 1913, 77–78). There were people around the house. Rave discovered that they were men from the German-organized Norwegian relief expedition led by Captain Staxrud, whom the ice pilot Stenersen had brought, together with the trapper Daniel Nøis and the physician Dr. Bøckmann. At the same time, the mining engineer Ellingsen waited with the two Sámi,[1] Hansen and Klemmetsen, and with the reindeer and sledges, in Mossel Bay, until the draft reindeer had recovered from exhaustion. Rave described Captain Staxrud as a gentleman of smaller stature

---

[1] Christopher Rave, in his book passage specifically referenced here, referred to the two Sámi, Per Hansen and Samuel Klemmetsen, as "Lappen" ("Lapps"), a term widely used at that time but no longer in use, as Sámi is the proper name and is therefore the name used in this book text. He also did not mention them by name, nor did he mention Stenersen by name, referring to him as his expedition's Ice Pilot. Rave also did not mention Daniel Nøis's first name, introducing him as the trapper Nøis.

**Fig. 12.7** Physician Dr. Bøckmann with his dog sledge. *Source* Rüdiger (1913, 176)

with a calm speaking manner and a clever appearance; and he described both Dr. Bøckmann and Daniel Nøis as physically being very large.

From Daniel Nøis's perspective, observing while standing outside the house, Rave[2] opened the door right away, and there followed an animated conversation among the men, and although Nøis did not comprehend what was being said, he did understand the passion and the sentiment of what the two rescued Germans were feeling—that their isolation and uncertainty experienced during this hapless trip were finally at an end (Nøis, 1929, 8).

Staxrud reported that as soon as the party "met the two Germans, Rüdiger and Rave," that "Dr. Rüdiger's leg was immediately seen to and treated by doctor Bøckmann," commenting also that "Rave's care for and nursing of Dr. Rüdiger has been utterly admirable" (Aftenposten, 1913, evening: 1–2). With the well-strategized presence of a capable doctor on the rescue expedition, Rüdiger was now promptly given medical attention (Staxrud, 1914, 30) (Fig. 12.7).

Rave reveled in the confirmations of the quality care he had been able to provide to his companion. He later reported that after Bøckmann immediately looked at

---

[2] In Daniel Nøis's first-hand account, throughout the article, Christopher Rave's name "Rave" is spelled as "Rove.".

Rüdiger's wounds, he praised Rave for his successful first aid and the many amputations, for the wounds were in very good condition. "The captain said that there were not more than one percent who could have done the same under the same circumstances," he wrote (Rave, 1913, 79).

Daniel Nøis, too, reported on the satisfactory condition of Rüdiger's frostbitten and operated-on foot, as examined by the doctor (Nøis, 1929, 8).

While Rave fed the guests with a pot of red cabbage that had just been cooked, he and Rüdiger leapt on the letters from home. They learned about the various rescue efforts and support from their hometown of Hamburg, as well as from "Count Zeppelin, Geheimrat Professor Miethe, Professor Hergesell, and [that] even Nansen was active for us" (Rave, 1913, 79).

Observing the rescued Germans reading the letters of correspondence and newspaper articles, Daniel Nøis would later record that the two men gained an appreciative understanding of the advantages of being part of a caring social structure (Nøis, 1929, 8).

## Moving from the Swedish House to the *Herzog Ernst*, and Chasing Bear Cubs

Staxrud's group, with the addition of Rave and Rüdiger and their two dogs Jule and Caesar, now moved from the Swedish hut to the iced-in ship *Herzog Ernst*, where the men previously had been trapped on board, and where Stenersen now would remain in order to look after the ship (Fig. 12.8).

When those at the hut had proceeded to the *Herzog Ernst*, Hansen, Klemmetsen, and Ellingsen appeared with the reindeer (Rave, 1913, 79; Nøis, 1929, 8). According to Daniel Nøis, it was a couple of days later that the three men and reindeer reached Treurenberg Bay, after which they returned to Mossel Bay for more moss. Meanwhile, Staxrud and his men considered making a sledging journey to the Northeast right away in order to search for the missing expeditioners Herbert Schröder-Stranz, Richard Schmidt, August Sandleben, and Max Mayr, but the conditions of the ice prohibited this action at this time. The wind had again pushed the ice in front of Treurenberg Bay, and thus Staxrud had to refrain from searching for Schröder-Stranz further to the east. They now remained in Treurenberg Bay, and Staxrud made an important decision.

Staxrud reported on these thwarted attempts and on his ensuing decision as follows (Aftenposten, 1913, evening: 1–2): "From the 20th to the 27th April there was continuously bad weather, likewise were the ice conditions unfavorable. I therefore considered it most propitious to postpone the Northeast Land trip, until Rüdiger and Rave were brought safely to Green Harbour. On 'Herzog Ernst' there turned out to be a lot of entirely excellent provisions, so that it looks like, that the leader was prepared for an overwintering."

**Fig. 12.8** *Herzog Ernst* iced in at Treurenberg Bay, seen here after the arrival of Staxrud's rescue expedition. *Photographer* Arve Staxrud; Source/Owner: Svalbard Museum

Rave now cooked diligently for everyone, while the others took care of driftwood to heat or helped wash the dishes (Rave, 1913, 79). Meanwhile, Staxrud and his men continued to check the ice. On April 25, following a storm that had occurred on the previous day, Staxrud and Daniel Nøis went up the mountain to view the situation of the ice below and to ascertain if it had altered (Nøis, 1929, 8–9). Finding it utterly useless to attempt to cross in a boat, they descended again and, while doing so, found a mother bear with her two cubs out in the distance on the ice. As Daniel tells the story, they were in need of fresh meat for the party, and so they enlisted some of the sledge dogs and engaged in another polar bear hunt. Again, they unleashed Suggen and Baysen, who helped them give chase, and eventually the female stopped, during which time the cubs were able to get away. Dr. Bøckmann shot and killed the mother. The dead female was skinned and taken back, and Daniel Nøis continued chasing the two cubs, capturing one of them and driving the cub back to the Swedish house in the midst of heavy snowfall that obscured him from the others. Upon returning to the ice, Nøis was asked by Staxrud if he had seen the cubs, to which Nøis replied in the affirmative and updated Staxrud. The second cub was later also captured in this same manner—a type of hunting that Daniel Nøis the trapper/hunter professed was unappealing to him (Fig. 12.9).

Rave and Rüdiger recount this tale of the hunt after the mother bear and the capture of the two baby bears, as well (Rave, 1913, 80–81; Rüdiger, 1913, 178–180). Rave's

85-13

**Fig. 12.9** Capture of two polar bear cubs. *Photographer* Arve Staxrud; Source/Owner: Svalbard Museum

account is particularly descriptive. According to Rave's report, the event took place on the 26th, and he himself hurried to participate, so as to observe. Upon being alerted by Staxrud, Rave quickly grasped his cinema camera, photo camera, and Canadian snowshoes. Riding on a sledge pulled by the sled dog Storm, one of Amundsen's South Pole Expedition dogs, and a second unnamed dog, Rave watched as the mother bear chased the cubs to run on ahead and escape as she herself was taken down with two gunshots. He then witnessed only some of the ensuing chase after the two cubs, as the two young bears were being encircled and driven from both sides, before the air in front of Rave became quite opaque with heavy snowfall and obstructed his view. During what was considered a successful hunt, recorded Rave, the physician killed the mother bear, while the party was able to catch her two cubs and bring them to the ship. Rave reported Rüdiger as being delighted to see the baby bear, who with his/her sibling was locked in a chamber for observation. The two little cubs were first placed in a dark room, then a light room, as they roared and attempted to bite and scratch, and then were made to be situated in permanent captivity. One of them consistently jumped at Rave whenever he approached to feed them, and according to Rave, the cub soon came to have a respect for the whip handle. The two men engaged in watching what they considered to be the bear cubs' droll behavior, through a hole in the door. One of the two cubs soon died, although, as was reported, they were both fed "with condensed milk, seal blubber and reindeer meat" (Villinger, 1929, 24). The other bear cub was later brought to Germany on the *Herzog Ernst*.

## Looking for Eberhard, Searching for Schröder-Stranz, and Preparing for Departure

On April 28, Staxrud and Stenersen traveled to Verlegen Hook to search for Wilhelm Eberhard, who had disappeared while heading north with Stenersen and Rotvold at Mossel Bay in late December; but after looking around the area, the rescuers were not able to find him (Aftenposten, 1913, evening: 1–2): "I found however no trace of the disappeared," reported Staxrud about the futile search. "Most likely the body has been covered by drifting snow."

Unfortunately, there was no clue visible as to the German's vanishing (Nøis, 1929, 8).

The men continued checking on the state of the ice from the vantage point of the mountains, to see if there was any chance they could proceed to Northeast Land to search for Schröder-Stranz and his three companions Schmidt, Sandleben, and Mayr, but they saw no change in the ice (Nøis, 1929, 9).

Thus, by the end of April, after bad weather and a still impassable Hinlopen Strait continued to prevent the search for Schröder-Stranz on Northeast Land from taking place, Staxrud had decided to first bring Rüdiger and Rave back across the inland ice to Advent Bay. Staxrud's targeted date of departure, according to Rüdiger, was the first of May (Rüdiger, 1913, 181). For Rüdiger's transport, the men constructed a kind of bed sledge, and for this purpose, they attached a bed from the Swedish station house, without feet, to a Nansen sledge (Rüdiger, 1913, 184). Two pairs of skis served as a slatted base. For insulation, life jackets made of kapok and a sleeping bag were placed in the bed. Rüdiger himself would lie in a double sleeping bag made of eider down and reindeer fur (Fig. 12.10).

Daniel Nøis reported that he built the bed sledge in order to accommodate Rüdiger during the long sledging journey south (Nøis, 1929, 9).

As of the end of the month of April, Ellingsen, Hansen, Klemmetsen, and the reindeer who had stayed at Mossel Bay had all arrived at Treurenberg Bay; the Staxrud party was fully reunited; and everyone had made the necessary preparations to depart south. But at that moment, they received an unexpected surprise.

Just as preparations for departure were completed on April 30, three skiers arrived at Treurenberg Bay (Staxrud, 1914, 33–36). They were from Theodor Lerner's relief expedition, and they had come from the motor-cutter *Loevenskjold*, in which they had just arrived that same day at Mossel Bay (Aftenposten, 1913, evening: 1–2). The three men on skis were Dr. Rudolf Biehler, Bernhard Villinger, and Gerhard Graetz. They informed the group that they were members of Lerner's group.

Lerner, who had previously desired to help but had been denied by the German government, had indeed managed to form a private relief expedition on his own. And

**Fig. 12.10** Rüdiger in his bed sledge on the inland ice at an altitude of 1200 m. *Source* Rüdiger (1913, 188)

at present, he had come to Spitsbergen to find the missing Germans and to rescue Rüdiger and Rave—who had just been rescued by Staxrud and his group.

Staxrud would now need to make a critical decision regarding his plan to convey the two rescued German expeditioners safely south.

## Notes on Original Material and Unpublished Sources

Hilmar Nøis's diary SVB is in the Historical Archive at the Svalbard Museum (SVB) in Svalbard. Hilmar Nøis's diary NPI is in the Library Archives at the Norwegian Polar Institute (NPI) in Tromsø. Einar Rotvold's diary is in the Library Archives at the Norwegian Polar Institute (NPI) in Tromsø.

1. H. Nøis diary SVB, April 1913, journal page 52, SVB-AP2.
2. H. Nøis diary NPI, notebook 3, April 1913, pdf page 11, (NPI), D-307 / D00307_3_0001.
3. E. Rotvold diary, 12 April 1913, (NPI), D00125.
4. H. Nøis diary SVB, April 1913, journal pages 54, SVB-AP2.

5.  H. Nøis diary NPI, notebook 3, April 1913, pdf page 12, (NPI), D-307 / D00307_
    3_0001.
6.  H. Nøis diary SVB, April 1913, journal pages 55, SVB-AP2.
7.  H. Nøis diary NPI, notebook 3, April 1913, pdf page 12–13, (NPI), D-307/
    D00307_3_0001.
8.  H. Nøis diary SVB, April 1913, journal pages 55, SVB-AP2.
9.  H. Nøis diary NPI, notebook 3, April 1913, pdf page 12–13, (NPI), D-307/
    D00307_3_0001.
10. H. Nøis diary SVB, April 1913, journal pages 55, SVB-AP2.
11. H. Nøis diary NPI, notebook 3, April 1913, pdf page 12–13, (NPI), D-307/
    D00307_3_0001.

## Unpublished Sources and Original Material

Nøis, H., Diary NPI, handwritten journal of recollections recounting the events of the 1913 Staxrud expedition, Hilmar Nøis, "Minder og Fortelinger fra fangsttiden på Spitsbergen, Klade 3, 7 Mars 1970" [Hilmar Nøis, "Memories and Stories from the trapping time on Spitsbergen, Notebook 3, 7 March 1970"], dated 7 March 1970, D-307, ("Minder og fortelinger, Nøis, Hilmar, hefte 3"), D00307_3_0001. Library Archives, Norwegian Polar Institute, Tromsø, https://brage.npolar.no/npolar-xmlui/handle/11250/274077. Accessed 20 July 2021.

Nøis, H., Diary SVB, handwritten journal of recollections including an account of the 1913 Staxrud expedition, ("Hilmar Nøis. Fangstmann. 1909–1923, dagbok med erindringer fra fangstlivet") ["Hilmar Nøis. Trapper. 1909–1923, diary with recollections from the trapping life"], not dated (latest year referenced is 1942 in journal pages and 1950 in inserted sheet), SVB-AP2-Hilmar-Nøis-dagbok-1909–1923-compressed. Historical Archive, Svalbard Museum, Svalbard, https://svalbardmuseum.no/no/samlingene/historisk-arkiv/arkivprosjekt-fangst-og-annen-overvintringsvirksomhet-fra-perioden-1910-til-1970/. Accessed 6 June 2021.

Rotvold, E., Diary, Einar Rotvold's expedition diary written during the Schröder-Stranz expedition of 1913, "Dagbok for August Stenersen og Einar Rotvold, Tromsø: Treurenberg Bay fra 1.januar–7.juni 1913: Schröder-Stranz-ekspedisjonen 1912–13", "Einar Rotvold har skrevet dagboken", ["Diary for August Stenersen and Einar Rotvold, Tromsø: Treurenberg Bay from 1 January–7 June 1913: The Schröder-Stranz Expedition 1912–13", "Einar Rotvold has written the diary"], dated 12 January 1913 through 7 June 1913, D00125. Library Archives, Norwegian Polar Institute, Tromsø, https://brage.npolar.no/npolar-xmlui/handle/11250/2426394. Accessed 24 July 2021.

# References

Aftenposten. (15 May 1913). Hjælpeexpeditionen paa Spitsbergen. Kaptein Staxruds beretning til "Aftenposten". [The rescue expedition on Spitsbergen. Captain Staxrud's account to "Aftenposten".]. *Aftenposten* (Kristiania [Oslo]), 15 May 1913, evening edition, pp. 1–2. National Library of Norway, Oslo, online archives, www.nb.no. Accessed 28 August 2012.

Der Zeitspiegel Berlin im Frühjahr. (1913). *Newspaper clipping.* Ritscher Estate, Cornelia Lüdecke, München.

Miethe, A. (ed.). (1914). *Die Expedition zur Rettung von Schröder-Stranz und seinen Begleitern - geschildert von ihren Führern Hauptmann A. Staxrud und Dr. K. Wegener.* Berlin: Dietrich Reimer.

Nøis, D. (1929). Med kaptein Staxrud på leiting efter Schrøder-Stranz og hans folk [With Captain Staxrud on the search for Schröder-Stranz and his people]. In: *And-Ungen*, July 1929, pp. 4–12. Andenes: Andøyposten. Library Archives, Norwegian Polar Institute, Tromsø. Received 20 July 2021.

Rave, C. (1913). *Tagebuch von der verunglückten Expedition Schröder-Stranz.* Schaffsteins Gründe Bändchen 49. Cöln: Schaffstein.

Rüdiger, H. (1913). *Die Sorge Bay. Aus den Schicksalstagen der Schröder-Stranz-Expedition.* Berlin: Georg Reimer.

Staxrud, A. (1914). Die Staxrudsche Hilfs-Expedition für Schröder-Stranz. In A. Miethe (ed.), *Die Expedition zur Rettung von Schröder-Stranz und seinen Begleitern - geschildert von ihren Führern Hauptmann A. Staxrud und Dr. K. Wegener* (pp. 1–68). Berlin: Dietrich Reimer.

Villinger, B. (1929). *Die Arktis ruft! Mit Hundeschlitten und Kamera durch Spitzbergen und Grönland.* Freiburg i. Br.: Herder.

# Chapter 13
# The Theodor Lerner Frankfurt Relief Expedition, and the Staxrud Group's Decision

## The Unofficial, Privately Organized, and Controversial Rescue Expedition of Theodor Lerner

Nearly four months prior to April 30, 1913, when Captain Alfred Ritscher's telegram from Advent Bay had reached Germany, Theodor Lerner had been among those who had wanted to answer the call for help to rescue the Schröder-Stranz expedition.

Over the following few months, he had eventually succeeded in organizing his own, privately-funded Frankfurt relief expedition (Villinger, 1929, 5–8). The expedition left Germany accompanied by ten dogs (Lerner, 2005, 231). According to Lerner, these dogs were "four more-or-less real [Saint] Bernards, one Newfoundland, four strong shepherd dogs, and one so-called police dog, but he died on arrival in Spitsbergen" (Lerner, 2005, 237–238). Villinger described some of them as "three German Shepherds, and two indeterminable mixed breeds" (Villinger, 1929, 8). The ten dogs, who had been raised in Frankfurt to pull the sledges, and three members of the Freiburg/Breisgau academic ski club, sailed by ship from Hamburg to Tromsø (Villinger, 1929, 5–8). They were the physician Dr. Rudolf Biehler (born in 1882, died in World War I), and the two medical students, candidate of medicine (cand. med.) Gerhard Graetz (1890–1977) and cand. med. Bernhard Villinger (1889–1967). The fourth skier was Sepp Allgeier (1895–1968), only 18 years old, who worked as a film operator for the Freiburg Express-Films Company for the first German daily cinematographic report, *Der Tag im Film* (*The Day on Film*) (Wikipedia: Sepp Allgeier). Lerner wanted to have Allgeier along as a documentary filmmaker, so to speak, and signed a contract to that effect with the Express-Films Company (Allgeier, 1931, 15).

In Tromsø, as reported by Bernhard Villinger, the Norwegian fishing vessel *Loevenskjold* still had to be modified for the expedition (Villinger, 1929, 8–30). It was 20 m long by 5 m wide and the space content was 40 net tons. The equipment also included a single-cylinder Boliden petroleum engine with 40 hp. When the ship was finally ready to go after a week's stay, according to Villinger, Lerner's group learned by radio telegram that the expedition organized by the Berlin Relief Committee and led by the Norwegian Arve Staxrud had gone ashore at the entrance to the Ice Fjord in Spitsbergen on April 3 and was on its way to Treurenberg Bay via Advent Bay and Wijde Fjord on March 8 [sic—in actuality it was April 12] with nine men, 22 sledges, 16 dogs, and 20 reindeer. Almost at the same time, four Norwegians had brought news of their wintering on the *Herzog Ernst*, so that Lerner could roughly imagine where to start his search. After thinking that Staxrud would probably succeed in searching Wijde Fjord for survivors and in advancing to *Herzog Ernst*, Lerner wanted to proceed with his ship from the west coast of Spitsbergen directly to the coast of Northeast Land. Both in Berlin and in Norway, his success at such an early time of the year was highly doubted. On April 24, Lerner's relief expedition on the *Loevenskjold* left the last Norwegian harbour. Unfortunately, along the way, the police dog named Alex died; the dog had hardly eaten anything since leaving Frankfurt. (The travel route is given in Fig. 10.2a, b in Chap. 10.)

At the west coast of Spitsbergen, the *Loevenskjold* encountered unusually good ice conditions for this time of year, so that the expedition was able to completely circumnavigate the northwest corner of Spitsbergen as early as the morning of April 30. Checking the view beyond Hinlopen Strait, the ship's lookout spotted a sledge track near Mossel Bay moving inland toward Treurenberg Bay, about 25 km away. Biehler, Graetz, and Villinger immediately chased after it on skis until they finally saw Staxrud's dog and reindeer sledges in front of the hut of the 1899 Swedish Arc-of-Meridian expedition, where Hermann Rüdiger and Christopher Rave had retreated on March 15 after having wintered on the *Herzog Ernst* (Fig. 13.1).

Staxrud had arrived there, at the hut in Treurenberg Bay, according to Villinger, on April 22, six days earlier than Villinger and his group [sic—in actuality it was April 21, nine days earlier] and was now preparing to march back to Advent Bay.

And hence, the two parties met.

## The Surprise Visit, the Invitation, and the Decision

The Staxrud rescue expedition was on the cusp of its planned departure south when it was visited by the three skiing medical personnel from the Lerner expedition.

According to Christopher Rave, the skiers were a most welcomed sight. Recalling that visit of April 30, which he attributed to May 1, Rave reported that three figures had appeared on the horizon—the physician Dr. Biehler, cand. med. Villinger, and cand. med. Grätz (Graetz) (Rave, 1913, 81–83). Such a surprise it was, he recorded, to have them all come to help within 24 h. Moreover, according to Villinger, Dr. P. W. K. Bøckmann and Villinger knew each other already, because they had met earlier in March, when Villinger had attended the Holmenkollen ski race in Kristiania (today Oslo) (Villinger, 1929, 23) (Fig. 13.2).

Villinger would later recall that the joy of the two German survivors was great, for the skiers brought them mail from home as well as cigars and cigarettes (Villinger, 1929, 23).

**Fig. 13.1** Rüdiger and Rave in front of the hut of the Swedish Arc-of-Meridian expedition (Rüdiger, 1913, 183)

Highly disputed in Germany, and declared an unofficial German expedition that did not represent the government but was a private enterprise, the Lerner expedition had nonetheless sailed to Spitsbergen.

As Daniel Nøis would later record, they had made it to Mossel Bay that day amidst the current conditions of the winds and the ice (Nøis, 1929, 9).

Upon receiving the three skiers, Staxrud immediately sent Dr. Bøckmann and the engineer Jakob Ellingsen to Lerner's ship at Mossel Bay with a letter (Aftenposten, 1913, evening: 1–2; Nøis, 1929, 9). According to Rave, the letter contained an invitation to Lerner to join them on the *Herzog Ernst* (Rave, 1913, 82–83). Lerner, however, remained aboard his own ship and put the vessel at their disposal for the trip home. Staxrud's two men were informed that Lerner wanted to travel to Northeast Land and take the two Germans with him in order to convey them to safety (Nøis, 1929, 9).

Thus, Rave and Rüdiger had been invited to board Lerner's ship *Loevenskjold* for their return. But a decision was made among the Staxrud group members to continue on land and ice as they had planned—a decision which would prove to be fortuitous.

As Staxrud reported in his telegram, "Rüdiger and Rave preferred to come with me overland rather than with Lerner's vessel, as the ice conditions at the north coast were poor" (Aftenposten, 1913, evening: 1–2). He would later expand on this, saying that the *Loevenskjold* was still locked in ice off Mossel Bay, and that Rüdiger and

**Fig. 13.2**   Three medics from Lerner's relief expedition. *Source* Rüdiger (1913, 182)

Rave did not want to wait until the ship was released and preferred to take the safer route over land (Staxrud, 1914, 34–35). On the *Loevenskjold* they would be exposed to the danger of being stuck even longer due to constantly changing ice conditions. As it turned out later, this was a wise decision.

Rave recorded this decision as well, saying that he and Rüdiger decided to take the land route with Staxrud (Rave, 1913, 82–83).

And so, after careful consideration, a second letter was dispatched from Staxrud to Lerner, via the three skiers, to this effect, in which Staxrud also agreed to combine search efforts upon his return to Treurenberg Bay after he had safely conveyed the two rescued Germans to Advent Bay (Staxrud, 1914, 35). For, in his reply to Staxrud's first letter, Lerner had "suggested cooperation in order to search for Schrøder-Stranz [sic] and his mates on the Northeast Land," reported Staxrud, and "I answered concurring with this and then joined the return trip to Green Harbour" (Aftenposten, 1913, evening: 1–2).

Staxrud had ruled out immediate cooperation with Lerner because he needed all the pulling animals for the return transport of Rüdiger and Rave (Staxrud, 1914, 35). As further help, however, in his letter to Lerner, Straxrud also shared the information he had gathered about the progress of the Schröder-Stranz expedition thus far.

According to Villinger, the news that Lerner's men received from Staxrud about the missing Erwin Detmers, Walter Moeser, and Wilhelm Eberhard suggested that those men were gone forever (Villinger, 1929, 23). So, the only thing left to do was to find Herbert Schröder-Stranz's sledge expedition on Northeast Land. Upon request, Ritscher had telegraphed them the information "Schröder-Stranz set down on pack ice 80°30″ north latitude, 22° east longitude. Intended direction Rijp Bay" (Villinger, 1929, 30).

Hence, Lerner and the expedition aboard the *Loevenskjold* prepared themselves to travel further northeast.

Meanwhile, Staxrud and his party proceeded to embark on the return journey south to Advent Bay.

But Staxrud had already determined that he must then again make a return trek north, the entire length of Spitsbergen, in order to find the other missing Schröder-Stranz expeditioners. For now, however, the best decision, he concluded, was to get the two Germans he had rescued back safely where they could be treated. And that meant sledging south across the ice with his local experts, his sledge dogs, and his reindeer.

Stenersen would remain on the *Herzog Ernst* to look after the ship, and the Staxrud party of men and animals would begin the southward sledging journey over the treacherous ice in order to safely convey the two rescued Germans to Advent Bay.

Thus far, the strategy and method of using Norwegian overwinterers, Sámi guides, sledge dogs, and draft reindeer, to make this rescue effort a success, were indeed working. They had made progress, had successfully reached the two stranded men, and had saved lives.

# References

Aftenposten. (1913, May 15). Hjælpeexpeditionen paa Spitsbergen. Kaptein Staxruds beretning til "Aftenposten". [The rescue expedition on Spitsbergen. Captain Staxrud's account to "Aftenposten".]. *Aftenposten* (Kristiania [Oslo]), 15 May 1913, evening edition, pp. 1–2. National Library of Norway, Oslo, online archives, www.nb.no. Accessed 28 August 2012.

Allgeier, S. (1931). *Die Jagd nach dem Bild. 18 Jahre Kameramann in Arktis und Hochgebirge.* Stuttgart: Engelhorn.

Lerner, T. (2005). *Polarfahrer: Im Banne der Arktis.* F. Berger (Ed.). Zürich: Oesch/Kontrapunkt.

Nøis, D. (1929). Med kaptein Staxrud på leiting efter Schrøder-Stranz og hans folk [With Captain Staxrud on the search for Schröder-Stranz and his people]. In: *And-Ungen*, July 1929, pp. 4–12. Andenes: Andøyposten. Library Archives, Norwegian Polar Institute, Tromsø. Received 20 July 2021.

Rave, C. (1913). *Tagebuch von der verunglückten Expedition Schröder-Stranz.* Schaffsteins Gründe Bändchen 49. Cöln: Schaffstein.

Rüdiger, H. (1913). *Die Sorge Bay. Aus den Schicksalstagen der Schröder-Stranz-Expedition.* Berlin: Georg Reimer.

Staxrud, A. (1914). Die Staxrudsche Hilfs-Expedition für Schröder-Stranz. In A. Miethe (ed.), *Die Expedition zur Rettung von Schröder-Stranz und seinen Begleitern - geschildert von ihren Führern Hauptmann A. Staxrud und Dr. K. Wegener* (pp. 1–68). Berlin: Dietrich Reimer.

Villinger, B. (1929). *Die Arktis ruft! Mit Hundeschlitten und Kamera durch Spitzbergen und Grönland*. Freiburg i. Br.: Herder.
Wikipedia. Sepp Allgeier. https://de.wikipedia.org/wiki/Sepp_Allgeier. Visited 12 August 2021.

# Chapter 14
# A Tale of Two Treks: Rotvold and the Jensen Brothers, and Hilmar Nøis and Johannes Kemi

## Simultaneous Excursions: The Einar Rotvold Group to Green Harbour, and the Hilmar Nøis Group to Advent Bay

While the Arve Staxrud rescue expedition had been making its way north toward Treurenberg Bay during the month of April 1913, two simultaneous excursions had been taking place—the procession of the remaining Schröder-Stranz Norwegian crewmembers from Advent Bay to Green Harbour, and the retreat of the early-returning Staxrud expeditioners from the north back to Advent Bay.

On April 13, the day after Staxrud departed northward from Advent Bay, the three remaining Norwegians who had rescued themselves made their own departure (Einar Rotvold Diary). They were Einar Rotvold and the brothers Julius and Jørgen Jensen, who, together with their fourth companion August Stenersen, had returned safely from Treurenberg Bay to Advent Bay on April 5 (Chap. 11). Stenersen had set off north again with Arve Staxrud, on April 12, to rescue the stranded Christopher Rave and Hermann Rüdiger. The four Norwegians had previously received a telegram from the Green Harbour telegraph station manager Olaf Henriksen, on April 7, welcoming them to Green Harbour, which was the center of activity on Spitsbergen at that time, and they had decided to take him up on his invitation.

Therefore, on that day of April 13, Rotvold and the two Jensens began their trek from Advent Bay to Green Harbour, leaving behind Johan Nilsen Nøis who was ill with the mumps, but taking along with them one reindeer (most likely from the rescue expedition) as they had been asked to do (Einar Rotvold Diary). Marching in "deep snow and hummocked ice" with the reindeer in tow proved to be "arduous" for Rotvold and his mates, compounded by the fact that the reindeer "lay down several times and refused to walk, so we had to chase it like any horse."[1] This description is in interesting contrast to Rotvold's previous verbal illustration of the herd of reindeer drawing onlookers while at rest at the Hotel-headland prior to drawing sledges on the rescue mission.

M. R. Tahan and C. Lüdecke, *Stranded at the Top of the World*,
https://doi.org/10.1007/978-3-031-56288-4_14

The Rotvold group arrived in Green Harbour early in the morning on April 14, first stopping in at the "Skrøder-cabin"—the Schröder House at Cape Heer that now served as the *Hertha* depot—where they were "well received" by Martin Pettersen Nøis and Søren Zachariassen and given food and a place to rest.[2] Here Rotvold observed that *Hertha* still lay at anchor and had not yet gone out. Later that same day, in the afternoon, he and his two companions proceeded to the telegraph station, where they were "wonderfully received" by the manager Henriksen and his staff, who gave them a nice welcoming. "We had hot supper, and afterwards we were served a glass of wine," wrote Rotvold in his diary. "It has been a long time since we have been together with such great people. Here it is quite like home for us. We got a good bed to sleep in, and a really nice as nice room, so we are now living well, after all that we have had to endure over the winter"[2] (Fig. 14.1).

From there the three men relocated to a room at the whaling station, on April 15, "as at the telegraph [station] is an English man as guest, and soon a wife is coming from one of the coal mines thereto, to give birth, so there will be no room for us there then, and we like it best when we can look after ourselves," wrote Rotvold.[3] They had the necessary supplies at the whaling station, and, once the weather cleared, they would be able to access provisions from the depot at the Schröder House. Indeed, on the following day, they went to the depot for supplies,[4] and on the day after that, Rotvold and Jørgen Jensen, besides fielding questions about the expedition, paid a

**Fig. 14.1** Telegraph station at Green Harbour. Photographer/Byline: Arve Staxrud; Source/Owner: Norwegian Polar Institute

visit onboard the *Hertha*, which had gone out briefly but had been forced back in by storm winds as well as "high sea and ice."[5]

On the very next day, April 18, Rotvold heard news about some of those rescue expeditioners who had arrived on the ship *Hertha* and who had now returned to Advent Bay from the trek north.

The Staxrud expedition members Hilmar Nøis, Johannes Kemi, and seven exhausted reindeer, who had been sent back by Staxrud from the West Fjord in Wijde Bay (Hilmar Nøis NPI), had begun their return to Green Harbour a few days prior.[6] They had left the expeditioners early, turning back after the grueling climb at Dicksonfjord on April 14, and proceeded to experience a rough return south. While Hilmar Nøis successfully learned to drive a pulk (reindeer sled), the going was quite difficult for both men and animals and further exhausted the reindeer (Hilmar Nøis SVB), as there was no sufficient food supply to be found for them, where they could graze and feed for several days and thus replenish themselves and regain enough strength to go on.[7] After driving for five (metric) miles, the first casualty occurred—one of the extremely exhausted reindeer lay down, and it was apparent that the animal could walk no further.[8] According to Nøis, in the face of this impossible situation, and with no other alternative, Kemi quickly wielded his wide knife, sat astride the animal, and dispatched the reindeer in an instant. It was a quick death for the toiling and depleted animal.

Unfortunately, the killing of a reindeer had to be repeated a second time over the course of two days spent in the ice during the return trip south toward Advent Bay (Nøis, 1929, 10). As Daniel Nøis would later recount, Hilmar reported to him that two of the seven reindeer had to be slaughtered on the way to Advent Bay and that the men themselves were fearful of the poor conditions, as there was a great risk of the ice breaking up beneath them.

Hilmar, Kemi, and the five remaining reindeer finally safely arrived in Advent Bay on April 17, for, on April 18, in Green Harbour, Rotvold (Einar Rotvold Diary) reported: "We have received [a] message from Advent Bay, that yesterday there returned from Dickson Bay, 2 men and 5 reindeer, but they cannot come hereto as there is open water all the way in to Rævnæs."[9]

## The Encounter of the Two Groups in Green Harbour

Upon reaching Advent Bay, Hilmar Nøis was able to reunite with his uncle Johan Nilsen Nøis, who had been too sick to travel with the Staxrud rescue expedition, but who was now recovered from the mumps and able to travel again (Hilmar Nøis Diary NPI). Thus Johan Nilsen Nøis joined Hilmar, Kemi, and the remaining reindeer (presumably five) in trekking to Green Harbour.[10] Rotvold (Einar Rotvold Diary) monitored their progress, writing on April 21: "There has arrived [a] message from Advent Bay, that the 2 men and the reindeer left Advent Bay yesterday at evening-tide, heading hereto, but they are still not here. They are coming the mountain-way over Coal Bay."[11] According to Hilmar, the trek was indeed made overland and lasted four days (Hilmar Nøis Diary SVB). This portion of the excursion, too, was extremely arduous, so that by the time they reached Green Harbour, only three of the reindeer remained alive.[12]

Whereas Hilmar Nøis reports that he arrived in Green Harbour with three of the originally seven reindeer still alive (Hilmar Nøis Diary SVB), thus indicating that four had been slaughtered along the way,[13] he reverses those numbers in another diary (Hilmar Nøis Diary NPI), saying that he had arrived with four reindeer and had been forced to slaughter three.[14] His previous reporting of three reindeer remaining alive (thus four being killed) is most likely the correct one, as this is also confirmed by Rotvold (Einar Rotvold Diary), who records in his diary on April 23, after Hilmar Nøis and Kemi had arrived in Advent Bay with five reindeer and had subsequently departed from Advent Bay with the reindeer, that "The two men with the reindeer have now arrived at the Skrøder-house [Schröder House in Green Harbour]. They have had to slaughter two reindeer, due to lack of reindeer moss."[15] Thus two reindeer slaughtered, of the five who had arrived in Advent Bay, would leave three who arrived in Green Harbour. This number is further reinforced by a letter Staxrud would later write on May 25, as shall be seen (Chap. 16).

Upon reaching Green Harbour with Johan Nilsen Nøis, Kemi, and the reindeer, Hilmar Nøis immediately met with Martin Pettersen Nøis and Søren Zachariassen (Hilmar Nøis Diary NPI), who had been guarding the depot of supplies unloaded from the ship *Hertha* and stored at the Schröder House near Cape Heer.[16] At this time, the returning Staxrud expeditioners also met with the returned Schröder-Stranz expeditioners (Einar Rotvold Diary), who by now—ever since April 19—were working at the telegraph station doing general painting and carpentry work as well as maintenance on Staxrud's motorboat, and who were receiving invitations to coffee and some much-needed clothing from the telegraph manager.[17] Evidently they were being given courteous attention by Henriksen. Some of the Staxrud expeditioners visited the telegraph station on the 23rd soon after their arrival, and various members helped Rotvold and Jørgen Jensen work on Staxrud's motorboat on the 24th, 25th, and 26th of April.[18] On April 25, apparently Rotvold's group connected with Kemi: Rotvold reported in his diary "The Sámi was here visiting today" and proceeded to describe Kemi as "a very well-mannered Sámi.[1]"[19]

Meanwhile, Rotvold documented the hunting excursions after adult and young ringed seal, conducted by the telegraph station assistants, including Øyvind Widding-Danielsen (1888–?), on April 25, 29, and 30.[20] As for the sealing vessel *Hertha*, which had transported the Staxrud expedition to Spitsbergen, and which still remained locked in the ice at Green Harbour, Rotvold and Jørgen Jensen delivered a telegram to its Captain on April 19, coming onboard for a visit and receiving "a case [of] cigars as [a] present from him."[21] The ship was finally released from the ice on April 22, according to Rotvold,[22] and thus presumably began to embark on its hunting

---

[1] Einar Rotvold, in his diary, referred to the Sámi expedition member, Johannes Kemi, as "Lappen" ("The Lapp") and "lap" ("Lapp"), a term widely used at that time but no longer in use, as Sámi is the proper name and is therefore the name used in this book text.

expedition (Hilmar Nøis SVB). It had been icebound for 14 days, according to Hilmar Nøis,[23] ever since Staxrud's departure from Green Harbour to Advent Bay.

Hilmar Nøis, it was reported, subsequently left for Advent Bay on April 27 (Olaf Henriksen Diary), according to the diary entry of Green Harbour telegraph manager Olaf Henriksen,[24] presumably to prepare for Staxrud's return from the north. Indeed, as of May 2, a weary Rotvold was telling his diary that he anxiously awaited Staxrud so as to know which step to take next (Einar Rotvold Diary): "We are waiting with longing for Staxrud to return, so that one is able to consult with him about what we should do."[25] On that very same day that Rotvold wrote this entry into his diary in Green Harbour, Staxrud and his expedition were leaving from Treurenberg Bay, beginning their journey south, headed toward Advent Bay.

# Notes on Original Material and Unpublished Sources

Einar Rotvold's diary is in the Library Archives at the Norwegian Polar Institute (NPI) in Tromsø. Hilmar Nøis's diary NPI is in the Library Archives at the Norwegian Polar Institute (NPI) in Tromsø. Hilmar Nøis's diary SVB is in the Historical Archive at the Svalbard Museum (SVB) in Svalbard. Olaf Henriksen's diary is in the Library Archives at the Norwegian Polar Institute (NPI) in Tromsø.

1. E. Rotvold diary, 13 April 1913, (NPI), D00125.
2. E. Rotvold diary, 14 April 1913, (NPI), D00125.
3. E. Rotvold diary, 15 April 1913, (NPI), D00125.
4. E. Rotvold diary, 16 April 1913, (NPI), D00125.
5. E. Rotvold diary, 17 April 1913, (NPI), D00125.
6. H. Nøis diary NPI, notebook 3, April 1913, pdf page 13, (NPI), D-307/D00307_3_0001.
7. H. Nøis diary SVB, April 1913, journal pages 55–56, SVB-AP2.
8. H. Nøis diary SVB, April 1913, journal page 56, SVB-AP2.
9. E. Rotvold diary, 18 April 1913, (NPI), D00125.
10. H. Nøis diary NPI, notebook 3, April 1913, pdf page 13, (NPI), D-307/D00307_3_0001.
11. E. Rotvold diary, 21 April 1913, (NPI), D00125.
12. H. Nøis diary SVB, April 1913, journal page 56, SVB-AP2.
13. H. Nøis diary SVB, April 1913, journal page 56, SVB-AP2.
14. H. Nøis diary NPI, notebook 3, April 1913, pdf page 13, (NPI), D-307/D00307_3_0001.
15. E. Rotvold diary, 23 April 1913, (NPI), D00125.
16. H. Nøis diary NPI, notebook 3, April 1913, pdf page 13, (NPI), D-307/D00307_3_0001.
17. E. Rotvold diary, 19 and 20 and 21 April 1913, (NPI), D00125.
18. E. Rotvold diary, 24 and 25 and 26 April 1913, (NPI), D00125.
19. E. Rotvold diary, 25 April 1913, (NPI), D00125.

20.  E. Rotvold diary, 25 and 29 and 30 April 1913, (NPI), D00125.
21.  E. Rotvold diary, 19 April 1913, (NPI), D00125.
22.  E. Rotvold diary, 22 April 1913, (NPI), D00125.
23.  H. Nøis diary SVB, April 1913, journal page 51, SVB-AP2.
24.  O. Henriksen diary, 27 April 1913, (NPI), D-305/D00305_0001.
25.  E. Rotvold diary, 2 May 1913, (NPI), D00125.

## Unpublished Sources and Original Material

Henriksen, O., Diary, Olaf Henriksen's diary written while he was the Green Harbour Telegraph Station Manager, "Dagbok for Olaf Henriksen, 1911–13" ["Diary for Olaf Henriksen, 1911–13"], dated 14 August 1911 to 2 May 1913, D-305, D00305_0001. Library Archives, Norwegian Polar Institute, Tromsø, https://brage.npolar.no/npo lar-xmlui/handle/11250/2600678. Accessed 24 July 2021.

Nøis, H., Diary NPI, handwritten journal of recollections recounting the events of the 1913 Staxrud expedition, Hilmar Nøis, "Minder og Fortelinger fra fangsttiden på Spitsbergen, Klade 3, 7 Mars 1970" [Hilmar Nøis, "Memories and Stories from the trapping time on Spitsbergen, Notebook 3, 7 March 1970"], dated 7 March 1970, D-307, ("Minder og fortelinger, Nøis, Hilmar, hefte 3"), D00307_3_0001. Library Archives, Norwegian Polar Institute, Tromsø, https://brage.npolar.no/npolar-xmlui/ handle/11250/274077. Accessed 20 July 2021.

Nøis, H., Diary SVB, handwritten journal of recollections including an account of the 1913 Staxrud expedition, ("Hilmar Nøis. Fangstmann. 1909–1923, dagbok med erindringer fra fangstlivet") ["Hilmar Nøis. Trapper. 1909–1923, diary with recollections from the trapping life"], not dated (latest year referenced is 1942 in journal pages and 1950 in inserted sheet), SVB-AP2-Hilmar-Nøis-dagbok-1909–1923-compressed. Historical Archive, Svalbard Museum, Svalbard, https://svalbardmuseum.no/no/samlingene/historisk-arkiv/arkivprosjekt-fangst-og-annen-overvintringsvirksomhet-fra-perioden-1910-til-1970/. Accessed 6 June 2021.

Rotvold, E., Diary, Einar Rotvold's expedition diary written during the Schröder-Stranz expedition of 1913, "Dagbok for August Stenersen og Einar Rotvold, Tromsø: Treurenberg Bay fra 1.januar–7.juni 1913: Schröder-Stranz-ekspedisjonen 1912–13", "Einar Rotvold har skrevet dagboken", ["Diary for August Stenersen and Einar Rotvold, Tromsø: Treurenberg Bay from 1 January–7 June 1913: The Schröder-Stranz Expedition 1912–13", "Einar Rotvold has written the diary"], dated 12 January 1913 through 7 June 1913, D00125. Library Archives, Norwegian Polar Institute, Tromsø, https://brage.npolar.no/npolar-xmlui/handle/11250/2426394. Accessed 24 July 2021.

# Reference

Nøis, D. (1929, July 1929). Med kaptein Staxrud på leiting efter Schrøder-Stranz og hans folk [With Captain Staxrud on the search for Schröder-Stranz and his people]. In: *And-Ungen*, pp. 4–12. Andenes: Andøyposten. Library Archives, Norwegian Polar Institute, Tromsø. Received 20 July 2021.

# Chapter 15
# The Staxrud Expedition's Journey South with the Saved Rave and Rüdiger

## The Southward Trek from Treurenberg Bay: Sled Dogs and Draft Reindeer

The Staxrud rescue expeditioners began their return journey south from Treurenberg Bay on May 2, 1913, bringing with them the two German Schröder-Stranz expedition members they had found and rescued—Christopher Rave and Hermann Rüdiger—and departing early at 2:00 am in relatively good snow conditions (Aftenposten, 1913, evening: 1–2; Staxrud, 1914, 36). The rescued Rave and Rüdiger were in tow, with the injured Rüdiger being pulled in the specially-made bed sledge constructed just for him. The returning men were Arve Staxrud, Daniel Nøis, Dr. P. W. K. Bøck-mann, Per Hansen, Samuel Klemmetsen, and the mining engineer Jakob Ellingsen. The only rescue expeditioner not returning south with Staxrud was the ice pilot August Stenersen—the formerly stranded Schröder-Stranz expeditioner—who now remained behind alone on the *Herzog Ernst* in order to re-float the iced-in ship for the later departure and to take care of the rescue expedition's equipment. An additional reason for Stenersen's staying, according to Daniel Nøis, was to secure the vessel in light of the difficult situation and possibly financial considerations of the Schröder-Stranz expedition (Nøis, 1929, 9).

"Rüdiger had to be transported in a bed on one of the sledges," reported Staxrud in his telegram to the *Aftenposten* newspaper that would be dispatched soon after his return (Aftenposten, 1913, evening: 1–2). While the invalid Rüdiger, reposing in his bed sledge, was being pulled by draft reindeer, the painter Rave was being pulled by his two faithful dogs Jule and Caesar (Rave, 1913, 83). The remaining reindeer, from the herd of reindeer who had boarded the *Hertha* in Bosekop, Alten (Chap. 9), pulled the men's loaded sledges along the ice.

The teams of dogs who helped pull the other sledges included Storm and possibly Lussi from Roald Amundsen's South Pole expedition, who, as reported by Hilmar Nøis, had both been loaned by Amundsen and worked on the rescue expedition (Hilmar Nøis Diary SVB; Hilmar Nøis Diary NPI), and of whom Lussi had at some point during the expedition given birth to puppies, some of whom would become future rescue dogs.[1, 2] (This new information will be further presented in this narrative.) In addition to the Greenland dogs Storm and Lussi (whose name Hilmar Nøis spells as Lussi in the first diary and Lusi in the second), both of whom were brought from Kristiania (Oslo) by Staxrud (Chap. 8), other driving dogs on the rescue expedition included those canines brought by Daniel Nøis (1929, 6) and those recruited in Tromsø and Green Harbour (Chaps. 9 and 11).

As perceived by Rüdiger, in addition to the reindeer whom the rescue expedition had brought, and Rüdiger and Rave's own dogs Jule and Caesar, the sledge dogs brought by the rescuers were diverse and experienced, as he wrote in the following statement: "The dogs of the relief expedition were mostly of Norwegian origin; some Eskimo and Samoyed dogs, as well as various mixed breeds were among them, as well as some specimens that had already participated in polar expeditions. For example, Storm, who was born on the 'Fram' during Amundsen's march to the South Pole; Amundsen's Lucie was left pregnant in Advent Bay. The oldest of the dogs had been on Wellman's ill-fated airship expedition. All Norwegian dogs were small and light, and the main advantage they had over our larger German dogs probably lay in the speed of their movement." (Rüdiger, 1913, 187) (Storm had actually been born during the voyage to Antarctica prior to the South Pole march; Lussi, states Rüdiger, *was left pregnant in Advent Bay*, but this statement is rather ambiguous, as he does not specify if this occurred prior to leaving on the trek north or later after returning from the trek; also, it is unclear which Walter Wellman expedition was being referenced— 1906, 1907, or 1909.) Rüdiger's statement presents two possibilities: That Lussi may have remained in Advent Bay in April when the expeditioners departed northward for Treurenberg Bay or that Lussi may have remained in Advent Bay in May after the expeditioners returned there with Rüdiger and Rave and subsequently departed again. Rüdiger's account seems to indicate that the birth of Lussi's puppies may have taken place in Advent Bay in April or May. Another newly found report seems to indicate that the births occurred earlier, in the month of March (Morgenbladet, 1913, evening: 1), as shall be discussed later in this narrative (Chap. 22). Given this new information regarding Lussi's pregnancy and her having puppies, there is a possibility that she may not have gone on this trek. There is no indication, however, from either Staxrud or Hilmar Nøis, that Lussi did not go on the trek to Treurenberg Bay, so at this point the timing is uncertain (Fig. 15.1).

## Mountains, Blizzards, and a Sad Loss

According to Rave, the rescue expedition party started out from Treurenberg Bay in fine weather, but, after two hours, the good conditions deteriorated into a heavy snowstorm (Rave, 1913, 83–84). With the strong headwind, Rave's sledge, although pulled by Jule and Caesar, could not keep up with the others. He did not arrive at the camp until two hours after the other expeditioners, where the storm held them for 43 h.

**Fig. 15.1** The famous dogs in front of *Herzog Ernst*. *Source* Allgeier (1931, 40)

Daniel Nøis sheds further light on this beginning portion of the trek southward during which the storm had continued to increase. According to his report (Nøis, 1929, 9), the sledges were completely loaded and quite heavy, and the incline the party encountered went on for some challenging distance. The men wanted to make the ascent as quickly as possible, and the sledge dogs, although typically needing some help up the steep incline, nonetheless worked hard and diligently to accomplish the goals that they understood the men were attempting to achieve. To Nøis, the dogs' recognition of the men's intent was a pleasure to witness and was a unique and special partnership between canine and human that employed wordless and silent communication. Daniel Nøis noticed, however, that Rave was lagging behind. According to Nøis, Rave had previously portrayed a rosy picture of his driving skills for the journey and had insisted on sledging with his two dogs Caesar and Jule (whose name Nøis spells Juli) as he viewed them equal to the rescuers' draft dogs. But both Rave and Dr. Bøckmann now remained quite a long distance behind the rest of the party; however, according to Nøis, he and the other men felt confident that the two would be able to see the tracks left by the sledging party. Nøis, Staxrud, and the rest— except for Rave and Bøckmann—reached a height at the bottom of a mountain range where they sought shelter from the blizzard winds and decided to camp. Finding that Rave and Bøckmann still were nowhere in sight, Daniel Nøis went back for them, taking with him three sledge dogs unharnessed from the sledges, and making good distance. He came upon Bøckmann traveling solo, having, he said, necessarily left Rave as Rave evidently could no longer go on. Nøis backtracked further, descending along the incline of the hill and reaching Rave as the latter tried working with his snowshoes, unable to ski. At this point, it was apparent Rave needed to ride on the sled and be pulled by Nøis and the dogs—a task that Nøis intimated was by no means an easy one for him and was an interesting representation of the role the artist Rave had aspired to and had painted for himself. According to Nøis, by the time he and Rave caught up to the others at the base of the mountain, with Rave's falling behind and his stated weak condition, the now very strong blizzard coming at them from the south caused a delay, forcing the men to remain in place until May 4.

Staxrud would later record that, on the very next day after their embarking south, the sledge group was held in a tent by a blizzard for two days (Staxrud, 1914, 36). Rave reported that at this time Staxrud, Bøckmann, Rüdiger, and Rave shared one of the small tents (Rave, 1913, 84) (Fig. 15.2).

While the sledge dogs survived this time well curled up under the snow cover, the reindeer unfortunately ate up all of the moss that had been brought along and that was actually intended as food for their entire journey south across the plateau (Staxrud, 1914, 36–37). When the storm ended, the party resumed their trek on May 4, and Daniel Nøis now observed Rave struggling on his snowshoes again and trampling on his dogs to vent his cross mood and frustration (Nøis, 1929, 9). Meanwhile, the reindeer and their herders walked a great distance to the rear of the other men and the dogs. By now the reindeer had become greatly weakened with no further nourishment to sustain them, as they continued trekking and pulling their sledges. Bøckmann came ahead and informed the rest of the party of the reindeer's plight and utter exhaustion. The remainder of the party raised the tents and awaited the

**Fig. 15.2** Camp in the blizzard. *Source* Staxrud (1914 Tf. 4)

reindeer sledges for more than an hour, freezing in the dampness created by their physical exertion and having no stove on which to cook food or heat a beverage as the stove was packed on the reindeer sledges. Finally, the two Sámi arrived with the extremely weary and toiling reindeer who had worked and given all their strength and who could truly go no further.

It had been a tremendous effort that they had made, and there had been much energy that they had dutifully expended. But the unfortunate result of the previous two days' delay now took effect.

After traveling 10 km that day, and after reaching an altitude of 700 m, the men had to shoot all but the strongest reindeer because of exhaustion (Staxrud, 1914, 37). It was a sad moment for everyone, according to Staxrud, because the reindeer had exceeded all their expectations.

It was the unplanned wait that had caused the reindeer to run out of food, reported Rave, and so now they had to be shot one by one (Rave, 1913, 84).

As the men unhappily began to slaughter the reindeer, they left only one alive to pull Rüdiger's sledge (Nøis, 1929, 9). According to Daniel Nøis, some of the meat from the slaughtered reindeer was given to the dogs to eat, but the men had no desire to partake in the flesh of their exhausted draft reindeer.

Ultimately, the men built a depot at this place, stocked with their reindeer meat, which they felt should secure their way back to Treurenberg Bay if necessary (Staxrud, 1914, 37). At 6:00 pm they began to resume their trek, but at this time they truly felt the absence of their hard-working and reliable reindeer (Nøis, 1929, 9). The one remaining reindeer toiled as long as possible until exhaustion was too overwhelming. In the evening, that last reindeer, who was still pulling Rüdiger's sledge, sadly also had to be shot (Staxrud, 1914, 37) (Fig. 15.3).

Staxrud summarized the events up to this point as follows (Aftenposten, 1913, evening: 1–2): "From Treurenberg Bay we went straight up onto the inland-ice and set the course toward Cape Petermann. There was a lot of bad weather, so that we often had to stay put for a longer period. The trip therefore lasted longer than expected and we ran out of moss for the reindeer, which we therefore had to shoot."

## Traveling to Wijde Bay and Through Dickson Bay: Fog, Absence, Rage, and Reliance

Due to the loss of the draft reindeer, reported Christopher Rave, the men had to harness themselves in front of the sledges in order to help the few dogs pull them (Rave, 1913, 84–85). According to Rave, in addition to the injured Rüdiger, Rave also was exempted from this pulling assignment, as he would otherwise experience heart problems at the pace presented. This time, in this portion of the march, dense fog forced them to stop for 30 h until 2 a.m. on May 9. When they left again, it was −20 °C and Rave froze his fingertips.

Die letzten Renntiere und der Bettschlitten.

**Fig. 15.3**   Last remaining reindeer and Rüdiger's bed sledge. *Source* Rüdiger (1913, 205)

Daniel Nøis also describes this leg of the trip as one that was hampered by frost fog and snowstorm, but additionally as one that was greatly troubled by Rave (Nøis, 1929, 9–10). According to Daniel Nøis, during this time he observed Rave trampling on his dogs again, wildly wielding himself in his snowshoes and pouncing in fury, with Caesar in particular appearing to receive most of the brunt of the force. Seeing this, Nøis, who was second-in-command, pointedly requested that expedition leader Staxrud issue Rave a command to cease stomping on his dogs, as there could be adverse consequences for Rave himself, and as Nøis was losing patience witnessing these actions. Simultaneously, Nøis gave Rave his own two sledge dogs and ordered him effectively to stay on his sledge and not go in front of the others. The result was that, from that point on, Rave remained seated and calm and evidently refrained from what Nøis seemingly deemed to be animal cruelty. In this way, despite the heaviness of Rave's sledge and the arduous journey, the men and dogs were able to work together smoothly and fairly, with no further tormenting of any dogs.

As the fog lifted sufficiently and the party was able to begin the narrow descent to Wijde Bay—a descent which they found quite rugged and steep at times—the expeditioners were beset with crevasses within the ice sheet, all of which necessitated taking extra care and additional time, while traveling on an estimated 10-day supply of provisions that just barely lasted after the ninth day as the party approached Wijde Bay (Nøis, 1929, 10). Daniel Nøis reported that, upon arrival at Wijde Bay, the men were able to kill two wild reindeer for needed meat. Unfortunately, Caesar—the dog who had helped pull Rave—was in such bad shape and was having such difficulty, hardly being able to walk, that he had to be euthanized.

As told by Rave (Rave, 1913, 86–87), when the party finally reached Wijde Bay, the men were able to successfully hunt several (wild) reindeer, thus supplementing their meager provisions (Fig. 15.4). On May 10, they spent the night in the small log hut at Cape Petermann where Captain Alfred Ritscher had spent so much time with his men. Rave reported that unlike Jule, who continued to pull the sledge tirelessly, Caesar was finished and Bøckmann gave him the coup de grace.

When the group reached Cross Point hut in Cape Petermann, where they spent the night, according to Daniel Nøis, they also accessed their depot there, which included a good supply of food (Nøis, 1929, 10). Although the journey across the ice thus far had been exhausting and cold, they did not spend further time at the hut, but left on the following day, Pentecost morning (May 11), as Staxrud wished to continue traveling on as quickly as possible following one good night's rest. Traveling inland on Vestfjorden (West Fjord), and stopping at the bottom of the fjord for coffee and to allow the dogs to stretch prior to the grueling ascent they were about to make, they ascended the mountain in two trips, taking a pemmican break after the first 18 h, then continuing to energetically complete the excruciating ascent. Immediately after, on this same second day of Pentecost (May 12), they continued to descend to

**Fig. 15.4** Reindeer hunting in Advent Valley. *Source* Rüdiger (1913, 191)

the bottom of Dickson Bay in Ice Fjord, where they paused to access their depot before continuing on over the Bay and setting up camp at the end of what had been a 12-h trek.

According to Rüdiger (1913, 195), fortunately, the way along Dickson Bay was level, so that he could also be pulled on the sledge. They were even able to set square sails on the sledges, using a tent roof pole as the yard, skis as the mast, and tarpaulin—that served as the floor of the tent—as the sails. With Rüdiger in his sleeping bag, his sledge became what he described as the first "sleeper" or "sleeping car" in Spitsbergen.

Staxrud's telegram, as published in the newspaper, describes the rescuers' march, from the time of the killing of the reindeer to the reaching of Dickson Bay, as follows (Aftenposten, 1913, evening: 1–2):

> We ourselves and the dogs pulled the toboggans [sledges] over the unknown inland-ice, which reached a height of 1200 meters above the sea. We found [a] descent to Wijdebay [Wijde Bay] through a narrow valley, but the descent was extremely arduous. The sledges and Rüdiger's bed had to be hoisted downward [lowered down] the precipitous hillsides [slopes] [with ropes].
>
> On the east side of Wijdebay [Wijde Bay] we found an abundance of wild reindeer. Arrived at the depot at the Korspynt-hut [Cross Point Hut] (no. 2) [on the map] on the eve of Pentecost [or the eve of Whit Sunday, thus Saturday, May 10]. At the watershed between Dickson Bay and Wijdebay [Wijde Bay], where we also had a depot, we arrived Whit Monday [May 12]. We now went quickly down to Dickson Bay and out to the Isfjord [Ice Fjord] itself, which however was found broken up, so the caravan had to follow the east side of the Nordfjord [North Fjord] and onwards around Cape Thordsen to Skansbay, where we at last succeeded in getting across the fjord to Hyperithatten, three hours travelling from Advent Bay.

On that following day (presumably May 13), according to Daniel Nøis, the group had arrived at Cape Delta located on the southern side of the Ice Fjord, and as the rest of the men arranged the tents for camp, Nøis followed Staxrud up to the apex of a precipitous transition point from where Staxrud could take a path toward Advent Bay (Nøis, 1929, 10).

From that point, Staxrud eventually left the rest of the party behind in order to make advance arrangements in Advent Bay, where he would arrive by May 14 (Aftenposten, 1913, evening: 1–2). "Here [Hyperithatten] I left the others in order to get to Green Harbour and arrange the northeast-land [Northeast Land] trip," reported Staxrud in his telegram to the *Aftenposten* newspaper sent on May 14. "The others will go to De Geer's valley in Advent Bay, where they probably will arrive Thursday [May 15]."

## The Reaching of Advent Bay: Staxrud's Telegrammed Reports, the Summoning of Expeditioners, and the Arrival of the Rescuers and the Rescued

According to Christopher Rave, as of May 14, the rescue expedition and rescued men had finally reached the Advent Bay vicinity, and Arve Staxrud had gone ahead to telegraph to Germany about his rescue mission (Rave, 1913, 90). Staxrud himself later reported that he had managed to travel on ahead to provide shelter for the two rescued German expeditioners Rave and Rüdiger, arriving in Advent Bay two days earlier than the rest of the group (Staxrud, 1914, 42).

Immediately upon arriving, Staxrud communicated the outcome of the expedition, telegraphing the Norwegian newspaper *Aftenposten* with his expedition report "Radio-special-telegram to 'Aftenposten' from Spitsbergen," sent from "Advent Bay, 14th May at 6 p.m." and "Received in Kristiania 15th May at 3:30 a.m." (Aftenposten, 1913, evening: 1–2). The front-page article was published that very day of receipt, on Thursday, May 15, 1913, and bore the headline "The Rescue Expedition on Spitsbergen", with large subheadings declaring "Captain Staxrud's Account to 'Aftenposten'" and announcing "Two Germans Brought to Safety at Advent Bay". Other subheads stated "No Trace of the Others.—The Investigations Continue," and "Encounter with Lerner's expedition in Treurenberg Bay.—Poor weather- and ice-conditions.—A painstaking 14 days' journey.—In a bed across Northwest-Spitsbergen." The prominent article featured the full telegram sent by Staxrud describing the entire expedition in his own words. His account of events ended with the positive news that "It is well with both Rüdiger and Rave and with all participants of the sledge expedition." The article included a map of the Spitsbergen area searched, as well as photographs of Staxrud, the expeditioners on board the ship *Hertha*, and the Swedish House. It is evident that this was all major news (Fig. 15.5).

On the same day of his arrival in Advent Bay, May 14, Staxrud also telegraphed his supporter and the person who had recommended him, Fridtjof Nansen, to report to him that the expedition had had a good outcome and to thank his mentor for his welcomed suggestions (Staxrud 14 May 1913). Addressing the telegram to "Professor Nansen" in Kristiania, Staxrud wrote: "Expedition successful[.] Thanking good advice."[3]

At this same time, Staxrud also telegrammed the other rescue expeditioners waiting in Green Harbour, summoning them to Advent Bay to help transport the Germans (Einar Rotvold Diary). According to Einar Rotvold, who was still in Green Harbour with the other Norwegian Schröder-Stranz expeditioners, word of Staxrud's arrival had been received, as had word of the rescue of the German survivors and the fate of the reindeer, and at that very moment, the other rescuers from the Staxrud expedition were preparing to leave Green Harbour to assist, while the Green Harbour telegraph station was spreading the word about the Staxrud expedition's progress. Rotvold's May 14th diary entry reads as follows: "At 4 o'clock afternoon one received message from Advent Bay, that Staxrud had arrived. He came alone. The others are going up De Geer Valley and down the Advent Bay valley, most likely arriving in

**Fig. 15.5** Front-page article published in the *Aftenposten* May 15, 1913, newspaper featuring Arve Staxrud's telegrammed report from Spitsbergen sent on May 14 (National Library of Norway)

Advent Bay tomorrow. The Nøis-guys received message to come to Advent Bay immediately, in order to help with getting the ailing here to Green Harbour. Staxrud has had to shoot down all the reindeer, due to lack of reindeer moss. Here it is now busy with sending telegrams to the press."[4]

The men summoned to Advent Bay to help would include Martin Pettersen Nøis, Hilmar Nøis, and Johannes Kemi (Nøis, 1929, 10). Previously, Pettersen Nøis had remained with the cases at the depot headquarters in Green Harbour, and Hilmar

Nøis and Kemi had returned early from the rescue journey north to Treurenberg Bay (Chap. 14).

Hilmar Nøis reported receiving the telegram from Advent Bay in May (Hilmar Nøis Diary NPI) advising of the arrival of Staxrud and his expedition with the two ill German expeditioners.[5] According to Hilmar (Hilmar Nøis Diary SVB), after receiving the communication, he met with his uncle Johan Nilsen Nøis in Advent Bay, where he then received further news about the rescue expedition as well as learned of the Lerner expedition on the *Loevenskjold*.[6] Evidently, it was from here in Advent Bay that Hilmar Nøis, together with Pettersen Nøis and Kemi, proceeded north to rendezvous with their fellow rescue expeditioners.

Staxrud himself, with Johan Nilsen Nøis, left Advent Bay on May 15, headed toward Green Harbour (Nøis, 1929, 10; Einar Rotvold Diary). He left in the evening, according to Einar Rotvold's May 15th diary entry, and was expected to arrive the following morning. Rotvold, who was still working at the Green Harbour telegraph station, also relayed in this same diary entry the other news Staxrud had disseminated—that "Stenersen stayed behind in Treurenberg Bay, to look after the ship" and that "Lerner lay in Mossel Bay with Løvenskjold."[7]

Staxrud was now working on preparing for the second northward sledging journey, as he had reported in his telegram that appeared in the newspaper that very day: "I intend to return to Treurenberg as soon as possible in order to undertake the search for Schrøder-Stranz in June" (Aftenposten, 1913, evening: 1–2).

Meanwhile, back at Cape Delta, on this same day of May 15, the rest of the Straxrud expedition—Daniel Nøis, Dr. Bøckmann, Hansen, Klemmetsen, and Ellingsen, with Rüdiger and Rave—departed from their location, traveling through De Geer Valley and making their exhausting ascent, and finally reaching a hunter's hut in Advent Valley, according to Daniel Nøis (Nøis, 1929, 10). There, the men were able to shoot two wild reindeer, which provided them with a meal they eagerly welcomed, and then proceeded to camp for the night. Just as the men were climbing into their sleeping bags, the dogs began barking a communication that someone was near and approaching. Quickly surveying the area, the men found that their colleagues and fellow expeditioners Martin Pettersen Nøis, Johannes Kemi, and Hilmar Nøis had just arrived from Advent Bay to rendezvous with them and to help them make the final stretch safely. It was at this time that Hilmar Nøis reported to Daniel Nøis about his and Kemi's rough trip south during their early return from the northward expedition and about their needing to slaughter some of their reindeer along the way. Hilmar also informed Daniel that Staxrud and Johan Nilsen Nøis had left Advent Bay for Green Harbour and gave Daniel a letter in which Staxrud stipulated to his deputy the arrangements he had made and further steps to be taken. After spending the night there together, the reunited Staxrud expeditioners and the two rescued German expeditioners, along with their dog teams, traveled on the following day, going down to the Arctic Coal Company camp in Advent Bay, and camping adjacent to the settlement's houses. Daniel Nøis quickly spoke with the American coal mining company's engineer Mangham (most likely Bert Mangham, winter superintendent for the Arctic Coal Company), who immediately provided Rave and Rüdiger with housing and attention. According to Daniel Nøis, Mangham was very hospitable, which he indicated was the norm for him, and the two Germans were looked after and taken care of in the best manner possible. Daniel Nøis himself and the rest of the group remained camped alongside the settlement and visited the coal company camp that night.

Christopher Rave echoes in his account (Rave, 1913, 90–92) that, two days after May 14, after Staxrud had gone ahead, they reached the American coal settlement of Longyear City in Advent Bay, where the two rescued Germans were assigned three rooms in a house under construction.

Hermann Rüdiger is reported to have further described the arrival in Advent Bay as follows, according to his book passage about the events that began on May 15, as quoted in the Arctic Coal Company's published account by Nathan Haskell Dole:

> "The water-shed between De Geer Valley and Advent Valley is only about 150 meters high. We easily surmounted the ridge and slid down the wonderful, long-stretching valley to the upper hunting hut of the Arctic Coal Company. In the hut were all sorts of provisions and utensils and the first indications of approaching civilization – empty beer bottles with the inscription, 'Bayrisk Øl.' [sic]. The hut itself was small, with three bunks, one above the other, fireplace, table, and chairs set out: it had as a tiny Vorraum [anteroom or outer room], a horse-stall, and outside a two-wheeled cart....
>
> "So in the early morning of May 16 we started on our three-hour journey to Advent Bay. One sledge always went more swiftly than the other; mine was drawn by four dogs, escorted by the three [Sami].[1] Before Longyear Valley we made a halt. Dr. Böckmann [sic] went ahead to announce our arrival to the little mining-settlement. Soon came a great sledge drawn by a horse. Our packs were loaded on it and my Schlafwagen [sleeping car or sleeper] attached behind. At twelve o'clock Rave and I made our entrance into Longyear City.["]
>
> (Rüdiger quoted in Dole, 1922, vol. 2: 194). [The passage appears to be from Rüdiger's, 1913 book, pages 198 and 199].

Based on this account, evidently, after Kemi's return to the rescue party, he had joined Hansen and Klemmetsen in supervising the pulling of Rüdiger's bed sledge, which, with the absence of the fallen reindeer, was subsequently performed solely by the sledge dogs. Upon arrival at Advent Bay, horses were further employed to help tow the injured Rüdiger.

And hence Rüdiger and Rave, accompanied by Bøckmann (Dole, 1922, vol. 2: 195), were able to settle into a house in Advent Bay on May 16th.

Thus, the first half of the rescue expedition had been accomplished. The two German expeditioners of the Schröder-Stranz expedition had been rescued. The return trek south had been an arduous excursion. As Staxrud would later describe in his published report (Staxrud, 1914, 36–42), in which he recounts the trek toward Longyear City: Again and again, snowfall and storms had interrupted the onward journey, but they had managed, over steep ups and downs, and with the additional provisions from the previously laid depots, to reach Advent Bay on May 16 and to bring Rüdiger and Rave to safety.

---

[1] Hermann Rüdiger, as quoted in the Dole 1922 book which features his book passage specifically referenced here, referred to the three Sámi, Per Hansen, Samuel Klemmetsen, and Johannes Kemi, as "the three Lapps," "Lapp" being a term widely used at that time but no longer in use, as Sámi is the proper name and is therefore the name used in this book text. Also as quoted in the Dole book, Rüdiger did not mention them by name in this passage.

## Regrouping in Green Harbour, Recuperating in Advent Bay, and Welcoming New Life on Spitsbergen

On the same day that the rescuers brought Rave and Rüdiger to the safe and warm housing in Advent Bay, May 16, Staxrud, who had left Advent Bay the previous day, arrived in Green Harbour, where he—remarkably, and arguably in true leadership form—took the symbolic time and effort to establish an additional special prize for a ski-jumping competition scheduled to take place the following day, as was reported in Einar Rotvold's May 16th diary entry (Einar Rotvold Diary). According to Rotvold, Staxrud arrived at 2:00 pm and was welcomed at the "Skrøder-cabin" (the Schröder House at Cape Heer) by the Green Harbour telegraph station manager Olaf Henriksen.[8] On the next day, Norway's National Day of Independence, May 17, Staxrud's prize and other prizes were distributed at the telegraph station, where, in addition to Henriksen, Staxrud himself delivered a holiday speech to those individuals gathered, including the rescue expedition's ice pilot Søren Zachariassen, all of which appear to have given a feeling of comradeship and encouragement to the Schröder-Stranz crewmembers Einar Rotvold and his Norwegian compatriots, as evidenced by Rotvold's words, "We have had a particularly enjoyable day here, which many in Norway maybe are longing for."[9]

In contrast to the ceremonies taking place in Green Harbour, there were no holiday festivities witnessed by Daniel Nøis among the working people in the coal mining town at Advent Bay, where the rest of the rescue expedition slept through most of that quiet May 17th (Nøis, 1929, 10). After resting, most of the expeditioners (all but Bøckmann) departed back to Green Harbour (Hilmar Nøis Diary SVB), leaving the two Germans in Advent Bay to recuperate for the time being until they could be transported.[10] That evening, Daniel Nøis and his group began to make their way to Green Harbour, traveling along the Longyear Valley to Coal Bay, where, during the night, they stopped in Daniel Nøis's cabin to prepare a celebratory holiday meal, cooking two wild reindeer that Johan Nilsen Nøis had previously shot and left there at the cabin (Nøis, 1929, 10). Following a dinner break, which lasted two hours, the group completed the trek to Green Harbour, arriving there at 8:00 am on the following morning of May 18, where they would continue to perform their work.

Einar Rotvold (Einar Rotvold Diary) recorded the arrival of Daniel Nøis and the rest of the Staxrud expeditioners in Green Harbour that day, reporting in his diary on May 18 that "Staxrud's people have now arrived at the Skrøder-cabin with the dogs."[11]

Rotvold and the Jensen brothers, Julius and Jørgen, were at this time still keeping themselves busy working at the telegraph station. Having previously rescued themselves from the frozen ship *Herzog Ernst* at Treurenberg Bay, they were now continuing to make an effort to contribute to the community in Green Harbour. Rotvold himself had helped a pregnant woman give birth on May 9 (Einar Rotvold Diary). It is uncertain if this was the same expectant mother whom the telegraph station personnel had been expecting, based on Rotvold's previous diary entry of April 15 (Chap. 14). At that time, Rotvold had reported that a spouse was "coming from one of the coal mines" to stay at the telegraph station in order "to give birth."[12] On May 9, a mother did indeed go into labor, and, according to his descriptive and substantial diary entry devoted to the momentous events of that day, Rotvold assisted during the delivery of the baby. "At 4 o'clock am there came message from a family, Glad, who works for Bjelland, Stavanger, here in Green Harbour, whether I would come out to them, as the wife was about to give birth," he wrote.[13] Immediately going to the family's aid, and stopping only to obtain "some medicine and cotton as well as a bottle of wine" from telegraph station manager Henriksen's location, Rotvold spent the day at Glad's place, aiding until "she gave birth to a pretty, healthy Boy" at 11:45 am, which "all went well and good," and then staying to help the mother with the afterbirth, as she was having difficulty with the placental expulsion. Rotvold tended to her during this time and seems to express concern for her in his diary. At 6:00 pm, Rotvold was relieved by another so that he could go back to his room to rest. He later that night indicated that the expulsion of the placenta had been completed, stating that "At 9 o'clock pm all was over and done with, and the wife and the child [were] in the best of health." Rotvold concluded his summary of events with the following positive statement: "We are all of us happy that everything went so well, hope she will soon get well again." On the following day, May 10, Rotvold, as well as others, including Henriksen, visited the Glads, and Rotvold reported that "there all is just well, both with the wife and the child."[14] He again checked in on the mother and baby on May 12, recording in his diary that "There all is just well" and commenting that he, Jørgen Jensen, and others had enjoyed going ski racing later that evening.[15]

Einar Rotvold, who had felt helpless on the iced-in ship, watching his mate Knut Stave grow ill and die while being able to do nothing to deter it, could now do something to help bring new life into the community on Spitsbergen. His diary entries about this event seem to express cheer and jovial relief. This childbirth and aiding in the childbearing, this bringing in of new life, as well as helping the mother who bore the child, was important, as lives had been lost as well as saved during this Schröder-Stranz expedition.

Another pregnancy event occurred around this time as well—it was Lussi, the Amundsen South Pole Expedition dog and Staxrud relief expeditioner, who gave birth to six puppies during the rescue expedition, according to Arve Staxrud and Hilmar Nøis (Staxrud 11 August 1913; Hilmar Nøis Diary SVB; Hilmar Nøis Diary NPI). While it is unknown at this time exactly when and where Lussi had her puppies,[16] Hilmar Nøis states in his diaries that most of her progeny were subsequently designated as rescuers and destined for life-saving efforts on Spitsbergen, and that the telegraph station would become home to these puppies and their offspring, resulting in a team of rescuer sled dogs who would reside in Green Harbour and be ready to respond to any tragic accidents, mine disasters, or shipwrecked expeditions similar to the Schröder-Stranz situation.[17, 18] Correspondence from Leon Amundsen (Amundsen 14 August 1913a, b) would later confirm the fact that some of Lussi's puppies would become rescuer dogs at the Green Harbour telegraph station on Spitsbergen.[19, 20]

Thus, a new pack of rescuer dogs would be begun by Lussi, taking after their mother and continuing the legacy and mission of rescuing stranded souls out on the ice who had become lost or injured or stranded either through illness, mishap, misfortune, or—as in the Schröder-Stranz case—lack of experience or preparation by some of the expedition's leaders.

Lussi may have been in Advent Bay in mid-May with Rave, Rüdiger, and Dr. Bøckmann, prior to their leaving.

As for Rave and Rüdiger, they remained in Advent Bay to recover and would end up staying through the end of the month. The rescue expedition's physician Dr. Bøckmann remained with them, according to Rüdiger's account as quoted by Dole:

> "Dr. Böckmann, Rave, and I – the other members of the help-expedition had gone on the 17th of May to Green Harbor [sic] – were housed in the highest up of the twenty houses in Longyear Valley. Just before our arrival three rooms in it had been made ready and they were working on the others: the house has four apartments, each for a family. Now Rave and I had a room to ourselves, and Dr. Böckmann had another, the third we used as a sitting-room in common. A young semi-invalid miner from Tromsø looked out for us.["]
>
> (Rüdiger quoted in Dole, 1922, vol. 2: 195). [The passage appears to be from Rüdiger's, 1913 book, page 201].

Dole's own account regarding Rave and Rüdiger's stay in Advent Bay, as presented in his book about the Arctic Coal Company, states that "As they were made quite welcome and hospitably treated[,] they enjoyed their visit, especially as they had various German countrymen with whom to associate, and others from the wireless station at Green Harbor [sic], who told them that a few weeks before, the first child ever born on Spitzbergen had arrived." (Dole, 1922, vol. 2: 196).

It is an apt coincidence that "the first child ever born on Spitzbergen," to whom the Dole account refers, most likely was the baby whom Rotvold had helped deliver on May 9. There is also the possibility that the person mentioned in Rotvold's April 15th diary entry was a different mother and that this was the baby to whom he refers. But, given the timing and dates, based on this account, it seems likely that Rotvold, the expeditioner who had saved himself from the Schröder-Stranz expedition, may have had a hand in helping the baby whom the Dole account states was the first born on the Arctic archipelago of Spitsbergen, news of whose birth had now reached fellow expeditioners Rüdiger and Rave, who had just been saved by the Staxrud expedition.

## Notes on Original Material and Unpublished Sources

Hilmar Nøis's diary SVB is in the Historical Archive at the Svalbard Museum (SVB) in Svalbard. Hilmar Nøis's diary NPI is in the Library Archives at the Norwegian Polar Institute (NPI) in Tromsø. Arve Staxrud letters of correspondence, written from and to Arve Staxrud, are in the Manuscripts Collection at the National Library of Norway (NB) in Oslo. Einar Rotvold's diary is in the Library Archives at the Norwegian Polar Institute (NPI) in Tromsø. Roald Amundsen letters of correspondence, written from and to Roald Amundsen and Leon Amundsen, are in the Manuscripts Collection at the National Library of Norway (NB) in Oslo.

1. H. Nøis diary SVB, 1913, journal page 65, SVB-AP2.

2. H. Nøis diary NPI, notebook 3, 1913, pdf page 39, (NPI), D-307/D00307_3_ 0001.
3. A. Staxrud to F. Nansen, telegram, 14 May 1913, NB Ms.fol. 1924.
4. E. Rotvold diary, 14 May 1913, (NPI), D00125.
5. H. Nøis diary NPI, notebook 3, 1913, pdf page 13, (NPI), D-307/D00307_3_ 0001.
6. H. Nøis diary SVB, 1913, journal page 56–57, SVB-AP2.
7. E. Rotvold diary, 15 May 1913, (NPI), D00125.
8. E. Rotvold diary, 16 May 1913, (NPI), D00125.
9. E. Rotvold diary, 17 May 1913, (NPI), D00125.
10. H. Nøis diary SVB, 1913, journal page 57, SVB-AP2.
11. E. Rotvold diary, 18 May 1913, (NPI), D00125.
12. E. Rotvold diary, 15 April 1913, (NPI), D00125.
13. E. Rotvold diary, 9 May 1913, (NPI), D00125.
14. E. Rotvold diary, 10 May 1913, (NPI), D00125.
15. E. Rotvold diary, 12 May 1913, (NPI), D00125.
16. A. Staxrud to R. Amundsen, telegram, 11 August 1913, NB Brevs. 812:1.
17. H. Nøis diary SVB, 1913, journal page 65, SVB-AP2.
18. H. Nøis diary NPI, notebook 3, 1913, pdf page 39, (NPI), D-307/D00307_3_ 0001.
19. L. Amundsen to Ø.W. Danielsen, telegram, 14 August 1913, NB Brevs. 812:3:8.
20. L. Amundsen to Foreign Minister of Norway, letter, 14 August 1913, NB Brevs. 812:3:8.

## Unpublished Sources and Original Material

Amundsen, L., 14 August 1913a, Telegram from Leon Amundsen to Øyvind Widding-Danielsen, NB Brevs. 812:3:8. Manuscripts Collection, National Library of Norway, Oslo. Received 9 June 2021.

Amundsen, L., 14 August 1913b, Letter from Leon Amundsen to Foreign Minister of Norway, NB Brevs. 812:3:8. Manuscripts Collection, National Library of Norway, Oslo. Received 23 June 2021.

Nøis, H., Diary NPI, handwritten journal of recollections recounting the events of the 1913 Staxrud expedition, Hilmar Nøis, "Minder og Fortelinger fra fangsttiden på Spitsbergen, Klade 3, 7 Mars 1970" [Hilmar Nøis, "Memories and Stories from the trapping time on Spitsbergen, Notebook 3, 7 March 1970"], dated 7 March 1970, D-307, ("Minder og fortelinger, Nøis, Hilmar, hefte 3"), D00307_3_0001. Library Archives, Norwegian Polar Institute, Tromsø, https://brage.npolar.no/npolar-xmlui/handle/11250/274077. Accessed 20 July 2021.

Nøis, H., Diary SVB, handwritten journal of recollections including an account of the 1913 Staxrud expedition, ("Hilmar Nøis. Fangstmann. 1909–1923, dagbok med erindringer fra fangstlivet") ["Hilmar Nøis, Trapper, 1909–1923, diary with recollections from the trapping life"], not dated (latest year referenced is 1942 in journal pages and 1950 in inserted sheet), SVB-AP2-Hilmar-Nøis-dagbok-1909–1923-compressed. Historical Archive, Svalbard Museum, Svalbard, https://svalbardmuseum.no/no/samlingene/historisk-arkiv/arkivprosjekt-fangst-og-annen-overvintringsvirksomhet-fra-perioden-1910-til-1970/. Accessed 6 June 2021.

Rotvold, E., Diary, Einar Rotvold's expedition diary written during the Schröder-Stranz expedition of 1913, "Dagbok for August Stenersen og Einar Rotvold, Tromsø: Treurenberg Bay fra 1.januar–7.juni 1913: Schröder-Stranz-ekspedisjonen 1912–13", "Einar Rotvold har skrevet dagboken", ["Diary for August Stenersen and Einar Rotvold, Tromsø: Treurenberg Bay from 1 January–7 June 1913: The Schröder-Stranz Expedition 1912–13", "Einar Rotvold has written the diary"], dated 12 January 1913 through 7 June 1913, D00125. Library Archives, Norwegian Polar Institute, Tromsø, https://brage.npolar.no/npolar-xmlui/handle/11250/2426394. Accessed 24 July 2021.

Staxrud, A., 14 May 1913, Telegram from Arve Staxrud to Fridtjof Nansen, NB Ms.fol. 1924. National Library of Norway, Oslo, online archives, https://www.nb.no. Retrieved 6 July 2021.

Staxrud, A., 11 August 1913, Telegram from Arve Staxrud to Roald Amundsen, NB Brevs. 812:1. National Library of Norway, Oslo, online archives, https://www.nb.no. Retrieved 6 July 2021.

# References

Aftenposten. (1913, May 15). Hjælpeexpeditionen paa Spitsbergen. Kaptein Staxruds beretning til "Aftenposten". [The rescue expedition on Spitsbergen. Captain Staxrud's account to "Aftenposten".]. *Aftenposten* (Kristiania [Oslo]), 1913, May 15, evening edition, pp. 1–2. National Library of Norway, Oslo, online archives: www.nb.no. Accessed August 28, 2012.

Allgeier, S. (1931). *Die Jagd nach dem Bild. 18 Jahre Kameramann in Arktis und Hochgebirge.* Stuttgart: Engelhorn.

Dole, N. H. (1922). *America in Spitsbergen: The romance of an Arctic coal-mine.* In Two Volumes. Boston: Marshall Jones Company. Volume II. Facsimile reprint by Scholar Select.

Morgenbladet. (1913, September 15). "Obersten"s Barn. [Children of "Obersten".]. *Morgenbladet* (Kristiania [Oslo]), 1913, September 15, evening edition, page 1. Digital Archive, National Library of Norway, Oslo, Received 30 March 2022; National Library of Norway, Oslo, online archives: https://www.nb.no. Retrieved April 2, 2022.

Nøis, D. (1929). Med kaptein Staxrud på leiting etter Schrøder-Stranz og hans folk [With Captain Staxrud on the search for Schröder-Stranz and his people]. In: *And-Ungen*, July 1929, pp. 4–12. Andenes: Andøyposten. Library Archives, Norwegian Polar Institute, Tromsø. Received July 20, 2021.

Rave, C. (1913). *Tagebuch von der verunglückten Expedition Schröder-Stranz.* Schaffsteins Gründe Bändchen 49. Cöln: Schaffstein.

Rüdiger, H. (1913). *Die Sorge Bay. Aus den Schicksalstagen der Schröder-Stranz-Expedition.* Berlin: Georg Reimer.

Staxrud, A. (1914). Die Staxrudsche Hilfs-Expedition für Schröder-Stranz. In A. Miethe (ed.), *Die Expedition zur Rettung von Schröder-Stranz und seinen Begleitern - geschildert von ihren Führern Hauptmann A. Staxrud und Dr. K. Wegener* (pp. 1–68). Berlin: Dietrich Reimer.

# Chapter 16
# The Reckoning and the Waiting

## The Difficult Task of Finding a Ship to Take the Rescued Expeditioners Home

When the returning members of the Arve Staxrud Norwegian rescue expedition (the official German rescue expedition) had arrived in Green Harbour on May 18, they had continued working, rotating assignments throughout the remainder of May, as reported by Daniel Nøis (1929, 10). Most likely, part of their work was making preparations for the second trek to Treurenberg Bay to search for Herbert Schröder-Stranz himself and the rest of the men who were still missing.

Meanwhile, the two German members of the Schröder-Stranz expedition who had been rescued and brought to Longyear City on May 16—Hermann Rüdiger and Christopher Rave—remained in Advent Bay with Dr. P. W. K. Bøckmann. It was not until May 18 that Rave and Rüdiger received news from home (Rave, 1913, 92–93). Among the items of news was a telegram from Prof. Adolf Miethe, who on the one hand congratulated them and thanked Rave for his prudence, but on the other hand also requested from Rave a detailed telegraphic report on the expedition up to their rescue.

The third German survivor of the expedition, Captain Alfred Ritscher (Chap. 11), evidently remained convalescing in Advent Bay at the same time that Rüdiger and Rave were there.

As for the surviving Norwegian crewmembers of the Schröder-Stranz expedition, who were still working in Green Harbour, they were apparently requested to provide further accounts of the failed expedition on the following day of May 19 (Einar Rotvold Diary). According to Einar Rotvold's diary entry, Rotvold went to Staxrud's place where Rotvold "submitted several explanations."[1] The day also included a visit from the engineer Jakob Ellingsen to Rotvold.

Staxrud himself, it seems, was making plans both for the Norwegian rescue expedition's second trip to Treurenberg Bay and for the Schröder-Stranz German and Norwegian expeditioners' transportation home from Spitsbergen; in addition to this,

© The Author(s), under exclusive license to Springer Nature Switzerland AG 2024    245
M. R. Tahan and C. Lüdecke, *Stranded at the Top of the World*,
https://doi.org/10.1007/978-3-031-56288-4_16

**Fig. 16.1** Arve Staxrud strategized and made plans (Photographer/Byline: Arve Staxrud; Source/
Owner: Norwegian Polar Institute)

in what appears to have been true multi-tasking, he was also planning for his scheduled surveying expedition in Spitsbergen on behalf of Norway (Staxrud 25 May
1913) (Fig. 16.1).

The task of arranging transportation for the rescued expedition members, however,
as well as for some of his own expedition members, was particularly challenging. In
a letter written to his surveying partner Adolf Hoel, dated May 25 in Green Harbour,
Staxrud relayed some of the complications he was encountering in his quest to bring
the men home and complained of the difficulties involved in making the arrangements
for the Germans and the Norwegians to be transported to Tromsø. In particular,
finding a ship that would be available, or that would have available space, was a
difficult matter. The letter reveals the frustrations that Staxrud was experiencing. "It
is a heck of a hassle with getting these [three] Germans and my surplus crew back to
Tromsø," he wrote to Hoel.[2] "First I was promised [a] place for them with Munroe
[the *Munroe*], then I get [a] message that there will not be [a] place for [them]."

The *Munroe*, or the *William D. Munroe*, was the ship of the American coal mining
firm, the Arctic Coal Company, that was owned by John Munro Longyear (Dole,
1922, vol. 2: 4, 466).

With the change in availability of the ship *Munroe*, Staxrud next attempted to hire
space on another vessel that would be affordable (Staxrud 25 May 1913). According

to his written correspondence to Hoel, he was met with exorbitant prices, all of which he listed in the letter and stated were "unreasonable,"[3] leading him to what evidently seemed to him the necessary decision that the expeditioners would need to "wait until second Munroe"—the following scheduled sailing of the *Munroe*—unless a more reasonably-priced ship passage that could be afforded were offered. By the end of his six-page letter, however, Staxrud had managed to identify one good possibility for travel. "There is now hope that I will get the Germans and the others home with 'Activ' to Norway," he wrote on the last page. "I have sent [a] letter to [Ernest] Mansfield today."

The *Activ*, on which Staxrud hoped to be able to send the expeditioners home, was the vessel provided by Ernest Mansfield, the director of the English firm, the Northern Exploration Company (Dole, 1922, vol. 2: 196).

While Staxrud worked to organize the transport for the rescued expeditioners to sail back to Norway and Europe, the Norwegians and Germans both felt the pull of home, especially after the tragic events of the Schröder-Stranz expedition.

Rotvold was eagerly awaiting a ship to collect him and his expedition mates so that they could go home to their families (Einar Rotvold Diary). He noted on May 21 that "One has heard that 'Moonroe' [*Munroe*] will be leaving from Tromsø at the end of next week"[4] and exclaimed on May 24 that "We are waiting impatiently for [Ernest] Mansfield's smack, which 17th May left from Tromsø."[5]

Rüdiger, too, waiting in Advent Bay, expressed the longing for home, according to his book passage, as quoted by Nathan Haskell Dole in the Arctic Coal Company's published account:

> "The little mining-settlement, Longyear City, at Advent Bay, was affected during the days and weeks of our stay by a peculiar fever. 'Munroe-fever,' as it is called, is the yearning expectation of the arrival of the first steamer so named. Here also we discovered the same phenomenon as we had observed in ourselves: the yearning for home is not strongest during the winter night, but rather in the transition-time between Winter and Summer. That it would show itself so violently we would not have believed possible. For the majority of the laborers – two hundred had wintered there; among them six women and two children – knocked off work toward the end of May, so as to have nothing to do but wait impatiently for the coming of the steamer.["] (Rüdiger quoted in Dole, 1922, vol. 2: 194–195). [The passage appears to be from Rüdiger's, 1913 book, page 200.]

Rave, also, would later report that they spent the next days and weeks waiting in Advent Bay and that, unfortunately, they were not allowed to travel to Tromsø on the next coal steamer of the Arctic Coal Company (Rave, 1913, 92–93). Almost every day, he stated, new telegrams came for them from home.

## Planning and Strategizing the Second Trek Northward
## to Treurenberg Bay

In Green Harbour, Staxrud's comings and goings during this time were documented, at least in part, by Rotvold (Einar Rotvold Diary), who reported that Staxrud, along with five men, traveled to Bell Sound on May 20 so as "out on the low-headland to erect a seamark,"[6] returned on May 22,[7] and left for a one-day stay at his "Skrøder-cabin"—the expedition's headquarters and depot in Schröder House at Cape Heer— on May 23 with plans to return the following day.[8]

Staxrud was apparently managing many elements, analyzing his just completed rescue trek to Treurenberg Bay, and planning and strategizing for his next foray north to search for the missing Schröder-Stranz expeditioners, as well as coordinating an upcoming topographical and hydrographical surveying expedition on Spitsbergen for Norway (Staxrud 25 May 1913). In his May 25th letter to Hoel, he provided his analysis that he "stayed longer here than estimated"[9] and that "The ice has shown its bad side," describing the impossibility of accessing the fjord via boat, and expressing that "to wear out the dogs with hauling through the valleys from here to Sassen Bay, there is little purpose in," adding that this would also require a (presumably additional) lengthy amount of time in the difficult conditions of the ice and snow. Thus previewing his plans for his second rescue expedition north to find the remaining German expeditioners, during which he planned to return south to Sassenfjorden, Staxrud mentioned that he hoped he would be able to return from the expedition in the beginning of July—a timeframe that, as will be seen later, would prove not possible to maintain. He revisits this timeline later in his letter, stating that, on his way back from the second trek, he hoped to return "in the first part of July" to the Sassen Bay cabin, for which he references Nøis (it is not clear which Nøis), and suggesting to Hoel that he bring the surveying motorboat to look for them there on July 15—the motorboat that Staxrud recommends Hoel use, as one of several vessels, for part of their summer surveying expedition that begins in the first half of July, and for which he outlines the locations, timing, and methodology of usage (possibly this is the same motorboat that Einar Rotvold and Jørgen Jensen, plus some of Staxrud's men, had been working on in Green Harbour [Chap. 14]), as well as outlines the work. "I assume you are able to organize the work thus, that the motorboat can come in there and look for us first time 15 July," he writes to Hoel in his letter. Very importantly, Staxrud stipulates that if he and his group arrive at the Sassen Bay cabin before July 15, then they "can probably struggle ourselves forward to Advent Bay by [our] own effort," thus indicating that, in that event, they would likely continue on their own. Seemingly trying to streamline their rescue efforts and make use of all time and resources for the rescue, and indicating that the motorboat would be working in the vicinity, Staxrud surmises that after the motorboat has been taken to investigate if they have arrived at Sassen Bay, it could possibly be used for a few days by the hydrographer in the surveying work in the area. Staxrud makes it a point to instruct Hoel, however, that the motorboat should not wait for them if the rescuers have not arrived by July 15, but should "go out to the field again immediately," and then could return to the Sassen Bay cabin again on July 30 to look once more to see if the rescue party had arrived there.

As part of his prescribed plan and strategy, Staxrud indicates the proposed routes to follow, including working off of Kap Staratschin. He also advises Hoel on where a harbour can be found for the motorboat, describing an area where he had assigned a spot for a sea marker the previous week while there with a few other men, and apparently mentioning Martin (possibly Pettersen Nøis) in conjunction with it. (Most likely, this is the trip that Rotvold had documented as taking place on May 20, during which, he reported, Staxrud had gone to Bell Sound to place a seamark along the headland.)

Interestingly, in his letter, Staxrud mentions his plan to use his four reindeer (most likely those remaining from the rescue expedition) in the upcoming summer surveying expedition, and to employ one of the Sámi to take care of the reindeer during the topographical and surveying work. He further proposes that the four reindeer be loaned to the "society"—likely the Norwegian Geographical Society—to help in making "the first overland trip from Van Mijen's Bay to Cole Bay [Coal Bay]," and that Hoel have the help of a "locally well-acquainted man" for that inland excursion.[10] Staxrud thus suggests that Hoel, who would be leaving from Tromsø to Spitsbergen, reiterate to the society's contact person Staxrud's proposal to have reindeer and an expert for the overland crossing from Van Mijenfjorden to Colesbukta.

The Sámi whom Staxrud had in mind to employ most likely was Johannes Kemi. This theory is borne out by the existence of a signed contract between Arve Staxrud and Johannes J. Kemi, dated May 31, 1913, in Green Harbour (Staxrud & Kemi 31 May 1913). In the handwritten document, Staxrud agrees to provide Kemi with 5 kroner per day and all necessary accommodations, plus free health care—medical doctor and medications, in return for Kemi's commitment to work as an assistant on Staxrud and Hoel's Spitsbergen expedition, and to work per the direction of Staxrud.[11]

In regard to the four reindeer referenced by Staxrud, most likely these were the one reindeer whom Rotvold had brought with him from Advent Bay to Green Harbour and the three surviving reindeer whom Hilmar Nøis and Johannes Kemi had brought back with them to Green Harbour from their early return from the northward trek to Treurenberg Bay (Chaps. 11 and 14).

In addition to several names listed as the hydrographical and topographical work team members, other proposed workers for the surveying expedition listed in Staxrud's letter (Staxrud 25 May 1913), who were also associated with previous rescue efforts, include "Martin [Pettersen] Nøis" (Pettersen Nøis guarded the rescue expedition depot at Green Harbour and helped the rescued at Advent Bay) and "Einar Tessem" (Tessem was part of the failed rescue effort by the Arctic Coal Company); and there is also a mention of "Ellingsen" (the engineer who was also a member of the rescue expedition).[12] A couple of names originally listed are then crossed out, including Hilmar Nøis, possibly due to the fact that Hilmar Nøis would make the second trip to Treurenberg Bay.

In this letter, Staxrud also congratulates Hoel for the funding recently secured for the topography and surveying work. Most likely this may refer to the same funding for which Hoel had written to the Ministry of Foreign Affairs on April 29, requesting financial support for the topographical and geological survey expedition to Spitsbergen planned to be undertaken with Staxrud during the summer season of 1913 (Hoel 29 April 1913). In his request letter, Hoel had cited the disastrous Schröder-Stranz expedition as a prime example for the necessity to have accurate maps,[13] and thus, by extension, seemed to indicate the importance of financing and conducting this surveying and mapping mission.

Evidently, based on Staxrud's letter to Hoel, Staxrud was possibly planning to segue the return from the rescue mission with the beginning portion of the surveying expedition that would be conducted by Hoel, strategizing how best he could complete the search and rescue thoroughly and then join the topographical and geological work already in progress (Staxrud 25 May 1913). This arrangement, like the July timeframe, also would end up necessarily having to be adjusted, as Staxrud would return much later from the relief expedition, as shall be seen later in this narrative.

## Preparations to Bring the Rescued Germans from Advent Bay to Green Harbour

The day after Staxrud wrote his letter to Hoel, on May 26, he sent a telegram to Hoel from the Green Harbour telegraph station (Staxrud 26 May 1913) referencing the hydrographical survey and notifying Hoel that additional information was contained in his letter.[14] On that same day, while the Norwegian and the German Schröder-Stranz expedition members were still waiting for a ship to take them home via Tromsø, Einar Rotvold and Jørgen Jensen went to Staxrud's "Skrøder-cabin [Schröder House]" (Einar Rotvold Diary), where they "fetched tobacco," while Julius Jensen and the telegraph station manager Olaf Henriksen traveled from Green Harbour to Advent Bay.[15] On the 29th of May, in Green Harbour, Rotvold reported that Mansfield's ship came in that morning at 7:00 am, that it disembarked five men, and that it brought correspondence—"One received quite a bit of mail with it," although the vessel then left again at 12:00 noon, returning to Bell Sound.[16] That diary entry is echoed by Christopher Rave, whose later account states that, during his and Rüdiger's stay in Advent Bay, another coal steamer brought them mail, but also sailed back to Tromsø without them (Rave, 1913, 94).

On the following day, May 30, Einar Rotvold and Jørgen Jensen went to Bell Sound (Einar Rotvold Diary), where they took part in goose hunting and seagull-egg collecting, and would not return to Green Harbour for several days.[17]

It was during this time, at the end of the month, that a transport ship had evidently finally been arranged by Staxrud for transporting the Germans and some of the Norwegians from Spitsbergen to Tromsø.

Staxrud would later summarize these events briefly in his published account, reporting that it was not until the end of May that the journey home could be organized for the six survivors of the Schröder-Stranz expedition (three Germans and three

Norwegians) and for five men from Staxrud's relief expedition (Staxrud, 1914, 42). This was not easy, he stated, for the American coal ships were not allowed to carry any passengers except their own officials. It was only through the concession of Mansfield, the director of the English Northern Exploration Company, that the men could be taken to Norway. (The name Northern Exploration Company appears as "Northern Exploiting Company" in Staxrud's published report.)

And so, on May 30, Staxrud wrote a telegram to Hoel (Staxrud 30 May 1913), in which he informed Hoel that the motorboat was now completed, outfitted with sails, and fit for seagoing, so that it could be used for the hydrographical work, and toward the end of which he announced that he would be "Traveling north tomorrow."[18]

Thus, indeed, sometime after signing his contract with Kemi on May 31, Staxrud left Green Harbour for Advent Bay, accompanied and followed by other members of his rescue expedition, who now made ready to bring the three surviving Germans and the relief expedition's physician from Advent Bay to Green Harbour. Now the first leg of the survivors' journey home could actually begin.

# Notes on Original Material and Unpublished Sources

Einar Rotvold's diary is in the Library Archives at the Norwegian Polar Institute (NPI) in Tromsø. Arve Staxrud letters of correspondence, written from and to Arve Staxrud, are in the Library Archives at the Norwegian Polar Institute (NPI) in Tromsø.

1. E. Rotvold diary, 19 May 1913, (NPI), D00125.
2. A. Staxrud to A. Hoel, letter, 25 May 1913, Norwegian Polar Institute (NPI), 1913_05_25_Brev fra Staxrud til Hoel.
3. A. Staxrud to A. Hoel, letter, 25 May 1913, Norwegian Polar Institute (NPI), 1913_05_25_Brev fra Staxrud til Hoel.
4. E. Rotvold diary, 21 May 1913, (NPI), D00125.
5. E. Rotvold diary, 24 May 1913, (NPI), D00125.
6. E. Rotvold diary, 20 May 1913, (NPI), D00125.
7. E. Rotvold diary, 22 May 1913, (NPI), D00125.
8. E. Rotvold diary, 23 May 1913, (NPI), D00125.
9. A. Staxrud to A. Hoel, letter, 25 May 1913, Norwegian Polar Institute (NPI), 1913_05_25_Brev fra Staxrud til Hoel.
10. A. Staxrud to A. Hoel, letter, 25 May 1913, Norwegian Polar Institute (NPI), 1913_05_25_Brev fra Staxrud til Hoel.
11. A. Staxrud and J. Kemi, contract signed in Green Harbour, 31 May 1913, Norwegian Polar Institute (NPI), Johannes_Kemi_Kontrakt_Green_Harbour_31_mai_1913.
12. A. Staxrud to A. Hoel, letter, 25 May 1913, Norwegian Polar Institute (NPI), 1913_05_25_Brev fra Staxrud til Hoel.
13. A. Hoel to The Royal Ministry of Foreign Affairs, letter, 29 April 1913, Norwegian Polar Institute (NPI), 1913_Brev fra Hoel til Utenriksdepartementet.

14.  A. Staxrud to A. Hoel, telegram, 26 May 1913, Norwegian Polar Institute (NPI), 1913_mai_Telegrams_fra_Staxrud.
15.  E. Rotvold diary, 26 May 1913, (NPI), D00125.
16.  E. Rotvold diary, 29 May 1913, (NPI), D00125.
17.  E. Rotvold diary, 30 May 1913, (NPI), D00125.
18.  A. Staxrud to A. Hoel, telegram, 30 May 1913, Norwegian Polar Institute (NPI), 1913_mai_Telegrams_fra_Staxrud.

## Unpublished Sources and Original Material

Hoel, A., 29 April 1913, Letter from Adolf Hoel to The Royal Ministry of Foreign Affairs, 1913_Brev fra Hoel til Utenriksdepartementet. Library Archives, Norwegian Polar Institute, Tromsø. Received 20 July 2021.

Rotvold, E., Diary, Einar Rotvold's expedition diary written during the Schröder-Stranz expedition of 1913, "Dagbok for August Stenersen og Einar Rotvold, Tromsø: Treurenberg Bay fra 1.januar—7.juni 1913: Schröder-Stranz-ekspedisjonen 1912–13", "Einar Rotvold har skrevet dagboken", ["Diary for August Stenersen and Einar Rotvold, Tromsø: Treurenberg Bay from 1 January—7 June 1913: The Schröder-Stranz Expedition 1912–13", "Einar Rotvold has written the diary"], dated 12 January 1913 through 7 June 1913, D00125. Library Archives, Norwegian Polar Institute, Tromsø, https://brage.npolar.no/npolar-xmlui/handle/11250/2426394. Accessed 24 July 2021.

Staxrud, A., 25 May 1913, Letter from Arve Staxrud to Adolf Hoel, 1913_05_25_Brev fra Staxrud til Hoel. Library Archives, Norwegian Polar Institute, Tromsø. Received 20 July 2021.

Staxrud, A., 26 May 1913 and 30 May 1913 (the latter "delayed due to line fault" and re-sent by the telegraph office on 6 June 1913), Telegrams from Arve Staxrud to Adolf Hoel, 1913_mai_Telegrams_fra_Staxrud. Library Archives, Norwegian Polar Institute, Tromsø. Received 20 July 2021.

Staxrud, A. and Kemi, J., 31 May 1913, Contract between Arve Staxrud and Johannes Kemi signed in Green Harbour, Johannes_Kemi_Kontrakt_Green_Harbour_31_mai_1913. Library Archives, Norwegian Polar Institute, Tromsø. Received 20 July 2021.

## References

Dole, N. H. (1922). *America in Spitsbergen: The Romance of an Arctic Coal-Mine*. In Two Volumes. Boston: Marshall Jones Company. Volume II. Facsimile reprint by Scholar Select.

Nøis, D. (1929). Med kaptein Staxrud på leiting efter Schrøder-Stranz og hans folk [With Captain Staxrud on the search for Schröder-Stranz and his people]. In: *And-Ungen*, July 1929, pp. 4–12. Andenes: Andøyposten. Library Archives, Norwegian Polar Institute, Tromsø. Received 20 July 2021.

Rave, C. (1913). *Tagebuch von der verunglückten Expedition Schröder-Stranz.* Schaffsteins Gründe Bändchen 49. Cöln: Schaffstein.

Rüdiger, H. (1913). *Die Sorge Bay. Aus den Schicksalstagen der Schröder-Stranz-Expedition.* Berlin: Georg Reimer.

Staxrud, A. (1914). Die Staxrudsche Hilfs-Expedition für Schröder-Stranz. In A. Miethe (Ed.), *Die Expedition zur Rettung von Schröder-Stranz und seinen Begleitern - geschildert von ihren Führern Hauptmann A. Staxrud und Dr. K. Wegener* (pp. 1–68). Berlin: Dietrich Reimer.

# Chapter 17
# The Long-Awaited Journey Home

## The Motorboat Ride: Bringing Rave, Rüdiger, and Ritscher from Advent Bay to Green Harbour

As of the end of May 1913, Arve Staxrud had finally been able to arrange a vessel for the surviving German and Norwegian members of the Schröder-Stranz expedition to leave Spitsbergen and sail to Tromsø.

Now, therefore, the time had come to transport the three German survivors—Christopher Rave, Hermann Rüdiger, and Captain Alfred Ritscher—from Advent Bay to Green Harbour, and this would be accomplished by utilizing the motorboat (Hilmar Nøis Diary SVB; Hilmar Nøis Diary NPI; Nøis, 1929, 10–11). (Most likely, this is the motorboat that Einar Rotvold, Jørgen Jensen, and some of Staxrud's men had worked on in Green Harbour [Chap. 14] and that Staxrud had suggested to Adolf Hoel to use to look for the returning rescuers in Sassen Bay [Chap. 16].) The Staxrud expeditioners were thus tasked with collecting the artist and geographer who had been rescued, and the captain who had walked back from his frozen ship, so as to bring them to where they could board a vessel to go home.[1, 2]

According to Daniel Nøis, he and the other Staxrud rescue expeditioners left Green Harbour on June 1, after the ice in the fjord had broken up to an extent, at which time they took the motorboat to Advent Bay, towing two small boats behind them (Nøis, 1929, 10). Hilmar Nøis reported that he and Martin Pettersen (Nøis) drove in with the motorboat to meet the three Germans,[3] and Daniel Nøis further reported that Rave, Rüdiger, and Ritscher were brought to the edge of the ice using horses from the Arctic Coal Company (Nøis, 1929, 10).

Rave recounts the events of meeting the boat, reporting that on June 2, he and Rüdiger were taken by horse-drawn sleigh to the ice edge of the Ice Fjord, where Staxrud was waiting for them in a motorboat (Rave, 1913, 93–94). Rave recalls further that Ritscher, who had been recovering at the sick camp in Advent Bay, was also on board.

© The Author(s), under exclusive license to Springer Nature Switzerland AG 2024
M. R. Tahan and C. Lüdecke, *Stranded at the Top of the World*,
https://doi.org/10.1007/978-3-031-56288-4_17

Witnessing this meeting among the surviving Germans, Daniel Nøis observed what he described to be seemingly tense relations and a heavy silence between the three men, reporting that Rave and Rüdiger did not speak at all to their former captain Ritscher, despite the fact that the three men sat closely together in the boat (Nøis, 1929, 10–11). Daniel Nøis further noted that he and his colleagues attributed this apparently frayed relationship to the sad events that had befallen the Germans during the failed expedition. Hence, Nøis's account seems to indicate a possible bitter feeling and strong tension between the two friends Rave and Rüdiger and the captain Ritscher which he indicates possibly had taken root during the tragic travels of the previous autumn.

As was their goal, the Norwegian rescue expeditioners immediately brought the three Germans from Advent Bay to Green Harbour, taking them to the Heer-headland (Hilmar Nøis Diary NPI), after which, according to Hilmar Nøis, the expeditioners picked up Captain Staxrud, the engineer Jakob Ellingsen, Dr. Bøckmann, and two of the Sámi (it is not clear if Per Hansen or Samuel Klemmetsen or Johannes Kemi had remained in Green Harbour).[4] Hilmar Nøis reports that he and Daniel Nøis proceeded to drive with the sledge dogs along the coast to Green Harbour.

Within five hours of boarding the motorboat, recounts Rave, the German expeditioners sailed to Green Harbour, where they were picked up on the ice by sledge and taken to the telegraph station (Rave, 1913, 94) (Fig. 17.1).

**Fig. 17.1**  "Back in culture." Gathered at the table of the telegraph station in Green Harbour are, from left to right: telegraph station director Henriksen, Dr. Rüdiger, Rave, telegraph station first assistant Danielsen, and Dr. Bøckmann. (*Source* Rüdiger, 1913, 204)

**Fig. 17.2** Draft reindeer helping to transport Rüdiger in Green Harbour (Photographer/Byline: Ernest Mansfield; Source/Owner: Norwegian Polar Institute)

They were warmly received in the house of the telegraph officials. According to Rüdiger, they spent the most beautiful five days of their stay in Spitsbergen there until their departure (Rüdiger, 1913, 205–206).

Summarizing this journey that the Germans made from Advent Bay, Staxrud would later state that, following their initial rescue from Treurenberg Bay, the men were taken further across the ice to Green Harbour (Staxrud, 1914, 42–44) (Fig. 17.2).

## Rotvold and the Jensens Return: Waiting for *Munroe*, and Boarding the *Activ*

In Green Harbour, news of the rescued Germans' arrival reached Einar Rotvold a few days later, when he and Jørgen Jensen returned from Bell Sound, having left there on June 4 and having arrived in Green Harbour in the early morning hours of June 5, just in time to see that the *Munroe* was departing from Spitsbergen for Norway (Einar Rotvold Diary). Rotvold and his mates were desperately waiting for a ship to take them home. "Mansfield's S/S 'Activ' is coming here one of the first days, in order to take us to Tromsø," he wrote in his diary. "The Germans came hereto the other day from Advent Bay."[5] Thus, the consolation of not being on the *Munroe* to Norway was that Ernest Mansfield's *Activ* would soon come and transport them to Tromsø. This diary entry—which uncharacteristically combines the news of several days into one entry, from the dates of May 30 through June 5, also is one of the only two entries during this specific time period that do not end with Rotvold's usual signature sign-off, "God with us. All well." The entry ends with only "All well." The missing words may be a reflection of the frenzied days of travel and anxious time of anticipation. Rotvold was still waiting for a ship on the next day, June 6, reporting that "From Mansfield not heard anything."[6] On the following day, June 7, he wrote in his diary, "We have heard that 'Activ' is in Bell Sound; expecting Mansfield any moment hereto in order to fetch us."[7] The excitement is palpable in this entry, which appears to be Rotvold's final entry in his diary.

Indeed, on the very next day, June 8, the steamship *Activ*, courtesy of the Englishman Ernest Mansfield, director of the English firm the Northern Exploration Company, picked up Rotvold and Rüdiger and all the expeditioners waiting to go home from Spitsbergen (Figs. 17.3, 17.4, and 17.5).

According to Rüdiger, both Rave and Rüdiger were personally taken to Tromsø on that day of June 8 by Mansfield himself, on the *Activ*, together with the other survivors of the Schröder-Stranz expedition, (some of) the men of the Norwegian relief expedition, and others (Rüdiger, 1913, 206–207).

Hilmar Nøis (Hilmar Nøis Diary SVB; Hilmar Nøis Diary NPI) reports that the expeditioners returning on the *Activ* to Tromsø were: the Norwegians Einar Rotvold, Julius Jensen, and Jørgen Jensen; the Germans Christopher Rave, Hermann Rüdiger,

**Fig. 17.3** Rüdiger, in his special sledge, is transported to the departing vessel in Green Harbour, by the three Sámi expeditioners, five other men, the four remaining draft reindeer, and two of the sledge dogs (Photographer/Byline: Ernest Mansfield; Source/Owner: Norwegian Polar Institute)

**Fig. 17.4** Rüdiger is carefully and painstakingly brought on board Ernest Mansfield's ship (Photographer/Byline: Ernest Mansfield; Source/Owner: Norwegian Polar Institute)

**Fig. 17.5** Waving goodbye to the ship as Rüdiger, the other German survivors, the Norwegian survivors, and some of the rescue expeditioners depart from Green Harbour aboard Mansfield's steamship *Activ* (Photographer/Byline: Ernest Mansfield; Source/Owner: Norwegian Polar Institute)

**Fig. 17.6** The three Sámi rescue expeditioners Hansen, Kemi, and Klemmetsen, together with the rescued Germans Rave and Rüdiger, who seem to be happily aboard the ship *Activ*, en route from Green Harbour to Tromsø (Photographer/Byline: Ernest Mansfield; Source/Owner: Norwegian Polar Institute)

and Alfred Ritscher; and the rescuers Dr. P. W. K. Bøckmann, Søren Zachariassen, Per Hansen, Samuel Klemmetsen, and Johannes Kemi[8, 9] (Fig. 17.6).

(Martin Pettersen Nøis is not mentioned by Hilmar Nøis at this time, although he presumably departed as well. Pettersen Nøis had been proposed by Staxrud for the upcoming topographical survey expedition; Kemi was contracted to work on Staxrud's topographical survey expedition [Chap. 16].)

According to Daniel Nøis, five rescue expeditioners—Arve Staxrud, Daniel Nøis, Hilmar Nøis, Johan Nilsen Nøis, and Jakob Ellingsen—remained on Spitsbergen so that they could make the second trek to Treurenberg Bay, and the other rescue expeditioners returned home on the ship provided by Mansfield (Nøis, 1929, 11).

Both Hilmar Nøis (Hilmar Nøis Diary SVB) and Daniel Nøis (Nøis, 1929, 11) report that the *Activ* was a Norwegian ship hired by Mansfield[10] (Fig. 17.7).

And so, rather than having to wait until the next voyage of the Arctic Coal Company's *Munroe*, the expeditioners (now, most likely, very grateful passengers) left on the steamship *Activ*, "which sailed from Green Harbor [sic] just before the drift-ice cut off communications" (Dole, 1922, vol. 2: 196).

The ship arrived in Tromsø on June 10, where Ritscher was immediately taken to the hospital for surgery on his frostbite injuries, while Rüdiger and Rave were taken on to Hamburg aboard the Norwegian tourist steamer *Neptun*, arriving there on June 21 (Rave, 1913, 95; Rüdiger, 1913, 207). Thus, the Schröder-Stranz expedition had come to its provisional end.

**Fig. 17.7** The hospitable Englishman Mansfield (in foreground) on the *Activ* (*Source* Rüdiger, 1913, 206)

With the arrival of the ship *Activ* in Tromsø, Rotvold and the two brothers Jensen had reached their home port (Hilmar Nøis Diary NPI). And the returning rescuers were now back in Norway as well.[11]

The two exhausted and injured German expeditioners, who had taken refuge in the huts near the frozen ship, had now finally been able to board a vessel home to Germany. The injured and weakened German captain, who had walked the length of Spitsbergen, was now finally hospitalized. And the three hard-working Norwegian expeditioners, who had endured the winter months stranded on the *Herzog Ernst*, were now finally home in Tromsø. Only August Stenersen currently remained on the frozen ship at Treurenberg Bay, from which the survivors had escaped.

It was later reported that, at the time of Rüdiger and Rave's arrival in Germany, a judicial inquiry was still taking place in Kristiania (Oslo) to resolve disagreements between the German expedition members and the ship's Norwegian crew aboard the *Herzog Ernst* (Anonym, 1913, 409).

It had been a set of complicated logistics and complex strategies that had been involved in bringing the expeditioners home after they were rescued—a rescue which had been fraught with complications in and of itself. Now, finally, the survivors were home.

## Notes on Original Material and Unpublished Sources

Hilmar Nøis's diary SVB is in the Historical Archive at the Svalbard Museum (SVB) in Svalbard. Hilmar Nøis's diary NPI is in the Library Archives at the Norwegian Polar Institute (NPI) in Tromsø. Einar Rotvold's diary is in the Library Archives at the Norwegian Polar Institute (NPI) in Tromsø.

1. H. Nøis diary SVB, 1913, journal page 57, SVB-AP2.
2. H. Nøis diary NPI, notebook 3, 1913, pdf pages 13–14, (NPI), D-307/D00307_3_0001.
3. H. Nøis diary NPI, notebook 3, 1913, pdf pages 13–14, (NPI), D-307/D00307_3_0001.
4. H. Nøis diary NPI, notebook 3, 1913, pdf page 14, (NPI), D-307/D00307_3_0001.
5. E. Rotvold diary, 30 May 1913—describing events of 30 May through 5 June, (NPI), D00125.
6. E. Rotvold diary, 6 June 1913, (NPI), D00125.
7. E. Rotvold diary, 7 June 1913, (NPI), D00125.
8. H. Nøis diary SVB, 1913, journal page 57, SVB-AP2.
9. H. Nøis diary NPI, notebook 3, 1913, pdf page 14, (NPI), D-307/D00307_3_0001.
10. H. Nøis diary SVB, 1913, journal page 57, SVB-AP2.
11. H. Nøis diary NPI, notebook 3, 1913, pdf page 14, (NPI), D-307/D00307_3_0001.

## Unpublished Sources and Original Material

Nøis, H., Diary NPI, handwritten journal of recollections recounting the events of the 1913 Staxrud expedition, Hilmar Nøis, "Minder og Fortelinger fra fangsttiden på Spitsbergen, Klade 3, 7 Mars 1970" [Hilmar Nøis, "Memories and Stories from the trapping time on Spitsbergen, Notebook 3, 7 March 1970"], dated 7 March 1970, D-307, ("Minder og fortelinger, Nøis, Hilmar, hefte 3"), D00307_3_0001. Library Archives, Norwegian Polar Institute, Tromsø, https://brage.npolar.no/npolar-xmlui/handle/11250/274077. Accessed 20 July 2021.

Nøis, H., Diary SVB, handwritten journal of recollections including an account of the 1913 Staxrud expedition, ("Hilmar Nøis. Fangstmann. 1909–1923, dagbok med erindringer fra fangstlivet") ["Hilmar Nøis. Trapper. 1909–1923, diary with recollections from the trapping life"], not dated (latest year referenced is 1942 in journal pages and 1950 in inserted sheet), SVB-AP2-Hilmar-Nøis-dagbok-1909–1923-compressed. Historical Archive, Svalbard Museum, Svalbard, https://svalbardmuseum.no/no/samlingene/historisk-arkiv/arkivprosjekt-fangst-og-annen-overvintringsvirksomhet-fra-perioden-1910-til-1970/. Accessed 6 June 2021.

Rotvold, E., Diary, Einar Rotvold's expedition diary written during the Schröder-Stranz expedition of 1913, "Dagbok for August Stenersen og Einar Rotvold, Tromsø: Treurenberg Bay fra 1.januar—7.juni 1913: Schröder-Stranz-ekspedisjonen 1912–13", "Einar Rotvold har skrevet dagboken", ["Diary for August Stenersen and Einar Rotvold, Tromsø: Treurenberg Bay from 1 January—7 June 1913: The Schröder-Stranz Expedition 1912–13", "Einar Rotvold has written the diary"], dated 12 January 1913 through 7 June 1913, D00125. Library Archives, Norwegian Polar Institute, Tromsø, https://brage.npolar.no/npolar-xmlui/handle/11250/2426394. Accessed 24 July 2021.

# References

Anonym. (1913). Nord-Polargegenden. *Geographische Zeitschrift*, Vol. XIX, pp. 408–409.

Dole, N. H. (1922). *America in Spitsbergen: The Romance of an Arctic Coal-Mine*. In Two Volumes. Boston: Marshall Jones Company. Volume II. Facsimile reprint by Scholar Select.

Nøis, D. (1929). Med kaptein Staxrud på leiting etter Schrøder-Stranz og hans folk [With Captain Staxrud on the search for Schröder-Stranz and his people]. In: *And-Ungen*, July 1929, pp. 4–12. Andenes: Andøyposten. Library Archives, Norwegian Polar Institute, Tromsø. Received 20 July 2021.

Rave, C. (1913). *Tagebuch von der verunglückten Expedition Schröder-Stranz*. Schaffsteins Gründe Bändchen 49. Cöln: Schaffstein.

Rüdiger, H. (1913). *Die Sorge Bay. Aus den Schicksalstagen der Schröder-Stranz-Expedition*. Berlin: Georg Reimer.

Staxrud, A. (1914). Die Staxrudsche Hilfs-Expedition für Schröder-Stranz. In A. Miethe (ed.), *Die Expedition zur Rettung von Schröder-Stranz und seinen Begleitern - geschildert von ihren Führern Hauptmann A. Staxrud und Dr. K. Wegener* (pp. 1–68). Berlin: Dietrich Reimer.

# Chapter 18
# The Shipwreck of the Theodor Lerner Rescue Expedition: *Loevenskjold* at North Cape, Northeast Land

## The Bear Hunt, the Hunter's Hut, and Reaching the North Cape of Northeast Land

Previously, at the beginning of May 1913, when Arve Staxrud and his expedition had begun their departure south from Treurenberg Bay with the rescued Christopher Rave and Hermann Rüdiger, setting out on the return trek to Advent Bay, the rescue-intent Theodor Lerner and his private relief expedition had begun preparing themselves for their own search for Herbert Schröder-Stranz and his lost comrades (Chaps. 12 and 13). Lerner's men had met with Staxrud and his group on April 30 and then had returned to their ship *Loevenskjold*, at Mossel Bay, where Lerner had remained during the encounter, after which, on May 2, the Staxrud expedition had left Treurenberg Bay.

Back on board the *Loevenskjold* after the Lerner men's return, and following a storm that occurred a few days later, the ship's lookout spotted a polar bear on the coast of Ryss Island (Russøyane), and the catch boats were immediately launched for the hunt (Villinger, 1929, 25–26). Then it was off on skis across the pack ice to the shore. Medical student Bernhard Villinger would later report that he "had received from Lerner, who now distributed his graces like a ruler, the distinction of being allowed to shoot the first bear for my [Villinger's] services in bringing about the expedition" (Villinger, 1929, 26). When a second bear was sighted, "Graetz was allowed to collect his reward for honorary service as expedition treasurer" (Villinger, 1929, 27).

Skier and filmmaker Sepp Allgeier had wanted to film the hunt and had positioned himself "ready to shoot" for it, when suddenly a female polar bear with two cubs approached him in close proximity (Allgeier, 1931, 19). Without a weapon, he stood on a block of ice and had only the tripod legs available for his own defense. Fortunately, she paid little attention to him and continued on her way. Allgeier was greatly reprimanded for not alerting the others to this hunting prey, but he "was glad that the mother bear had not been shot away from her two cubs" (Allgeier, 1931, 19).

M. R. Tahan and C. Lüdecke, *Stranded at the Top of the World*, https://doi.org/10.1007/978-3-031-56288-4_18

The vessel's trip progressed unexpectedly well northward to the ice edge at the North Cape of Northeast Land (Villinger, 1929, 28–29). Lerner initially planned to set up headquarters for two to three months at the hunter's hut, which was four km from the entrance to Beverly Sound, and from there to search the surrounding area on several sledge trips. In the evening, however, masses of ice drifted in, completely enclosing their ship within a very short time, so that Lerner immediately wanted to bring ashore all the things that they needed (Fig. 18.1).

After finding no traces of Schröder-Stranz in the hut, they repaired the most necessary items so that the hut could perhaps serve as an emergency shelter one day. They also established a depot there with a portion of their provisions and equipment. During this time, the *Loevenskjold* was trapped by drift ice and was moored.

**Fig. 18.1** *Loevenskjold* trapped in ice at the North Cape of the Northeast Land. *Source* Allgeier (1931, 19)

## Lerner, Villinger, Graetz, Biehler, and Allgeier: The Search Parties and the Sad Determination

Finally, at 2 a.m. on May 9, Villinger and candidate of medicine Gerhard Graetz left the ship *Loevenskjold* for Rijp Bay, and then Lerner, the physician Dr. Rudolf Biehler, and Allgeier set out the following midnight with heavily loaded sledges pulled by nine dogs headed for Extreme Huk, where they all met up again (Allgeier, 1931, 20–21; Villinger, 1929, 33–39). Along the way, however, Lerner complained of heart trouble and wanted to remain behind at the tent together with Allgeier and to wait until the three medics returned after exploring Rijp Bay. So, Villinger's dog team was joined by Lea, a female Saint Bernard, and Sultan, a male Newfoundland, from Lerner's team, who immediately fought for the lead as a rival to Villinger's Barry. Only the smallest dog, Bobby, mostly stayed out of the fray. It was Pentecost Sunday, May 11, when the three expeditioners parted and left Lerner with his companion near the location where Schröder-Stranz and his comrades had disembarked and had last been seen (Figs. 18.2 and 18.3).

The three skiers searched but found no markings or tracks that would indicate the presence of Schröder-Stranz's group. Thus, with no success, the five men reunited and returned to the ship on May 17 (which was Norway's National Day of Independence, the day that the returned Norwegian expeditioners and rescuers were celebrating in Green Harbour). On the next sledge trip, Biehler and Villinger were to search the

**Fig. 18.2** Lerner's sick camp at Scoresby Island. *Source* Allgeier (1931, 25)

**Fig. 18.3**  Sultan, Lerner's
Newfoundland dog, who
vied for the lead in
Villinger's team. *Source*
Lerner (2005, 290)

area southward to Wahlenberg Bay, checking in on the huts on the islands of Lavö
(Lågøya) and Ryss Island (Russøyane) along the way (Fig. 18.4).

In the meantime, Graetz and Allgeier were to advance eastward to Cape Wrede
beginning on May 27 (Allgeier, 1931, 21–29). Their sledge was pulled by the
dogs Senta, Barry, Bobby, and Fuchs. En route they visited Scoresby Island. Here,

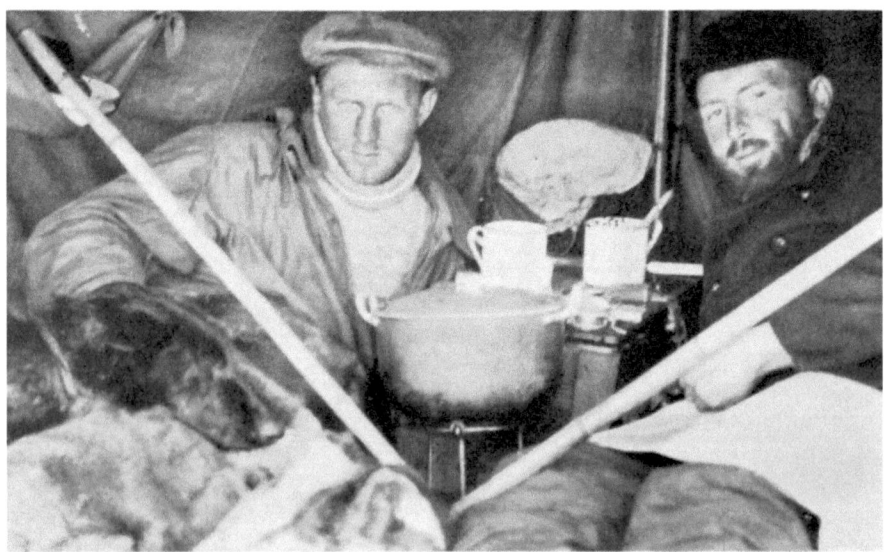

**Fig. 18.4**  Villinger (left) and Biehler (right) in their tent. *Source* Villinger (1929, 33)

**Fig. 18.5** This is how Allgeier's snow-blind Graetz was driven around Scoresby Island. *Source* Allgeier (1931, 26)

Allgeier's "favorite dog Senta proves [proved] to be particularly persistent and tough" (Allgeier, 1931, 24). On the third day of the sledge journey, Graetz became snow-blind and his comrade had no choice but to nurse him and pull him on the sledge in a sleeping bag (Fig. 18.5).

At one point, Allgeier thought that he spotted a sign of the lost Schröder-Stranz group in the distance over the ice, and he kept toward it as fast as he could until the dogs could hardly go on. Unfortunately, the mark turned out to be driftwood that had been straightened by the ice pressures. At another point, the men came near a polar bear, whom they wanted to hunt. To prevent their pulling dogs from interfering, the men tied the dogs to the sledge and left them behind temporarily. After hours of the men unsuccessfully stalking and chasing the bear while they were on skis, their dogs had by that time bitten their way through the bindings and were now running loose, away from the sledge. The expedition dog Bobby was the first dog to appear on the scene and promptly brought the hunt to a close with his barking. He had already several times before bitten through the pulling ropes and now acted as though he were completely innocent. The men at this point had to look for the sledge and the other dogs, which was rather time-consuming. On the way to Cape Platen, piled up press ice made a further advance impossible. So, the expeditioners returned to the ship. Within eight days, they had covered 160 km. Both search groups returned almost simultaneously on June 2 without having achieved anything. (This was on the same day that the surviving Germans were brought by the Staxrud rescuers from Advent Bay to Green Harbour).

A final reconnaissance of Walden Island north of Beverly Sound also proved fruitless (Villinger, 1929, 42). In their 600-km search, Lerner's men could only determine that Schröder-Stranz's sledge expedition had not reached the coast of Northeast Land. It was suspected, according to Villinger, that "a sudden storm had set the ice off the land and driven it north with the sledge detachment. ... Perhaps their boat was crushed by the floes and their equipment was lost, so that they perished helplessly in the Arctic Ocean" (Villinger, 1929, 42). On June 21, the last search was completed, and the Lerner expeditioners retreated back to their ship, empty-handed.

## The Unstoppable Ice, the Loss of the *Loevenskjold*, and the Offering of Thanks

Starting in mid-June, the ice began moving and began to threaten the *Loevenskjold* with huge ice pressures (Villinger, 1929, 42–46; Allgeier, 1931, 30–35). The expedition members now slept at night fully clothed so that they could leave the ship head over heels if necessary. On June 25, Bottolfsen, the young Norwegian captain, discovered a leak in the keel, and in consequence pumping was now performed on a permanent basis. After the shocks from the ice pressures became more and more violent, the men immediately began packing the bare necessities. During a storm on June 28, the pressings became worse and worse, until a thunderous crash indicated the end of the *Loevenskjold*. The oak planks had given way and the engine room was already under water when the catch boats were set on the ice, and the men left the expedition ship with their dogs and their belongings, the sledge-pulling dogs now pulling the small boats that were loaded with the expeditioners' gear, and hauling them from the ship to the coast. Allgeier's greatest concern was saving his exposed films, and then afterward, he began filming the sinking of the *Loevenskjold* (Fig. 18.6).

**Fig. 18.6** Sepp Allgeier filmed the *Loevenskjold* during the ice pressing. *Source* Allgeier (1931, 33)

**Fig. 18.7** Prayer of thanksgiving offered by the shipwrecked men of the *Loevenskjold* after they rescued themselves onto the ice. *Source* Allgeier (1931, 35)

Hastily, the men salvaged as much as they could from the wreck before it sank completely. When there was nothing left to do, they said a prayer of thanksgiving (Fig. 18.7).

Fortunately, the men were able to save themselves in the nearby hunter's hut on Beverly Sound, which had already been prepared for emergencies. Unfortunately, the hard bread had been destroyed by the water inrush, so that they had to share the provisions of dog biscuits and dried fish intended for the dogs.

Thus, the expeditioners were summarily shipwrecked. The aspiring rescuers now needed to be rescued. And so concluded the attempted relief expedition led by Theodor Lerner, which ended in the loss of his expedition ship *Loevenskjold* and resulted in the need for Lerner and his men to be rescued in turn. They had managed to sail from Mossel Bay to North Cape and had covered much distance on ice toward Cape Platen. But they had not been able to find any of the missing men, and in the process, they had lost their ship. The tenuous situation and harsh conditions in Spitsbergen had led to these aspiring rescuers, who were intent upon rescuing members of the Schröder-Stranz expedition, becoming stranded expeditioners themselves. As shall be seen, the Lerner expeditioners would remain on the ice for the next three-and-a-half weeks, until they would once again have a chance—and this time quite fortuitous—meeting with Staxrud.

# References

Allgeier, S. (1931). *Die Jagd nach dem Bild. 18 Jahre Kameramann in Arktis und Hochgebirge.* Stuttgart: Engelhorn.

Lerner, T. (2005). *Polarfahrer: Im Banne der Arktis.* F. Berger (ed.). Zürich: Oesch/Kontrapunkt.

Villinger, B. (1929). *Die Arktis ruft! Mit Hundeschlitten und Kamera durch Spitzbergen und Grönland.* Freiburg i. Br.: Herder.

# Chapter 19
# The Staxrud Rescue Expedition's Second Sledge Journey North Across the Ice to Northeast Land, in Search of the Remaining Seven Schröder-Stranz Members

## The Men and Dogs' Journey from Green Harbour to Advent Bay to Sassen Bay

On the day after the Arve Straxrud rescue expedition members had brought the rescued Christopher Rave and Dr. Hermann Rüdiger of the Schröder-Stranz expedition, and the third surviving German expeditioner Captain Alfred Ritscher, from Advent Bay to Green Harbour to board their ship that would take them home, the remaining rescue expeditioners themselves immediately began their second sledging journey north to Treurenberg Bay and Northeast Land. Their mission was to reach the iced-in *Herzog Ernst* ship once more and conduct another search for the remaining seven men of the Schröder-Stranz expedition who were still missing—Herbert Schröder-Stranz, Richard Schmidt, August Sandleben, Dr. Max Mayr, Dr. Erwin Detmers, Dr. Walter Moeser, and Wilhelm Eberhard. And so, while Rave and Rüdiger and Ritscher, and the other surviving Schröder-Stranz expedition members—Einar Rotvold, Julius Jensen, and Jørgen Jensen—as well as the returning relief expedition members—Dr. P. W. K. Bøckmann, Søren Zachariassen, Per Hansen, Samuel Klemmetsen, Johannes Kemi, and presumably Martin Pettersen Nøis—were anticipating the arrival of the *Activ* in Green Harbour, Staxrud and his now smaller group of rescuers were preparing to set out to accomplish their second trek across Spitsbergen, which they would now undertake.

Thus, on June 3, 1913, the second overland crossing of the ice northward toward Treurenberg Bay began for the expedition leader Captain Arve Staxrud and the four remaining rescuers (Staxrud, 1914, 44). Those four rescuers were Staxrud's deputy Daniel Nøis, Daniel's brother Johan Nilsen Nøis, Daniel's nephew Hilmar Nøis, and the mining engineer Jakob Ellingsen (Nøis, 1929, 11; Hilmar Nøis Diary SVB; Hilmar Nøis Diary NPI), who now made ready to leave Green Harbour for Advent Bay and then on to Treurenberg Bay.[1, 2] The five men were accompanied by four sledge teams of four dogs each (indicating 16 sledge dogs), according to Staxrud (Staxrud, 1914, 44), although one of Hilmar Nøis's accounts reports 15 dogs going from Green Harbour to Advent Bay, and the other account reports 17 dogs going from Advent Bay to Treurenberg Bay.[3, 4] According to Hilmar Nøis, in order to have with them all the provisions and equipment that they would need for a long overland trek north, they had to transfer all the food and supplies from Green Harbour to Advent Bay, with Staxrud, Daniel Nøis, and Ellingsen transporting it all via the motorboat on June 5, while Hilmar Nøis and Johan Nilsen Nøis drove with 15 dogs along the coast to Coal Bay then up Far Valley and down Longyear Valley.[5, 6] (It is possible that the men perhaps picked up the by now formerly pregnant sled dog Lussi in Green Harbour or Advent Bay, if she had had her puppies prior to May, and thus the 15 dogs became 16 or 17, although this is not known for certain at this time. The sled dog Storm presumably was on this trek. The motorboat which was used to transport the equipment and provisions from Green Harbour to Advent Bay most likely is the one that some of Staxrud's expeditioners and Einar Rotvold et al. had worked on, and about which Staxrud had written to Adolf Hoel—the same motorboat that they had used to transport Rave and Rüdiger from Advent Bay to Green Harbour.)

Daniel Nøis reports that the lengthy trip began on June 4 with a drive up Advent Bay Valley that was heavy-going and included a break at a hunter's hut followed by a continuation toward Sassen Bay the next day with two shot reindeer, and an additional day's stay due to there being foggy conditions on the ice sheet (Nøis, 1929, 11) (Fig. 19.1).

Hilmar Nøis further details this first portion of the trek, reporting that, on June 6, the five men left Advent Bay, traveling up the Advent Valley to the Bren crevasse, then down the Esker valley to Sassen Fjord, as the ice sheet cover was broken open along the Ice Fjord beyond Tempel Mountain.[7] After the party left Advent Bay on the evening of the 6th with 17 sledge dogs arranged in three sledge teams and entered the Advent Valley, one of the draft dogs worked their way out of their harness and chased after a wild reindeer who was, as Hilmar Nøis describes in his story, fortuitously a female reindeer heavily pregnant with a calf, and who was then necessarily shot by the men in order to retrieve the desperately-needed sledge dog; the pregnant reindeer's meat was then carried to Sassen Bay and consumed by the men as welcomed top-grade steaks, with the rest of the meat being given to the dogs to eat, as the men did not want to add to the weight of their already-heavy sledges.[8]

## Climbing the Von Post Glacier and Trekking Toward Kvit Glacier: Treacherous Crevasses and a Steep Descent

The trek to search for the seven Schröder-Stranz expedition members this time necessitated driving up the Von Post Glacier—this was the beginning of what would constitute the third lengthwise crossing of Spitsbergen by Hilmar Nøis and the rescue expeditioners.[9] This time, Staxrud and the rescue party would take an easterly route across the inland ice toward Lomme Bay (Staxrud, 1914, 45–48) (Fig. 19.2).

**Fig. 19.1** One of the huts along the way—this hut was located near the river crossing between Advent Valley and De Geer Valley. The inscription on the annexation board states that the hut is owned by the Spitzbergen Coal and Trading Company Limited. Photo date given is 1913, as part of Staxrud's Rescue Expedition. (Photographer/Byline: Arve Staxrud; Source/Owner: Norwegian Polar Institute)

Due to the uncertainty of the ice (that lay below along the fjords) this late into spring, the strategy was to summit the Von Post Glacier located off of Tempel Bay (Tempelfjorden) and follow the soaring mountains toward Treurenberg Bay (Nøis, 1929, 11). Thus, on the following day, the climbing commenced, with the group beginning the ascent to the Von Post Glacier from the bottom of Temple Bay, using all the dogs to pull one sledge at a time up the incline, thus making three trips up the rise until the glacier leveled out and the dogs could be harnessed in their normal formation again, by which time both the dogs and the men were in extreme need of a recuperative rest.[10] But what they had gained in terms of a now flat terrain, they lost in terms of unpredictability, for here the glacier was riddled with a plethora of cavernous crevasses, each deep and yawning and waiting to claim a victim. The danger and potential for death surrounded them, with fathomless crevasses that would cause a long and unfathomable death (Nøis, 1929, 11). At one point, the good draft dog Viggo suddenly fell down into one of the cracks, where he was hanging by only one paw in the harness. Dangling helplessly, the dog remained still and did not move. Immediately, according to Daniel Nøis, Daniel helped Staxrud lower himself down into the crevasse with a rope, and Staxrud proceeded to tie the rope around the dog, after which the two—Staxrud and Viggo—were pulled up out of the crevasse and back to safety (Fig. 19.3).

**Fig. 19.2** Surveying the route and checking the ice—both the men and dogs participate (Photographer/Byline: Arve Staxrud; Source/Owner: Norwegian Polar Institute)

This unnerving scenario played out with a second dog, according to Hilmar Nøis, who describes two dogs hanging in their harness as one man was carefully lowered down into the crevasse in ropes and in turn fastened sturdy ropes around the dogs so that the dogs could be pulled up successfully.[11] After this harrowing event, reports Hilmar Nøis, there was always one man walking in the front of the party with a rope tied onto himself, and one man walking in the back at the end of the rope, in the event that any expeditioner fell down an opening. In addition, they made sure to trek along the mountain side of the glacier, as they understood that the crevasses there were narrower and not as deep, and so they managed to circumvent the steeper cracks (Fig. 19.4). Hilmar describes the crevasses as best detected in bright sunlight, wherein one could spot the snow-covered indentations. The men had also had the foresight to equip themselves with a pole to help detect this type of danger on the glacier should it arise.

Up and down the terrain the group went, reaching, according to Daniel Nøis, an altitude of 1,600 m near Cydenihøiden on June 10, and then plunging down to 700 or 800 m, leaving them with no other choice but to climb up again with the terrain, and leaving the group disenchanted with what Daniel Nøis described as (presumably Adolf Erik) Nordenskiöld's map, which he laments gave an impression of a consistent land altitude of 2,000 feet (Nøis, 1929, 11). Daniel Nøis also reports that the party encountered fog during this tiring stretch of the trek but that it did not stop them

**Fig. 19.3** Traversing and falling into the treacherous ice crevasses—the sledge dogs' and men's lives hanging in the balance (Photographer: Arve Staxrud; Source/Owner: Svalbard Museum)

from making progress (although, based on Hilmar Nøis's account regarding needing sunlight, the fog would have also made it more difficult for them to see crevasses) (Fig. 19.5).

At last, the party found themselves on the ice plateau and proceeded on a northly route traveling west of the Kvit Glacier which, according to Hilmar Nøis, connected with all the glaciers that ran toward Stor Fjord or Hinlopen Strait and passed along many mountain peaks including Svanberg Mountain and Newton Peak.[12] Approaching the area in between Wijde Bay and Lomme Bay, the group suddenly was faced with an extremely steep and near-vertical descent which they negotiated by carefully lowering the dogs and sledges down, one-by-one, using ropes (Nøis, 1929, 11; Hilmar Nøis Diary SVB)[13] (Fig. 19.6).

Both Daniel and Hilmar Nøis report on the severity of the descent, and, in contrast, the evenness of the glacier below, where they landed, which afforded them smooth travel along a wide valley sandwiched on each side by mountains, and where they spent the night at the foot of the mountain along the opposite side (from whence they had come). Thus, the party traveled along the accommodating glacier, gradually ascending as it gently sloped upward, until they reached an altitude of 1,585 feet at the zenith of the glacier and were met by an approaching thick snowfall that forced them to make camp at the height of the ice. The date of their ascent was June 15,

**Fig. 19.4** Example of ascending the glacier close to the side of the mountain (Photographer/Byline: Arve Staxrud; Source/Owner: Norwegian Polar Institute)

and they were now visited by a storm with gales so strong it would soon make its presence known to both the men and the dogs.

## Stormy Weather Atop the Plateau, and the Snowy Way to Lomme Bay

Daniel Nøis remembers taking the usual precautions and securing everything at camp so as to be prepared in case of a storm (Nøis, 1929, 11). This included turning the sledges over so as to give added stability to the tents and tying all the skis together using the foot straps so as to keep them in a safe bundle. He lay himself down to sleep in his full outfit, presumably ready to spring up at any moment, and it was not too long afterward that a violent storm came upon them from the west. Soon after that, Daniel Nøis heard Johan Nilsen Nøis yell out to him for help. Daniel rushed out to see that the tent where Johan and Staxrud lay had been blown down by the storm, leaving the two sleeping partners trapped in the tent. Daniel Nøis and the other men found stones on a mountain pile nearby and later used them to weigh down the edges of the tent. According to Hilmar Nøis (Hilmar Nøis Diary SVB), the men secured the other tent's guy ropes so that it could hold up against this strong gale coming from

**Fig. 19.5** Pulling the sledges along the ice and toward the plateau (Photographer/Byline: Arve Staxrud; Source/Owner: Norwegian Polar Institute)

the west.[14] The sledge dogs, however, fared worse, according to Daniel Nøis, who reported that the hardness of the glacier ice prevented the dogs from digging down low enough into the snow to seek insulation and shelter (Nøis, 1929, 11). Hilmar Nøis, in addition, reports that the dogs were blanketed in a layer of snow so thick that it gave the appearance of an even surface where the individual dogs actually lay, thus concealing their shapes completely.[15] On the following day, the group was greeted with fine weather and a beautiful sight along the glacier, which now presented nice conditions after the passing of the storm (Nøis, 1929, 11). Everything around them was covered in fine snow whose bright whiteness competed with the sunshine.[16] According to Hilmar Nøis, three of the men indulged in refreshing snow-baths, and the dogs arose from their snow-caverns with large clumps of snow frozen in their hair, requiring the men to remove as much of the ice clumps as they could, and the sun to melt away what was still left. With the sunshine warming the air and causing the fine snow still falling to quickly evaporate, and with the mountain-scape around them now completely covered in ice and snow in every direction, the view from the height of the glacier was a dazzling one (Fig. 19.7). The expeditioners broke camp on this lofty glacier and followed its ice plateau, which, as reported by Hilmar Nøis, continued to stretch out evenly for another 50 km before any downward inclination was perceived, at which point a descent was felt. After going down to 800 m, the

**Fig. 19.6** Example of negotiating a sheer ice ridge (Photographer/Byline: Arve Staxrud; Source/ Owner: Norwegian Polar Institute)

incline became increasingly steep. Fortunately, the glacier merged into stable terrain in an area which was not filled with crevasses.

Thus, as Staxrud reported, the expeditioners had crossed the Lomme Bay Glacier and, despite further snowstorms that had interrupted their progress, they reached their destination on June 16 to now search the coast of Northeast Land (Staxrud, 1914, 49) (Figs. 19.8 and 19.9).

## Travails at Treurenberg Bay, Stenersen on the Ship, and the Catch Boat to Cap Foster

The party arrived at the foot of Treurenberg Bay's inland ice on June 17, according to Hilmar Nøis, encountering an excessive amount of thawed snow in that lower area along the fjord ice, which caused them great difficulty in driving the sledges full of their supplies over the ice, and necessitated that they exert tremendous effort and go to great lengths to successfully transport everything with them (Hilmar Nøis Diary SVB). They managed, however, to safely reach the frozen *Herzog Ernst* ship, which lay locked in the ice and standing on the clay situated on the west side of Treurenberg Bay, also called Sorge Bay (which means Mourning Bay)[17] (Fig. 19.10).

**Fig. 19.7**  Sledging across the ice and over the glacier plateau (Photographer/Byline: Arve Staxrud; Source/Owner: Norwegian Polar Institute)

As reported by Daniel Nøis, they found the whole bay covered in a layer of ice that looked like a cover or a sheet of armor and found August Stenersen, who had decided to remain on the *Herzog Ernst* (when they had first reached the ship on April 21, and who had remained behind when they had departed south on May 2) now seemingly somewhat nervous on the frozen ship, although he made no complaints to the men (Nøis, 1929, 11). Hilmar Nøis makes it a point to say that the ice pilot Stenersen had remained on the vessel in the event that any of the missing Schröder-Stranz expedition members returned.[18] Although he was the only man on the ship, Stenersen was by no means alone, for he had been in the company of the two polar bear cubs whom the men had previously captured during the first trek, after they had rescued Christopher Rave and Hermann Rüdiger. Unfortunately, according to Hilmar Nøis, one of the bear cubs by now had died, but the other cub had grown to the size of a large dog. The two cubs had been slated to be taken home aboard the ship. At the ship, both the sled-pulling dogs and the men, who had all just arrived and who had had a difficult journey, now were able to have a much-needed rest.

But the well-deserved rest would not last long, for soon after, the party prepared to search the coast of Northeast Land for any Schröder-Stranz survivors. According to Staxrud, for this purpose, two sledges pulled by five dogs each (indicating 10 dogs), and loaded with sufficient provisions each, were prepared for a depot at Cap Foster

**Fig. 19.8** View of the pass at an altitude of about 1400 m, according to Staxrud (Staxrud, 1914, Tf. 7) (Photographer/Byline: Arve Staxrud; Source/Owner: Norwegian Polar Institute)

**Fig. 19.9** On Lomme Bay glacier (*Source* Staxrud, 1914, Tf. 10)

**Fig. 19.10** Expedition ship *Herzog Ernst* frozen in the ice at Treurenberg Bay, surrounded by men and dogs from Staxrud's rescue expedition (Photographer/Byline: Arve Staxrud; Source/Owner: Norwegian Polar Institute)

(also Cape Foster) at the eastern entrance to Treurenberg Bay (Staxrud, 1914, 50). The men also took a catch boat with them to cross Hinlopen Strait (Fig. 19.11).

Daniel Nøis explains that, over those next few days, the men worked at transporting the fishing boat to Cap Foster in order to possibly be able to row across the strait and thus examine Northeast Land (Nøis, 1929, 11). According to Hilmar Nøis, it was planned that four men (presumably including Staxrud) and by his count six dogs would attempt to traverse Hinlopen Strait in the catch boat that had been driven to the ice edge at the Cap Foster headland, in order to land on Northeast Land and search for Herbert Schröder-Stranz and his three companions.[19] The pack ice, however, was still so thick that they had to wait a few more days (Staxrud, 1914, 50–51). And in addition to the thick ice preventing the boat from reaching the mainland on the other side of the strait, storm after storm raged and presented further difficulties that made the journey not possible at this time (Nøis, 1929, 11). Due to this, they explored the surrounding area, but found no sign of Schröder-Stranz (Staxrud, 1914, 51). They even made a detour to Mossel Bay in hopes of finding the expedition of Theodor Lerner, who Staxrud knew was attempting to conduct his own search for Schröder-Stranz, but this trip, too, brought no further news or evidence. Over the next few days, the rescuers tried again and again to get a little farther ahead by boat, to no avail.

**Fig. 19.11**  Transport of the catch boat to Cap Foster, according to Arve Staxrud (Staxrud, 1914: Tf. 11) (Photographer/Byline: Arve Staxrud; Source/Owner: Norwegian Polar Institute)

## Hampered at Hinlopen Strait

June became July, and additional challenges presented themselves. Daniel Nøis reports that during the first days of July, the party waited 10 days at the strait, along a headland located between Cap Foster and Lomme Bay, but their repeated attempts to travel were deterred by the uncooperative weather and the severity of the ice (Nøis, 1929, 11). On one day, they managed to travel a bit of a distance out along Hinlopen Strait but were forced to return soon after due to the heavy weight of the boat, which carried their supplies, food, equipment, and ten sledge dogs, in addition to themselves.

For 14 days, reports Hilmar Nøis, the four men remained at their camp, in their tents, as the thick drift ice thwarted their attempts to cross.[20]

According to Staxrud, even at the narrowest part of the Hinlopen Strait they waited in vain from July 8 to July 19 for a chance to cross (Staxrud, 1914, 52). "It had now become so late in the year that a search of the Northeast Land had become impossible even by land," concluded Staxrud (Staxrud, 1914, 52).

Thus hampered from crossing the Hinlopen Strait, and frustratingly finding his search for Schröder-Stranz fruitless, Staxrud, with his men and sledge dogs, determined that it was time to return to the ship and free it from the frozen bay.

# Notes on Original Material and Unpublished Sources

Hilmar Nøis's diary SVB is in the Historical Archive at the Svalbard Museum (SVB) in Svalbard. Hilmar Nøis's diary NPI is in the Library Archives at the Norwegian Polar Institute (NPI) in Tromsø.

1. H. Nøis diary SVB, 1913, journal page 57, SVB-AP2.
2. H. Nøis diary NPI, notebook 3, 1913, pdf pages 14–15, (NPI), D-307/D00307_3_0001.
3. H. Nøis diary NPI, notebook 3, 1913, pdf page 15, (NPI), D-307/D00307_3_0001.
4. H. Nøis diary SVB, 1913, journal page 57, SVB-AP2.
5. H. Nøis diary NPI, notebook 3, 1913, pdf pages 14–15, (NPI), D-307/D00307_3_0001.
6. H. Nøis diary SVB, 1913, journal page 57, SVB-AP2.
7. H. Nøis diary NPI, notebook 3, 1913, pdf page 15, (NPI), D-307/D00307_3_0001.
8. H. Nøis diary SVB, 1913, journal pages 57–58, SVB-AP2.
9. H. Nøis diary NPI, notebook 3, 1913, pdf page 15, (NPI), D-307/D00307_3_0001.
10. H. Nøis diary SVB, 1913, journal page 58, SVB-AP2.
11. H. Nøis diary SVB, 1913, journal page 58, SVB-AP2.
12. H. Nøis diary SVB, 1913, journal page 59, SVB-AP2.
13. H. Nøis diary SVB, 1913, journal page 59, SVB-AP2.
14. H. Nøis diary SVB, 1913, journal page 59, SVB-AP2.
15. H. Nøis diary SVB, 1913, journal page 59, SVB-AP2.
16. H. Nøis diary SVB, 1913, journal pages 59–60, SVB-AP2.
17. H. Nøis diary SVB, 1913, journal page 61, SVB-AP2.
18. H. Nøis diary SVB, 1913, journal page 61, SVB-AP2.
19. H. Nøis diary SVB, 1913, journal pages 61–62, SVB-AP2.
20. H. Nøis diary SVB, 1913, journal page 62, SVB-AP2.

# Unpublished Sources and Original Material

Nøis, H., Diary NPI, handwritten journal of recollections recounting the events of the 1913 Staxrud expedition, Hilmar Nøis, "Minder og Fortelinger fra fangsttiden på Spitsbergen, Klade 3, 7 Mars 1970" [Hilmar Nøis, "Memories and Stories from the trapping time on Spitsbergen, Notebook 3, 7 March 1970"], dated 7 March 1970, D-307, ("Minder og fortelinger, Nøis, Hilmar, hefte 3"), D00307_3_0001. Library Archives, Norwegian Polar Institute, Tromsø, https://brage.npolar.no/npolar-xmlui/handle/11250/274077. Accessed 20 July 2021.

Nøis, H., Diary SVB, handwritten journal of recollections including an account of the 1913 Staxrud expedition, ("Hilmar Nøis. Fangstmann. 1909–1923, dagbok med erindringer fra fangstlivet") ["Hilmar Nøis. Trapper. 1909–1923, diary with recollections from the trapping life"], not dated (latest year referenced is 1942 in journal pages and 1950 in inserted sheet), SVB-AP2-Hilmar-Nøis-dagbok-1909–1923-compressed. Historical Archive, Svalbard Museum, Svalbard, https://svalbardmuseum.no/no/samlingene/historisk-arkiv/arkivprosjekt-fangst-og-annen-overvintringsvirksomhet-fra-perioden-1910-til-1970/. Accessed 6 June 2021.

# References

Nøis, D. (1929). Med kaptein Staxrud på leiting efter Schrøder-Stranz og hans folk [With Captain Staxrud on the search for Schröder-Stranz and his people]. In: *And-Ungen*, July 1929, pp. 4–12. Andenes: Andøyposten. Library Archives, Norwegian Polar Institute, Tromsø. Received 20 July 2021.

Staxrud, A. (1914). Die Staxrudsche Hilfs-Expedition für Schröder-Stranz. In A. Miethe (ed.), *Die Expedition zur Rettung von Schröder-Stranz und seinen Begleitern - geschildert von ihren Führern Hauptmann A. Staxrud und Dr. K. Wegener* (pp. 1–68). Berlin: Dietrich Reimer.

# Chapter 20
# Returning to Treurenberg Bay, Releasing the *Herzog Ernst* from the Ice, and Rescuing the Lerner Expedition

## The Freeing of the Ship and the Final Farewell to Knut Stave

On July 19, 1913, having been unsuccessful in their long-lasting and repeated attempts to cross the ice-blocked Hinlopen Strait in order to search Northeast Land for the missing Schröder-Stranz expeditioners, the Arve Staxrud rescue expedition's men and dogs returned to Treurenberg Bay in order to free the German expedition's frozen ship *Herzog Ernst* from the ice (Staxrud, 1914, 52; Nøis, 1929, 11; Hilmar Nøis Diary SVB). It was time to go back to the ship, reported Staxrud, and, according to Hilmar Nøis, it was imperative to free and re-float the vessel while still in the height of summer,[1] as presumably this would be the best time to literally break the ice. And so, as Daniel Nøis reports, the Staxrud expeditioners sailed back from Hinlopen Strait to the ship at Treurenberg Bay.

The rescuers Captain Staxrud, Daniel Nøis, Hilmar Nøis, Johan Nilsen Nøis, Jakob Ellingsen, and four sledge teams of four dogs each –16 sledge dogs according to Staxrud, and 17 sledge dogs according to Hilmar Nøis (Chap. 19)—regrouped at the ship, where the former Schröder-Stranz expeditioner, now Staxrud rescue expeditioner, August Stenersen, had been holding down the fort ever since April 21 (Staxrud, 1914, 52) when the rescue party had arrived, and where he had remained behind when the other rescuers had left on May 2 (Chap. 15).

There at the *Herzog Ernst*, according to Staxrud (Staxrud, 1914, 52–54), Ellingsen and Hilmar Nøis helped to drain the ship and to carry out—using buckets—the water that had penetrated the vessel, as well as helped to finally blast the ship free from the ice in preparation for the return trip (Fig. 20.1).

As reported by Daniel Nøis, who also worked on the ship's motor, the rescuers found that they had to employ a combination of hand-sawing the thin ice in the vicinity of the vessel and dynamiting the extremely thick ice that had accumulated immediately surrounding the ship (Nøis, 1929, 11). They additionally unloaded some of the stone ballast and kerosene, as the ship had by now gone aground on a shallow area of clay due to the spring tides and storms. After this, they used six empty barrels

**Fig. 20.1**   Ice blasting to re-float the *Herzog Ernst* in Treurenberg Bay (*Source* Staxrud, 1914, Tf. 13)

which they placed underneath the stern during low tide and in between the ice, thus tilting the ship somewhat, and worked patiently together to successfully re-float the vessel.

As of July 23, the ship was blasted out of the ice and back afloat, according to Hilmar Nøis (Hilmar Nøis Diary SVB). It was finally released from its winter prison. Everything was organized, and a depot was left behind, near the site, for any possible survivors who might return.[2]

During this time of mid-July, as reported by Hilmar Nøis, the men also carried out the sad duty of burying Knut Stave, the Norwegian cook and crewmember who had died on the ship during its imprisonment in the winter. Stave was interred in a grave near the Swedish House where his body had been resting ever since his death in February.[1] A cross was placed on the grave, and Stave's name, year of birth, and date of death were engraved at the site (Fig. 20.2).

---

[1] While Hilmar Nøis seems to give the date of Knut Stave's death in this passage of his account as being February 12, 1913, the date that Stave died, as recorded in Einar Rotvold's diary, was February 24, 1913 (Chap. 7).

**Fig. 20.2** The Swedish scientific expedition house at Treurenberg Bay, near where, according to Hilmar Nøis (Hilmar Nøis Diary SVB)[2], Knut Stave was buried, and where, according to Daniel Nøis (Nøis, 1929, 10), Hermann Rüdiger and Christopher Rave stayed (*Photographer/Byline* Arve Staxrud; *Source/Owner* Norwegian Polar Institute)

## A Fortunate Second Encounter with the Now-Shipwrecked Lerner Group, and an Unfortunate Determination About Schröder-Stranz

Meanwhile, during the very same time that the *Herzog Ernst* was being freed from the ice, and that the late crewmember Stave was being prepared for burial, Theodor Lerner's attempted rescue expedition began to be on the move again. The shipwrecked group of men and dogs, who had conducted a fruitless search for the Schröder-Stranz expeditioners, and who had been sheltering at the hunter's hut on Beverly Sound and camping on the ice ever since losing their ship the *Loevenskjold* on June 25 at the North Cape of Northeast Land (Chap. 18), now started to travel toward Treurenberg Bay, where they would have yet another fortuitous—and this time life-saving—chance meeting with Staxrud.

Stranded on the ice for nearly a month, it was only on July 22, when the water channel at the horizon was ice-free far out enough, that the 13 men and nine dogs, with three fully loaded catch boats and three packed sledges, including "tents, sleeping bags, instruments, provisions...." (Villinger, 1928, 37), could leave and cross over the sea ice to the south (Villinger, 1929, 42–52; Allgeier, 1931, 30–40). On the way to the open water, they set up an emergency camp on the ice for the night (Fig. 20.3).

They then continued with the boats on the open water. Completely overtired and exhausted, because they had not had any sleep on the way, they reached the east entrance of Treurenberg Bay, where they met Staxrud again on the way from the Swedish House to the *Herzog Ernst*.

According to Staxrud, the fortunate meeting occurred as Staxrud and Johan Nilsen Nøis were returning their fishing boat to the depot at Cap Foster (Staxrud, 1914, 54). At this time, reported Staxrud, he and Johan Nilsen Nøis encountered Lerner's ski party that included Bernhard Villinger, Gerhard Graetz, and Dr. Rudolf Biehler, at the Swedish House, on July 25. Shortly thereafter, Lerner himself arrived with Sepp Allgeier. Now Staxrud and his men heard the entire story about the Lerner expedition's unsuccessful search efforts, the sinking of their ship *Loevenskjold*, and their own rescue ashore. At this point, it was determined that there was probably no chance of finding any of the missing members of the Schröder-Stranz expedition alive. As Staxrud stated (Staxrud, 1914, 55), "In all probability Schröder and his three comrades had already perished on the drift ice that separated him from land while attempting to reach it."

Daniel Nøis, in his account, reiterates the events and outcome of the meeting of July 25 with the Lerner expeditioners, mentioning that a portion of Lerner's crew

**Fig. 20.3** Emergency camp on the ice; left: Allgeier; right: two Norwegians (*Source* Allgeier, 1931, 36)

visited the Staxrud expeditioners on board the *Herzog Ernst* ship, wherein information was relayed regarding the shipwreck of the *Loevenskjold* at the North Cape and
the narrow escape of the expeditioners, the unsuccessful search for Schröder-Stranz
was parlayed to Staxrud, and, after conferring with Lerner himself, an agreement
was arrived at (and presumably an invitation was made) that the entire Lerner group
relocate to the *Herzog Ernst* as soon as possible (Nøis, 1929, 11–12).

Hilmar Nøis reported that on July 25 the ice along the way toward the Hinlopen
Strait was now open, and subsequently a decision was made regarding trying to
travel east on the strait, at the same time that the Staxrud expeditioners saw Lerner's
tents along the shore at Cap Foster and his men rowing toward them in sealing
boats (Hilmar Nøis Diary SVB). The former *Loevenskjold* crew told the Staxrud
expeditioners the tale of the demise of their ship and their own hard-won survival
crossing the hummocked ice while hauling their boats (through the help and efforts
of their dogs) and rowing their boats across the freezing waters.[3] They also shared
the information of not finding any traces of Schröder-Stranz, leaving the impression
that there was little hope that any further tracks would be found.

After that, according to Hilmar Nøis, the Staxrud expeditioners brought all the
Lerner men on board.[4] And, according to Daniel Nøis, the ship was ready on July
26; a storm from the northwest, however, brought the ice back into the strait so that
the expeditioners had to wait (Nøis, 1929, 12).

As summarized by Lerner's two men Allgeier and Villinger in their own accounts,
regarding Staxrud's and Lerner's expeditions coming together, after Lerner's auxil-
iary expedition had searched Northeast Land without success, and Lerner's people
had found no traces of Schröder-Stranz, it was decided that, according to Allgeier,
"all 19 men and 46 dogs"—some of the dogs having been together with Amundsen
(presumably referring to Storm and Lussi) (Allgeier, 1931, 38), and, according to
Villinger, "36 dogs and a young polar bear" (Villinger, 1928, 38; Villinger, 1929, 47),
should all return to Norway on the *Herzog Ernst*. For this, however, the ship had to
be blasted out of the ice first, while Lerner's people could rest from their exertions.
On July 29, they left Treurenberg Bay (Fig. 20.4).

Staxrud had fortuitously encountered and promptly taken the Lerner men to the
now-freed *Herzog Ernst* ship, on which both groups would return. Now the two
expeditions had combined on the freed ship, and all the men and dogs were brought
on board. The ship that had been iced-in all year and that had imprisoned some
of the Schröder-Stranz expeditioners was now the vessel that would transport two
rescue expeditions back south—the rescuers who had successfully rescued stranded
survivors, and the rescuers who had come to need to be rescued themselves. It must be
noted that it is fortunate that Staxrud and his expeditioners were there at that moment
and that place to also rescue the members of the Lerner expedition—13 men and nine
dogs were safely retrieved. With the consensus at this time that there probably was no
further chance of finding any of the missing members of the Schröder-Stranz expedition alive, the Staxrud rescuers and the Lerner expeditioners (perhaps reluctantly)
made ready to set sail southward.

**Fig. 20.4**  *Herzog Ernst* after over-wintering in Treurenberg Bay, in spring 1913 (*Source* Allgeier, 1931, 39)

Herbert Schröder-Stranz, Richard Schmidt, August Sandleben, Dr. Max Mayr, Dr. Erwin Detmers, Dr. Walter Moeser, and Wilhelm Eberhard, would forever remain lost on Spitsbergen.

## Notes on Original Material and Unpublished Sources

Hilmar Nøis's diary SVB is in the Historical Archive at the Svalbard Museum (SVB) in Svalbard.

1. H. Nøis diary SVB, 1913, journal page 62, SVB-AP2.
2. H. Nøis diary SVB, 1913, journal page 62, SVB-AP2.
3. H. Nøis diary SVB, 1913, journal page 62, SVB-AP2.
4. H. Nøis diary SVB, 1913, journal page 63, SVB-AP2.

## Unpublished Sources and Original Material

Nøis, H., Diary SVB, handwritten journal of recollections including an account of the 1913 Staxrud expedition, ("Hilmar Nøis. Fangstmann. 1909–1923, dagbok med erindringer fra fangstlivet") ["Hilmar Nøis. Trapper. 1909–1923, diary with recollections from the trapping life"], not dated (latest year referenced is 1942 in journal pages and 1950 in inserted sheet), SVB-AP2-Hilmar-Nøis-dagbok-1909–1923-compressed. Historical Archive, Svalbard Museum, Svalbard, https://svalbardmuseum.no/no/samlingene/historisk-arkiv/arkivprosjekt-fangst-og-annen-overvintringsvirksomhet-fra-perioden-1910-til-1970/. Accessed 6 June 2021.

## References

Allgeier, S. (1931). *Die Jagd nach dem Bild. 18 Jahre Kameramann in Arktis und Hochgebirge.* Stuttgart: Engelhorn.

Lerner, T. (2005). *Polarfahrer: Im Banne der Arktis.* F. Berger (ed.). Zürich: Oesch/Kontrapunkt.

Nøis, D. (1929). Med kaptein Staxrud på leiting etter Schrøder-Stranz og hans folk [With Captain Staxrud on the search for Schröder-Stranz and his people]. In: *And-Ungen*, July 1929, pp. 4–12. Andenes: Andøyposten. Library Archives, Norwegian Polar Institute, Tromsø. Received 20 July 2021.

Staxrud, A. (1914). Die Staxrudsche Hilfs-Expedition für Schröder-Stranz. In A. Miethe (ed.), *Die Expedition zur Rettung von Schröder-Stranz und seinen Begleitern - geschildert von ihren Führern Hauptmann A. Staxrud und Dr. K. Wegener* (pp. 1–68). Berlin: Dietrich Reimer.

Villinger, B. (1928). Erlebnisse auf der deutschen Arktischen Hilfsexpedition für Schröder-Stranz. In: *Akademischer Skiclub Freiburg im Breisgau 1908–1928.* C.A. Wagner, Freiburg, i. B. pp. 20–43.

Villinger, B. (1929). *Die Arktis ruft! Mit Hundeschlitten und Kamera durch Spitzbergen und Grönland.* Freiburg i. Br.: Herder.

# Chapter 21
# Sailing South Over the Ice-Filled Water, and Sledging South Over the Ice-Covered Land

## Heading West, then East, and then West Again

The Arve Staxrud rescue expedition of six men and 16 or 17 sledge dogs, and the Theodor Lerner *rescued* expedition of 13 men and nine sledge dogs, as well as the captive polar bear cub captured during a previous hunt, were now all joined together aboard the freed *Herzog Ernst* ship, and the vessel was ready for departure (Chap. 20).

On July 30, 1913, when the ice conditions finally appeared favorable for sea travel, the expeditioners began their difficult return journey south from Treurenberg Bay toward Green Harbour, attempting at first to sail via the west coast of Spitsbergen, and indeed trying to pass Verlegenhuken (Forlegenhuk, Verlegenhuk, Verlegen Hook, or Vrangneset),[1] but, finding that they were unable to travel through the ice-filled water there, cruising along the east coast instead, through the Hinlopen Strait, where fog and ice once again impeded their way and caused further delay (Staxrud, 1914, 55–56; Nøis, 1929, 12; Hilmar Nøis Diary SVB). The journey along the Hinlopen Strait had begun well, according to Daniel Nøis, but soon the dense fog enveloped the ship and caused them to stop their advancing along the water, while the copious ice locked them in overnight, so that, even after the fog dissipated on the following day, the thickened ice kept them from proceeding on their way. Based on Hilmar Nøis's report, the compressed and obstructive ice encountered by the ship was located near the Bastian Islands,[1] and, according to Staxrud's report, it was there, on August 1, that the fog and ice effectively prevented the expeditioners from continuing their journey. Thus hindered on the Hinlopen Strait, the ship approached Wilhelm Island (Wilhelmøya)[2] where it would presumably reverse course and sail northward again to attempt to access the western coast. According to Daniel Nøis, on the snow-filled Wilhelm Island were two very large polar bears whom the men interestedly and easily

---

[1] Both Daniel Nøis and Hilmar Nøis, in their first-hand accounts referenced here, name this place Forlegenhuk.

[2] Daniel Nøis, in his first-hand account, names this place Williamsøya (possibly as in King William Island); Staxrud, in his published 1914 account, names it Williamsinsel (as in William's Island).

© The Author(s), under exclusive license to Springer Nature Switzerland AG 2024
M. R. Tahan and C. Lüdecke, *Stranded at the Top of the World*,
https://doi.org/10.1007/978-3-031-56288-4_21

observed strolling along the island with a rather calm demeanor; after a while, one of the bears began to stroll in the direction of the ship, and Daniel Nøis went on shore and killed the animal.

Because it was uncertain if the ship would make quick progress back via the western route, Staxrud decided that he and the non-seafarers of the party would disembark from the vessel here, on the west side of Bismarck Strait (today: Bjørn-sundet—"The Bear Sound"), which separated Wilhelm Island from the mainland of Spitsbergen, and would sledge up the glacier and proceed overland to Advent Bay. The reason for this separation into two parties was that the sailors remaining on board would be much better off if the men not needed for navigation left the ship in time, so that those who stayed behind would have more provisions and could save themselves with the boats in case of an emergency. One of the Lerner expedition's skiers, Dr. Rudolf Biehler, was eager to join the land party.

"Since disagreements had developed between Lerner and his men and agreement could not be reached specifically about our plans," Arve Staxrud later reported, Staxrud drafted a document specifying that ice pilot August Stenersen was transferred to the role of captain, with whom Lerner's crew would sign a contract (Staxrud, 1914, 57–60). Daniel Nøis was to remain aboard with the sledging equipment to lead the party across the inland ice to Advent Bay in case they encountered an emergency situation. In his instructions document, Staxrud also specified the distribution of the salvage money for the *Herzog Ernst*. In addition, the men were instructed to establish five depots en route during the return trip, just in case these were needed.

At the Bismarck Strait, according to Hilmar Nøis (Hilmar Nøis Diary SVB), the men left behind a catch boat that could be used by the land party in the event that they were forced to turn back due to finding any rivers they were unable to cross along the way.[2] Furthermore, according to Daniel Nøis (1929, 12), Staxrud intended to reach Ice Fjord and there, on the northern side of the fjord, provide row boats for the sea party to use to cross the fjord in the event that they were unable to make the sea journey and were thus forced to travel over land.

It seems that Staxrud's reasoning and strategy attempted to take into account all possible outcomes and increase chances for the parties' reaching their destination.

And so, with Captain Staxrud leading the overland sledging party, and ice pilot Stenersen and skipper Daniel Nøis leading the ship's party, the five land party members left the *Herzog Ernst*—these were Staxrud, Johan Nilsen Nøis, Hilmar Nøis, Jakob Ellingsen, and Dr. Biehler—along with 12 dogs pulling three sledges (per Staxrud) or 10 dogs pulling two sledges (per Hilmar)[3], going ashore from the Bismarck Strait (Bjørnsundet) on August 1, and planning to sledge over the ice to Tempel Bay and on to Advent Bay, while the seafarers remaining on board would attempt to sail through the ice at Verlegenhuken and cruise down the western side of Spitsbergen (Staxrud, 1914, 61–62; Nøis, 1929, 12; Hilmar Nøis Diary SVB).

# Navigating the Ice by Land and Sea

According to Arve Staxrud's report (Staxrud, 1914, 61–62), he and his land party traveled all night across the ice and indeed did not set up camp until the morning of August 3 (Fig. 21.1). They came full circle as they returned to their old route.

Hilmar Nøis's report (Hilmar Nøis Diary SVB) lends further detail about the land party's perilous journey, which presented many pools of deep open water to negotiate, including a large glacier river that the sledgers encountered east of Svanberg Mountain and that posed severe problems for crossing.[4] After a long search for a means by which to traverse the water, the men located a snow bridge, and one of the men proceeded to risk his life to test its solidity, in order to see if it would be sturdy enough to bear his weight and not collapse below him into nothingness. The test was performed to ascertain if the men and dogs could safely cross over to the other side. Seeing that it held the weight, the men crossed individually, each making the crossing one at a time, looking at the fluid scene below them and, according to Hilmar, fearfully beholding the blue-green colored water as it cascaded down in the glacier, apparently many meters deep. As for the sledge dogs, they had to run across the snow bridge in teams, at full speed, so as not to fall through the snow and into the water.

**Fig. 21.1** Camp on the inland ice. (*Source* Staxrud, 1914, Tf. 14)

Once on the Plateau, the expeditioners found dry snow that allowed their skis and sledges to slide smoothly over the surface.[5] While descending the Von Post Glacier, however, they discovered a great volume of rocks and sand that created obstructions along the glacier's surface, so much so that, at approximately 500 m away from the Tempel Bay shoreline, they found it necessary to carry everything on their backs—with the exception of the dogs, who by now were running along unharnessed and free (Figs. 21.2 and 21.3).

Upon reaching the shoreline at Tempel Bay (Tempelfjorden) and making camp, Staxrud and his land party prepared to separate their party into two groups (Staxrud, 1914, 61–62; Hilmar Nøis Diary SVB). Three men—Staxrud, Johan Nilsen Nøis, and Dr. Biehler—proceeded to continue overland toward Advent Bay, while the other two men—Hilmar Nøis and Ellingsen—and the sledge dogs, along with the sledges, were left behind in Tempel Bay to wait at the shore.[6] The plan, according to Hilmar Nøis, was that Staxrud's group would arrange for a vessel to retrieve the remaining men and dogs and equipment at Tempel Bay. Therefore, following some much-needed rest, Hilmar Nøis and Ellingsen brought the sledges and gear to the sea's edge and made ready for their group to be picked up by boat.

Simultaneously, Staxrud and his two companions negotiated the icy terrain south in a continuation of the inland ice journey undertaken by the land party—a journey

**Fig. 21.2**  Dog-sledging over the snow. (*Photographer/Byline* Arve Staxrud; *Source/Owner* Norwegian Polar Institute)

**Fig. 21.3** Traveling across the snowy surface with intermittent rocky terrain. (*Photographer/Byline* Arve Staxrud; *Source/Owner* Norwegian Polar Institute)

that would later be described by Daniel Nøis as a quite strenuous trip (Nøis, 1929, 12).

Meanwhile, the *Herzog Ernst*, with Daniel Nøis, August Stenersen, and the joint crew of Staxrud and Lerner expeditioners, was trying to make its way westward (Hilmar Nøis Diary SVB).

The ship headed toward Verlegenhuken immediately following the land party's departure, finding the pack ice now loosened enough that the vessel could skirt the coast, but being promptly stopped again by heavy fog that completely surrounded the ship (Nøis, 1929, 12). After a time, the fog thinned out and the vessel was able to work its way through the ice to beyond Mossel Bay and to the entrance of Wijde Bay. In small increments the *Herzog Ernst* inched its way through the ice toward Grå Huk (Gråhukflya, Gråhuken, or Grey Hook). At last, slowly but steadily, the ship came out of the ice at Norskøyane, and, with the north wind at its back, sailed on to Green Harbour. The journey took four days.

## The End of the Expedition

The *Herzog Ernst* finally arrived in Green Harbour near the Norwegian radio tele-graph station on August 5 (Staxrud, 1914, 62). (Ironically, this was a year to the day from the date that the ship had first set sail under the flag of the Schröder-Stranz expedition in 1912.)

On the following day, August 6, at 14:30 (2:30 p.m.), the land party vanguard consisting of Staxrud, Dr. Biehler, and Johan Nilsen Nøis, reached the coal mine in Advent Bay, as Staxrud would later report.

According to Hilmar Nøis (Hilmar Nøis Diary SVB), on their third day of waiting at the Tempel Bay shoreline, the remaining members of the rescue expe-dition party—Hilmar Nøis, Ellingsen, and the sledge dogs—were picked up by the Tromsø steamship *Victoria* (Hilmar reports this as August 8) and subsequently were informed of the arrival of the *Herzog Ernst* at Green Harbour.[7] To Hilmar, this arrival signaled the end of the rescue expedition and the beginning of the return to Norway.

After the *Victoria* picked up Ellingsen, Hilmar Nøis, the sledge dogs, and the sledges in Temple Bay, it went on to Advent Bay, where the *Victoria* also picked up the recently-arrived Staxrud, Johan Nilsen Nøis, and Biehler group, and brought them all together, with Hilmar Nøis and Ellingsen, from Advent Bay to Green Harbour (Staxrud, 1914, 62).

In Green Harbour, according to Hilmar Nøis (Hilmar Nøis Diary SVB), he and the men disembarked from the *Victoria* and stayed in tents along the shore, as there was not enough room aboard the ship for all the men. There the rescue expeditioners remained to prepare for their journey to Tromsø, Norway.

## The Long View of Longyear—A Sharp Outlook on Schröder-Stranz

On the night of the day that the *Herzog Ernst* arrived in Green Harbour, the American mining businessman and owner of the Arctic Coal Company, John Munro Longyear, happened to be leaving from Advent Bay aboard his ship, *Munroe*, when he spied the beleaguered vessel, the formerly iced-in *Herzog Ernst*, which prompted a withering comment from Longyear in one of his letters to his family, as quoted by Nathan Haskell Dole in his book (Dole, 1922, vol. 2: 229):

> "About ten o'clock in the evening the *Munroe* was loaded with coal and ready to start; so I bade the folks all good-bye and sailed away on a smooth and glassy sea. Just as we left the dock we saw a boat coming around Advent Point and it came into the bay to anchor not far from the dock. It was the *Herzog Ernst*, the auxiliary schooner of the foolish, ill-fated so-called German Scientific Expedition of last year, part of whom wintered at our camp, at our expense. . . . I wondered what kind of dead-beat scheme they may be up to now. ..." (Dole, 1922, vol. 2: 229, quoting from Longyear's letter to his family).

The comment is rather striking, coming from Longyear as a business owner and bystander, whose company had extended assistance as well as demanded reimbursement (Chap. 5), and bearing in mind all that had transpired among the rescuers and the rescued expeditioners over the previous several months—their intense activities, their risk-taking to save lives, and their intricate strategy to travel twice-over to the North and back. Indeed, the appearance of the once-imprisoned *Herzog Ernst* reflected the hard-won efforts of the Staxrud rescue expeditioners and marked the sad conclusion for those Schröder-Stranz expeditioners who remained lost.

Longyear as primarily a business person, it seems, had a propensity to be carefully possessive about his property on "No-man's Land" (Dole, 1922, vol. 2: 223, quoting Longyear's letter to the American Minister to the Court of Russia) and adamant about his coal mining claims on Spitsbergen, as well as somewhat distrustful of others who came to Spitsbergen, especially those on scientific ventures, as proposed in P.J. Capelotti's paper "The Train Has Left the Station" (Capelotti, 2002).

Immediately after spotting the *Herzog Ernst*, Longyear, also by chance at this same time, happened to spot the *Victoria* heading toward what he thought would be Sassen Bay:

> "…We also saw the *Victoria*, an old whaler, now harboring some of the trespassers on our Green Harbor [sic] lands, steaming toward Sassen Bay. There are some Russians on board, who are said to be looking at the claims of our trespassers. Our guess as to this ship and its errand is that they are taking the Russians to Sassen Bay for reindeer-hunting." (Dole, 1922, vol. 2: 229; quoting from Longyear's letter to his family).

Of course, unbeknownst to Longyear at that time on that day of August 5, the *Victoria* would eventually go on to Tempel Bay to pick up Hilmar Nøis's group at the shoreline and then back to Advent Bay to pick up Staxrud's group after they arrived there on August 6.

Longyear himself had just concluded a summer tour of his mining fields domain and was departing from Spitsbergen when he by chance came across the recently freed Schröder-Stranz ship (Dole, 1922, vol. 2: 227–230). His own workers in Advent Bay had made two attempts to rescue the Schröder-Stranz expeditioners in January and February (Chap. 5), and his company and camp had served as hosts to the rescued German survivors Hermann Rüdiger and Christopher Rave (Chaps. 15 and 16). Longyear's last-minute encounter with the *Herzog Ernst* was among the last in a round-up of familiar faces and places he had visited during his summer tour—characters and locations that had interweaved and intersected with the Staxrud rescue expedition's sequence of actions and events.

Several days prior, apparently on August 1, Longyear had arrived with his ship *Munroe* at the Green Harbour telegraph station where he had met the "Government operator" who was soon to resign his post following two years of work in order to take a position at the headquarters in Norway, and who "showed the Americans over the admirably-kept station (Dole, 1922, vol. 2: 224)—presumably this was telegraph station manager Olaf Henriksen, who had documented some of the rescue efforts to find the Schröder-Stranz expedition and who had given Einar Rotvold, Julius Jensen, and Jørgen Jensen shelter and work upon their safe return from the frozen *Herzog*

*Ernst* (Chaps. 6 and 14, for example). (Henriksen is unnamed in this book passage but referenced for this page in the index as "Hendriksen, Mr., Norwegian wireless operator" [Dole, 1922, vol. 2: 456]). According to Dole, Henriksen "was very courteous" and Longyear was duly impressed by the "neatness" of the telegraph station as opposed to what Longyear described in his own words as "the very nasty wreck of the old whaling-station" (Dole, 1922, vol. 2: 224, including quote from Longyear from a letter). Longyear then had proceeded to survey the Ayer and Longyear camp at the mine in Green Harbour, which he called "our Green Harbor [sic] property," and severely criticized those "Norwegians [who] have filed claims four and five deep" as well as those "Russian trespassers" whom he alleged had "selected [property] to try and steal from us" (Dole, 1922, vol. 2: 225, quoting from Longyear's letter for home). At this time, Longyear also happened to see that "An old whaler, the *Victoria*, was at anchor in Green Harbor [sic]" and he made it a point to state that the *Victoria* "was said to have some Russian engineers aboard who are examining some of the properties claimed by some of the trespassers on our lands," to whom he then served notice by issuing a letter of warning to the leader of the group of engineers. Interestingly, and apparently unknown to Longyear at the time, this sighting occurred during the days just prior to the *Victoria*'s retrieving the Staxrud rescue expeditioners at Tempel Bay and Advent Bay.

Soon after, on August 2, Longyear had proceeded to take the telegraph operator Henriksen (now referring to him by last name, although it is spelled Hendriksen in the Dole book) with him on the *Munroe* from Green Harbour to Kings Bay to visit the camp of the Englishman Ernest Mansfield's Northern Exploration Company, as relayed in Longyear's account, according to Dole (1922, vol. 2: 226). It will be remembered that Mansfield's workers had participated in an attempted rescue mission, along with meteorological station director Kurt Wegener and others from the German Geophysical Observatory in Ebeltofthamna at Cross Bay, in February (Chap. 6). From Kings Bay, Longyear had proceeded on the *Munroe* to the western side of Cross Bay to see the "German Scientific Station" and to be shown its weather-monitoring equipment and earthquake-recording seismograph, during which time the station's director advised Longyear that he believed that falling ice from surrounding glaciers was the cause of most of the vibrations detected by the apparatus at the station (Dole, 1922, vol. 2: 227). Though he does not mention Kurt Wegener by name, Longyear referred to the Wegener expedition group's previous rescue efforts, saying that "The director and some of his men were out last winter on a rescue expedition after the members of that foolish so-called German Scientific Expedition which published so much 'guff' in the newspapers for more than a month" (Dole, 1922, vol. 2: 227, quoting from Longyear's letter account). This statement by Longyear, as quoted in Dole's book, is followed by a reference to other pages in Dole's book that describe the Schröder-Stranz expedition, one of which bears the following statement: "The Director of the German Scientific Station [thus presumably Kurt Wegener] told Mr. Longyear the next Summer that the Schröeder-Stranz [sic] party was better supplied with provisions and had a more complete equipment than the rescuers had. Their report of the deaths and hardships were in reality cases of pure 'funk'" (Dole, 1922, vol. 2: 199; Chap. 5). It will be remembered that the Wegener group's failed rescue efforts at times interfered with and outright obstructed Einar Rotvold's, August Stenersen's, and the Jensen brothers' efforts to save themselves (Chap. 10). Dole capped off the description of the visit to Wegener with the following assertion: "Mr. Longyear considered it a great pleasure and privilege to call on people whom they 'did not have to warn off as trespassers'" (Dole, 1922, vol. 2: 227, including quote from Longyear's letter account).

Longyear returned on the *Munroe* to Advent Bay on August 3 (the same day that Staxrud and his land party had reached Tempel Bay), and spent the next two days conducting business and overseeing the work at his mine, as well as observing his workers, prior to departing from Advent Bay (Dole, 1922, vol. 2: 227–229). According to Dole, Longyear wrote "in his diary-letter for his family" (Dole, 1922, vol. 2: 228) the following rather severe and startlingly disturbing description and comment about his labor force:

> "In the afternoon I spent some time watching the crude ways in which the Norwegian workmen do things. It is curious but exasperating when the spectator is paying their wages. If any one shows them a better way they will do it that way as long as they are watched, but they are soon doing the old way again. It seems to be too much trouble to do any thinking. Most of them are just a body, two feet and two hands; nothing else.["] (Dole, 1922, vol. 2: 228, quoting from Longyear's letter to his family)

It will be remembered that some of these Norwegian workers had mounted strong efforts to rescue the Schröder-Stranz expedition at considerable risk to their own persons and personal safety.

On that same night of August 5, after watching his workers, Longyear boarded his ship *Munroe* for his late-night departure from Advent Bay, and, after spying and verbally maligning the *Herzog Ernst*, he proceeded to Green Harbour, where he arrived in the very early hours of the following day, and where the telegraph operator Henriksen came on board the ship (Dole, 1922, vol. 2: 229). Indeed, a handwritten telegraph station staff record (Spitsbergen Radio) shows that Henriksen, after initially arriving August 14, 1911 to work as an assistant, and serving as manager since July 1, 1912, departed for good from the Green Harbour telegraph station on August 6, 1913.[8] Evidently, he left with Longyear on the *Munroe*.

Longyear thus took his leave of Spitsbergen, for the time being, and reached Tromsø on August 9 (Dole, 1922, vol. 2: 230).

But his encounter with the *Herzog Ernst* had left a sharply negative and lasting impression.

## Captain Ritscher Resumes His Command

Another important encounter occurred on this same day that the now ice-free *Herzog Ernst* arrived in Green Harbour.

On that day, August 5, the *Herzog Ernst*'s former captain and now frostbite-injured patient, Alfred Ritscher, currently being treated and hospitalized in the Catholic hospital in Tromsø, after sustaining severe physical loss and frostbite injuries during his walk from Treurenberg Bay, received a special visitor in his hospital room (Nansen, 1914, 1–14; Chap. 1). The young captain lay in his hospital bed, having lost half of his right foot, the large toe of his left foot, and the two leading joints of his small finger on his right hand (Ritscher, 1916, 28). The visitor who had come to see him that day was Fridtjof Nansen, ally of Roald Amundsen, and advisor to Staxrud, who had helped coordinate the formation of the Norwegian rescue mission and who had recommended Staxrud as its leader (Tahan, 2019, 3–4, 6–8; Tahan, 2021, 45–51; Chap. 6). Nansen was in the Arctic capital on other business, and decided to take the time and opportunity to visit the impaired captain of the failed Schröder-Stranz expedition, who told him "of all the difficulties they had met with" and who informed Nansen "that none of them had any previous experience of voyages of this kind" and "None of them had been in the Arctic or in the ice" (Nansen, 1914, 8).

The admired explorer questioned Captain Ritscher about the preparedness—or, rather, lack thereof—of the Schröder-Stranz expedition, querying him about "whether it was really possible, as I [he] had read in the paper, that when Schröder-Stranz, the leader of the expedition, with three others left the ship in the ice to the north of North-East Land to make for that island, the arrangement was that the ship should wait for them at Spitsbergen till as late as December 15?," adding that he "imagined it might be a misprint for September 15" (Nansen, 1914, 9). The convalescing captain replied "No... the arrangement was December 15" (Nansen, 1914, 9, quoting Ritscher's conversation with him). To that reply, Nansen "then asked him if they did not see that this was an impossible date for returning from Spitsbergen, as the sea was covered with ice at that time, and besides it was dark winter with no daylight," to which Ritscher answered "Oh no... it must seem strange to you, but none of us knew anything about it, for, you see, we had no experience; it is only now, when it is over, that I can see how foolish it all was" (Nansen, 1914, 9, including quote from Ritscher's conversation).

Nansen seemed to empathize with the young captain's earnest demeanor, writing that "He made a frank and winning impression, and seemed to be a keen and capable seaman," and reporting that "His one desire now was to get out of bed and go north again to Spitsbergen to fetch back his ship; for that was what he ought to do, there was no question about it, he said" (Nansen, 1914, 9–10). Nansen urged the injured skipper not to hurry back to Spitsbergen to retake his freed ship and thus risk his health. "I thought that might very well be left to others, and that he had better see about getting quite well first, as it was to be hoped he had a long life before him yet," Nansen wrote (Nansen, 1914, 10).

But despite Nansen's urging and efforts to talk the young German out of it, Ritscher traveled to Green Harbour. Lerner had informed Ritscher, while he was still in the hospital in Tromsø, that the *Herzog Ernst* had reached Green Harbour, and Ritscher indeed made the journey to Advent City to meet the vessel (Lerner, 2005, 288–289) (Fig. 21.4).

According to Staxrud, there, Captain Ritscher, having in the meantime recovered to some extent in Tromsø, took over the repatriation of the *Herzog Ernst* (Staxrud, 1914, 62). This action, reported Staxrud, allowed him to pay off the crew, "since

**Fig. 21.4** Ritscher (left) with Lerner aboard the *Herzog Ernst* after it had reached Green Harbour. (*Source* Lerner, 2005, 286)

according to the regulations of the Maritime Professional Association, the ship is liable for wages" (Staxrud, 1914, 67).

And, so, Ritscher resumed command of his ship, so as to part with it. According to Daniel Nøis, the *Herzog Ernst* stood out the Ice Fjord (Isfjorden) bound for Tromsø, approximately two days after it had arrived, and was sold in order to pay for its expenses incurred in relation to the Schröder-Stranz expedition (Nøis, 1929, 12).

## The Safe Return of the Rescued Expeditioners and the Rescuers

All the Schröder-Stranz expedition men who had been found had now returned from the icy grip of Treurenberg Bay. All the Lerner expeditioners had been rescued and returned. And all the Staxrud rescue expeditioners, with the exception of most of the draft reindeer—who had worked hard and toiled to the very end in order to retrieve the rescued men on the first journey north, and who had perished due to severe exhaustion or depletion of food—had now returned safely from the northern ice. The human expeditioners had made the trek there and back safely, twice; the sledge dogs who had transported their fellow rescuers both times had arrived successfully in Green Harbour; and a few of the reindeer who had returned early from the first trek had remained alive.

In analyzing Staxrud's leadership during this rescue mission, and the expeditioners' own execution of the rescue strategy, one must assess the fact that Staxrud had returned from this second expedition journey with the German ship *Herzog Ernst* freed from the ice, with Lerner's German rescue expedition safely rescued after the *Loevenskjold*'s sinking, with his own Norwegian expeditioners safe and unharmed, and with his sledge dogs safe and sound—including the famous South Pole Expedition sled dogs Lussi and Storm, who had safely returned just as Staxrud had promised to Leon and Roald Amundsen prior to his departure, and who had indeed returned with the addition of Lussi's puppies born during the trip. Most of these puppies, as shall be seen later in this narrative (Chap. 22), would prove to be instrumental in future rescue expeditions on Spitsbergen and in the continued saving of lives in the Arctic.

## Notes on Original Material and Unpublished Sources

Hilmar Nøis's diary SVB is in the Historical Archive at the Svalbard Museum (SVB) in Svalbard. The Spitsbergen Radio Personnel List is in the National Archives in Tromsø (received from NPI).

1. H. Nøis diary SVB, 1913, journal page 63, SVB-AP2
2. H. Nøis diary SVB, 1913, journal page 63, SVB-AP2
3. H. Nøis diary SVB, 1913, journal page 63, SVB-AP2
4. H. Nøis diary SVB, 1913, journal pages 63–64, SVB-AP2
5. H. Nøis diary SVB, 1913, journal page 64, SVB-AP2
6. H. Nøis diary SVB, 1913, journal page 64, SVB-AP2
7. H. Nøis diary SVB, 1913, journal page 64, SVB-AP2
8. Spitsbergen Radio Personnel List, 1 July 1911–17 August 1914, National Archives Tromsø (Spi radio065 received from NPI).

## Unpublished Sources and Original Material

Nøis, H., Diary SVB, handwritten journal of recollections including an account of the 1913 Staxrud expedition, ("Hilmar Nøis. Fangstmann. 1909–1923, dagbok med erindringer fra fangstlivet") ["Hilmar Nøis. Trapper. 1909–1923, diary with recollections from the trapping life"], not dated (latest year referenced is 1942 in journal pages and 1950 in inserted sheet), SVB-AP2-Hilmar-Nøis-dagbok-1909–1923-compressed. Historical Archive, Svalbard Museum, Svalbard, https://svalbardmuseum.no/no/samlingene/historisk-arkiv/arkivprosjekt-fangst-og-annen-overvintringsvirksomhet-fra-perioden-1910-til-1970/. Accessed 6 June 2021.

Spitsbergen Radio, Personnel List, "Personalfortegnelse", handwritten Green Harbour telegraph station staff record 1 July 1911–17 August 1914, courtesy of Per Kyrre Reymert, from the National Archives in Tromsø (Arkivverket Statsarkivet i Tromsø), File Spi radio065, received from Petr Masat, Library, Norwegian Polar Institute, Tromsø (Bibliotek, Norsk Polarinstitutt). Received 31 March 2022.

# References

Capelotti, P. J. (2002). The train has left the station: Archaeological formation processes and the abandoned coal mining village of Pyramiden, Svalbard. *International scientific cooperation in the Arctic: Proceedings of a conference devoted to the centenary of the Swedish-Russian Arc-of-Meridian expedition to Spitsbergen and 125 years of V.A. Rusanov's Birth*, ed. Eugene Bouzney (pp. 116–127). Moscow: Scientific World.

Dole, N. H. (1922). *America in Spitsbergen: The romance of an Arctic Coal-Mine*. In Two Volumes. Boston: Marshall Jones Company. Volume II. Facsimile reprint by Scholar Select.

Lerner, T. (2005). *Polarfahrer: Im Banne der Arktis*. F. Berger (Ed.), Zürich: Oesch/ Kontrapunkt.

Nansen, F. (1914). *Through Siberia, the land of the future*, (A. G. Chater, Trans.). New York: Frederick A. Stokes Company; London: William Heinemann. Digitized by Google. https://books.google.com. Accessed 4 August 2014.

Nøis, D. (1929). Med kaptein Staxrud på leiting etter Schrøder-Stranz og hans folk [With Captain Staxrud on the search for Schröder-Stranz and his people]. In *And-Ungen*, July 1929, (pp. 4–12). Andenes: Andøyposten. Library Archives, Norwegian Polar Institute, Tromsø. Received 20 July 2021.

Staxrud, A. (1914). Die Staxrudsche Hilfs-Expedition für Schröder-Stranz. In A. Miethe (Ed.), *Die Expedition zur Rettung von Schröder-Stranz und seinen Begleitern - geschildert von ihren Führern Hauptmann A. Staxrud und Dr. K. Wegener* (pp. 1–68). Berlin: Dietrich Reimer.

Tahan, M. R. (2019). *Roald Amundsen's sled dogs: The sledge dogs who helped discover the South Pole*. Cham: Springer International Publishing.

Tahan, M. R. (2021). *The return of the South Pole sled dogs: With Amundsen's and Mawson's Antarctic expeditions*. Cham: Springer International Publishing.

# Chapter 22
# The Closing of the Expedition, the Going Home of the Rescuers, and the Inception of the Spitsbergen Rescuer Dogs

## The Way Home Again

On August 7, 1913, prior to its departure to Norway, the *Herzog Ernst* was seen at anchor in Green Harbour, observed by visitors to Spitsbergen who had just arrived aboard the mail boat *Sirius* from Tromsø and who took notice of the Herbert Schröder-Stranz ship that had recently arrived from the northern coast of Spitsbergen with the Theodor Lerner expedition members (and some of the Arve Staxrud rescue expedition members) on board (*Fredrikshalds Avis* 15 September 1913, 2). According to an article by Jakob Moe recording the trip and published in the Fredrikshald (Halden) newspaper, the visitors also spied there, in the harbor, the *Victoria*, which was reported to be chartered by a Russian expedition (and which had recently brought Arve Staxrud and his land party rescue expeditioners from Tempel Bay and Advent Bay). The visiting group had anchored here, near the Green Harbour telegraph station, after passing Cape Heer where they had found that a large tent had been erected at the Schröder House. Disembarking at the radio telegraph station, they were informed that Captain Staxrud himself had arrived back in Green Harbour on the previous day and that he was, at the moment, staying in the telegraph station with his people (Figs. 22.1 and 22.2). Adolf Hoel (Staxrud's surveying partner) and his survey expedition group also were there.

Evidently, based on this article and others, the expeditions had all garnered much attention in Norway as well as in Germany. Seven Schröder-Stranz expeditioners had been returned safely, one had sadly died, and valiant efforts had been made to find the seven who remained lost—intensive efforts put forth by the Arve Staxrud, Theodor Lerner, Kurt Wegener, and Arctic Coal Company-sponsored expeditions, with only Staxrud's rescue expedition being fruitful.

In Germany, the person who had worked with Fridtjof Nansen to form a committee that would result in the selection of Arve Staxrud as rescue expedition leader,

© The Author(s), under exclusive license to Springer Nature Switzerland AG 2024
M. R. Tahan and C. Lüdecke, *Stranded at the Top of the World*,
https://doi.org/10.1007/978-3-031-56288-4_22

**Fig. 22.1** Possibly Hilmar Nøis (far left, partial image), then one of the sledge dogs (dark image), Daniel Nøis (holding accordion), Johan Nilsen Nøis with another dog, and Arve Staxrud with one of Nøis's draft dogs. (*Photographer/Byline* Arve Staxrud; *Source/Owner* Norwegian Polar Institute)

Professor Adolf Miethe, evidently agreed with the determination that the chance of finding any of the missing expeditioners still alive was highly unlikely. "Further searches must be considered futile," he wrote (Miethe, 1914, XIV)—it was the final statement with which he finished his book about the relief expeditions of Wegener and Staxrud. One could say that this conclusion literally and metaphorically closed the chapter on the tragic Schröder-Stranz expedition.

Now, shortly after the *Herzog Ernst* was sighted in the harbor at Green Harbour on August 7, the time had come for the closing of the rescue mission that had searched for the stranded and lost German and Norwegian expeditioners. Thus, the ship of the unfortunate Schröder-Stranz expedition, *Herzog Ernst* (Nøis, 1929, 12; Chap. 21), along with the successful Arve Staxrud rescue expedition and the rescued Theodor Lerner expedition, proceeded to return from Spitsbergen to Norway.

On August 16, the participants of the two relief expeditions (with the exception of Staxrud and two others) reached Tromsø, where they were greeted by countless small and large boats and a large crowd of curious onlookers (Villinger, 1929, 52).

Only Dr. Rudolf Biehler of the Lerner expedition and Staxrud himself stayed behind on Spitsbergen for a few more days (Staxrud, 1914, 62). The mining engineer Jakob Ellingsen, too, who had served on the Staxrud expedition, remained on Spitsbergen, as requested, to work with Staxrud's partner Adolf Hoel on the topographical

**Fig. 22.2** Daniel Nøis (left), Johan Nilsen Nøis (center), and most likely Jakob Ellingsen (right), along with four of the sledge-pulling dogs. (*Photographer/Byline* Arve Staxrud; *Source/Owner* Norwegian Polar Institute)

and hydrographical surveying expedition (*Aftenposten* 3 September 1913, evening: 4). He would participate briefly in the survey expedition as assistant geologist (Hoel, 1929, 22).

After arriving in Tromsø in mid-August, the Norwegian rescue expeditioners were later informed that the Lerner expeditioners of the shipwrecked *Loevenskjold* had indeed all arrived home to Germany, and that the Norwegian ice pilot August Stenersen of the *Herzog Ernst* also had arrived home in Tromsø (Hilmar Nøis Diary SVB). According to Hilmar Nøis, in Tromsø, the Norwegian rescuers comprising second-in-command Daniel Nøis, his brother Johan Nilsen Nøis, and their nephew Hilmar Nøis, received payment for the expedition, and made the journey home to Risøyhavn (Risøyhamn) with all their sledge dogs.[1]

The men and dogs indeed had worked exhaustively on the Arctic rescue mission.

## The Sled Dogs Lussi and Storm, the Six New Puppies, and the Spitsbergen Rescuer Dogs

Not all the sled dogs, however, had returned home just yet. During Staxrud's additional days on Spitsbergen, he set into motion a flurry of activity on behalf of the two most famous furry members of the rescue expedition—Lussi and Storm, the sled dogs from Roald Amundsen's South Pole expedition—and the puppies who had been born to Lussi during the rescue expedition. On August 11, from Spitsbergen, Staxrud sent a telegram to Roald Amundsen in Kristiania thanking him for the contribution of his two sled dogs to the rescue expedition, announcing that Lussi had had six puppies during the Arctic rescue mission, and inquiring as to where the two celebrated sled dogs and six new puppies should be sent (Staxrud 11 August 1913). "Thanking heartfelt the loan, Storm [and] Lucie [Lussi]," Staxrud wrote. "Where wishing these plus six puppies brought[?]."[2]

The telegram was received on August 12, and a handwritten notation on the bottom of it, presumably made after Leon Amundsen had conferred with his brother Roald (Fig. 22.3), indicates their preference that the dogs be sent back via Tromsø, Norway: "The dogs wishing preliminary returned Tromsø," the notation reads.

Leon indeed replied to Staxrud with a telegram sent on the following day of August 13 (Amundsen 13 August 1913) bearing those very same words, stating that "The dogs wishing preliminary returned [to] Tromsø" and signed off with his initials "L.A."[3]

At this point, at the Green Harbour Norwegian radio telegraph station, a new telegraph station manager had taken over the reins as a result of Olaf Henriksen's end of tenure and subsequent departure on August 6 (Chap. 21). Øyvind Widding-Danielsen, who had previously worked as an assistant under Henriksen (Chap. 14), had assumed the role of manager on August 4 (Spitsbergen Radio) and begun his one-year service in that position, as appears to be shown in a handwritten telegraph station staff record.[4] At this time, evidently manager Danielsen made a special request of Roald and Leon Amundsen to acquire four of Lussi's puppies for the telegraph station where they would be of use on Spitsbergen (Amundsen 14 August 1913b).

Leon Amundsen promptly replied to telegraph station manager Danielsen on August 14, sending a telegram in which he gave his consent that four of the puppies be kept on Spitsbergen, and in which he asked that the two adults Lussi and Storm as well as the other two puppies—at least one of whom it appears he wanted to be a male—be brought back to Tromsø (Amundsen 14 August 1913a). "Entrust four puppies[,] while Storm [and] Lucie [Lussi] plus two puppies[,] of these at least [one] male puppy[,] wishing [wanted] returned [to] Tromsø."[5] There is also a partially illegible sentence that seems to include a mention of "never onboard Hurtigruten [the public coastal route]"—possibly this was an instruction regarding whether to send the sled dogs via public transport, although this portion is not readable.

Leon Amundsen then took it upon himself on that day of August 14 to write a formal letter to the Foreign Minister of Norway saying that he had received a telegram that very same day from the telegraph station manager in Spitsbergen inquiring about

**Fig. 22.3** Leon Amundsen (center) surrounded by (left to right) Gustav, young Nicolay, and Roald Amundsen (far right). Photographed in Svartskog later in 1918. (Photo slightly cropped from original.). (*Photographer* unidentified; *Source/Owner* National Library of Norway)

buying four puppies, and that he had replied yes and had instructed the manager to return Lussi and Storm and keep four of the puppies at the telegraph station, as he believed those four Polar puppies would help Spitsbergen immensely (Amundsen 14 August 1913b). The letter emphasized the importance of the dogs to Norway and to the Arctic Archipelago:

Mister Foreign Minister,

Today arrived radio-telegram from the manager of the telegraph station at Spitsbergen with enquiry about whether there was opportunity to purchase four of the Polar-dogs, puppies under [from] the South Pole dogs "Storm" and "Lucie" [Lussi], about which I talked with Mister Amundsen a couple of days ago. As I did not succeed in getting to speak with You in time this morning I dared not wait longer with answering the telegram and found I should agree with the request as I asked to have "Storm" and "Lucie" [Lussi] returned to Tromsø in the instance that the government possibly would wish to have a smaller stock of these dogs at disposal for later contingencies. The four puppies will be used at the radio [telegraph] station and could probably be of great benefit—I therefore hope that the Mister Minister finds my course of action correct. My brother is for the moment absent.

Most sincerely

Leon Amundsen[6]

Thus the tale unfolds, through the information contained in these telegrams and letter of correspondence, of the inception and start of the Spitsbergen rescuer dog teams, created through the progeny of the South Pole expedition sled dog Lussi, as a result of Lussi and Storm's participation in the Staxrud Rescue Expedition, and resulting in future search-and-rescue capabilities and safety for the workers, explorers, trappers, hunters, scientists, and sailors in Spitsbergen.

And so it came to pass that four of Lussi's six puppies were transferred from the Staxrud Rescue Expedition to the telegraph station in Green Harbour (Hilmar Nøis Diary SVB; Hilmar Nøis Diary NPI). Hilmar Nøis confirms in his own recollection accounts that, of the six puppies Lussi had, Amundsen kept two puppies—one male and one female—and the telegraph station was given four puppies—three males and one female—who became the first generation of Spitsbergen rescuer dogs, and who would be bred to provide assistance in the event of any future emergencies, shipwrecks, or accidents.[7,8] The goal was for the telegraph station to have a team of 12 to 14 sled dogs to serve as Spitsbergen's official rescuer dogs and to help in any future tragic circumstances such as those which had recently occurred with the Schröder-Stranz expedition or any that might occur at the coal mining companies. These sled dogs came to be called the Spitsbergen Dogs. And Lussi, who along with Storm had comprised two of the rescuers on the Staxrud Rescue Expedition, was the one to provide the progeny for these Spitsbergen rescuer dogs—this special pedigree of dog—who would go on to assist future travel and rescue expeditions and to help all the coal companies on Spitsbergen.

## Leaving Spitsbergen, and Coming Home

After the telegrams were sent and received regarding returning the dogs and granting some of the puppies, thus helping ensure the future of rescue expeditions in Spitsbergen, and during the time that Staxrud, along with Dr. Biehler, remained for a few additional days on Spitsbergen, Staxrud met with his surveying partner Adolf Hoel in Green Harbour (*Aftenposten* 3 September 1913, evening: 4). Most likely it was during this time that Staxrud made logistical arrangements with Hoel to bring Storm and Lussi, along with two of the puppies, back to Norway and to the Amundsens, and that he bestowed upon Hoel the significant responsibility of transporting home safely the four highly valued working canines.

Once Staxrud had sorted out his last remaining tasks and organized things in Green Harbour, he left Spitsbergen on the *Kong Harold* and arrived in Tromsø on August 25 (Staxrud, 1914, 62).

On August 30, Fritz G. Zapffe, Roald Amundsen's great friend and trusted expedition agent who was a pharmacist in Tromsø (Tahan, 2019, 20, 496–497; Tahan, 2021, 232), sent a telegram to Leon Amundsen in Kristiania from Tromsø (Zapffe 30 August 1913) confirming that the sled dogs were scheduled to leave from Spitsbergen on September 6 and informing him that Staxrud would be sending a telegram to Leon.[9]

Staxrud himself arrived in Kristiania (Oslo) on the morning of September 3, returning on the train from Trondhjem (Trondheim) after being away from home for nearly six months (*Aftenposten* 3 September 1913, evening: 4). He gave an interview to the newspaper that very same day, summarizing the results of the rescue expedition on Spitsbergen and confirming the whereabouts of the German and Norwegian survivors and rescuers. According to Staxrud, the Schröder-Stranz expedition survivors (presumably he was referring only to the Germans and not also the Norwegians) were back in Hamburg, with the exception of Captain Alfred Ritscher, who remained hospitalized in Tromsø and who was in good condition but still unable to walk due to sustaining frostbite injuries and subsequent amputations to his feet and finger, as had Hermann Rüdiger also sustained these same types of injuries and surgeries. Lerner's expeditioners had all arrived in Germany, except for the leader Theodor Lerner, who remained in Tromsø. And the members of Staxrud's own rescue expedition (Daniel Nøis, Dr. P. W. K. Bøckmann, Per Hansen, et al.), whom Staxrud at this time counted as 10, also were all back home, with the exception of Ellingsen. (Perhaps Staxrud was counting the Schröder-Stranz-survivor-turned-rescuer August Stenersen as one of the 10; he did not specifically mention the Norwegian survivors Einar Rotvold and Julius and Jørgen Jensen although he did refer to all the Schröder-Stranz survivors being brought back safely). Staxrud made sure to mention that Adolf Hoel and his people would be leaving from Spitsbergen to travel to Tromsø on the mail boat scheduled for one of the next departures.

Indeed, a few days later, on September 6, Adolf Hoel, along with Jakob Ellingsen and some other members of the Norwegian Spitsbergen surveying expedition, left Spitsbergen on the mail boat, departing from Green Harbour and destined for Norway (*Trondhjems Adresseavis* 6 September 1913, 1). Reported in the same newspaper article that announced Hoel's departure was a mention that, at the meteorological station in Ebeltofthamna at Cross Bay, Kurt Wegener and Max Robitzsch (of the attempted rescue expedition) had resigned their posts. Not reported in this news article, however, was the fact that Hoel had brought with him on the mail boat from Spitsbergen the two rescuer sled dogs Storm and Lussi as well as the two puppies—a son and a daughter—born on Spitsbergen who had been selected—out of the six puppies born to Lussi—to return home with their mother to Norway.

The fact of Lussi and Storm's inclusion in the group that sailed from Spitsbergen was reported four days later, after the group's arrival in Tromsø via the mail boat on the night of September 9 (*Morgenbladet* 10 September 1913, evening: 2). In documenting Hoel and his group's arrival in Tromsø, the *Morgenbladet* newspaper

**Fig. 22.4** This photo from Adolf Hoel's 1913 topographical and hydrographical surveying expedition on Spitsbergen—the Norwegian Svalbard Expedition 1913—most likely was taken during his return to Tromsø on the postal ship with the Staxrud rescue expedition sled dogs Lussi (possibly the darker-color dog on the left) and Storm (possibly the lighter-color dog on the right) and Lussi's two puppies (one puppy appears to be visible in the foreground). (*Photographer/Byline* Adolf Hoel; *Source/Owner* Norwegian Polar Institute)

made sure to mention that "Amundsen's 2 dogs 'Lucie' [Lussi] and 'Storm', who had been loaned to Staxrud" were also part of Hoel's expedition group that had arrived safely (Figs. 22.4 and 22.5).

The postal ship that brought Hoel and Lussi and Storm presumably was the *Sirius*, which would make its last run for the season, from Tromsø to Green Harbour, on September 11, reportedly taking with it, on this final round trip of the year, another person to work at the telegraph station as well as two men who would, according to a newspaper article, overwinter on Spitsbergen on assignment from Staxrud (*Aftenposten* 12 September 1913, morning: 1).

Lussi and Storm and the two puppies subsequently arrived back in Kristiania on September 14 (*Morgenbladet* 15 September 1913, evening: 1). They returned there most likely with Hoel, who was reported to have arrived in Kristiania on the very same day along with the engineer Ellingsen and three others from the surveying expedition (*Aftenposten* 15 September 1913, evening: 1). The arrival of Lussi, Storm, and the two puppies generated much attention as the four made their way from "Østbanen" (Østbanestasjonen—East Railway Station) to the premises of the veterinarian

**Fig. 22.5** This photo, also from Adolf Hoel's 1913 topographical and hydrographical surveying expedition on Spitsbergen—the Norwegian Svalbard Expedition 1913—appears to feature possibly the engineer Jakob Ellingsen (seated on the left) with possibly one of Lussi's two puppies (appearing to be playing below him in the foreground) and possibly Storm (seemingly sleeping in the background), most likely en route to Tromsø, sailing back from Spitsbergen on the postal ship. (*Photographer/Byline* Adolf Hoel; *Source/Owner* Norwegian Polar Institute)

"Anker-Nielsen" (Bernt Anker Nielsen), where they would be taken in, presumably to be checked and tended to by the vet (*Morgenbladet* 15 September 1913, evening: 1). This attention paid by spectators along the way must have been reminiscent of the parade that had spontaneously formed upon the arrival of Lussi, Storm, and Obersten from Argentina seven months prior—after they had returned from Antarctica and the South Pole expedition—wherein a gathering of fascinated individuals, young and old, had followed the celebrated sled dogs through the streets from the Kristiania Harbor to the trusted veterinarian's clinic (Tahan, 2019, 609; Tahan, 2021, 279–280). Thus the dogs' arrival from the Arctic also was met with a receptive welcome, nearly a February 10th reception redux.

Now, on this day of September 14, the two additional puppies—"a Son and a Daughter"—garnered attention in their own right (*Morgenbladet* 15 September 1913, evening: 1). The newspaper *Morgenbladet* featured a front-page photograph of the two puppies, which appeared in a September 15 article announcing their arrival with the famed sled dogs Lussi and Storm in Kristiania from Spitsbergen (Fig. 22.6). The puppies were heralded as the son and daughter of Lussi and the South Pole dog

Obersten, and were described as being six months of age, with the daughter looking like her father, and sporting brown fur, and the son looking like his mother, who is described as being black.

This, then, brings into question the age as well as the paternity of the puppies. If they were indeed six months old in mid-September, as the article states, then this would mean that they would have been born in mid-March, which seems unlikely, as Lussi and Storm had left Leon Amundsen and Kristiania on March 5 and were en route to Spitsbergen, arriving on March 6 and March 9 in Trondheim and Tromsø, respectively, and sailing on March 19 toward Spitsbergen via Hammerfest, arriving in Green Harbour on April 3 (see Chaps. 8 and 9). Based on the Amundsens' letters regarding the importance of the surviving Antarctic expedition dogs (Chap. 7), Leon would have had a difficult time parting with a visibly pregnant Lussi, and no mention is made of any birthing event taking place during the journey from Kristiania to

„Obersten"s Barn.

**Fig. 22.6** The *Morgenbladet* article on September 15, 1913, announcing the return of Lussi and Storm from Spitsbergen and featuring a photo of Lussi's two puppies—one male and one female—who had arrived in Kristiania with Lussi and Storm on September 14. (National Library of Norway)

Spitsbergen. It seems more likely that the puppies were born in April, which would make them five months old, or in May, which would make them four months old (see Chaps. 11 and 15)—recall that Hermann Rüdiger had claimed that Lussi "was left pregnant in Advent Bay" (Rüdiger 1913, 187; see Chap. 15). With puppies' weaning normally occurring at five to six weeks,[10] Lussi could have made the second trek north in June if the puppies were born in April (Chap. 19). She had been recommended by Leon for mating with Oversten as early as February (Tahan, 2021, 292–294; Chap. 7), so, with the average gestation period for dogs being approximately 60–65 days,[11] Oversten could be the father if the puppies were born in April. But Lussi was with both Oversten and Storm in March—with Oversten until March 5, and with Storm thereafter, including in mid-March during the journey to Spitsbergen—so, if the puppies were born in May, it seems there would be a possibility that either of them—Oversten or Storm—could be the father. Based on the appearance of the two puppies in the photograph featured in the newspaper article (*Morgenbladet* 15 September 1913, evening: 1), as inspected and analyzed by an experienced Doctor of Veterinary Medicine, the age of the puppies could very well be four months or five months indeed, with a possibility of six months.[12]

Another puzzle to ponder regarding the paternity of the pups is the claims from the different sources. While the *Morgenbladet* newspaper states that Oversten is the father (*Morgenbladet* 15 September 1913, evening: 1), Leon Amundsen's August 14 letter to the Foreign Minister of Norway (Amundsen 14 August 1913b) seemed to indicate that he understood the puppies to be Lussi and Storm's, referring to them as "*hvalper under Sydpolhundene 'Storm' og 'Lucie',*" meaning "puppies under [or from] the South Pole dogs 'Storm' and 'Lucie' [Lussi],"[13] and thus it is implied that his initial thought was that Storm was the father. While in the Kristiania newspaper the spotlight was on Oversten as the father, in Spitsbergen the perception seems to have been that Storm was the father. Hilmar Nøis, in one of his accounts (Hilmar Nøis Diary SVB), would later say that the four puppies who had remained at the telegraph station on Spitsbergen were the product of parents Lussi and Storm.[14] One of his colleagues, Rolf S. Tandberg, would also later write that the puppies came from the couple Lussi and Storm (Tandberg, 1928, 176).

As for the puppies' appearance as described by the newspaper (*Morgenbladet* 15 September 1913, evening: 1), the female puppy's brown coloring may have originated from Oversten, who was large and "reddish-brown" (Amundsen, 1912, vol. 1: 308; Tahan, 2019, 39, 66, 85; Tahan, 2021, 438, 443), and the male's similarity to the mother Lussi who is described in the article as black is close to her dark coloring as described in an advertisement taken out presumably by the Amundsens themselves the following month which stated that "The Fram-dog 'Lussi'" was "Steel-grey" (*Aftenposten* 8 October 1913, morning: 12; Tahan, 2019, 610–611, 621; Tahan, 2021, 322, 327). Tandberg, too, years later, describes Lussi as having been black in color, which is close to the Amundsens' own ad description of Lussi as being steel-grey; and Tandberg describes Storm as having been yellow in color (Tandberg, 1928, 176)—to date, no color description of Storm given by Amundsen has yet been found.

The birth date and paternity aside, both puppies are described by the newspaper as being well advanced for their age, in both appearance and ability (*Morgenbladet*

15 September 1913, evening: 1). The article's description depicts them as strong and stable on their feet, with great stamina and a strength that requires a firm grasp on their leash. It paints a picture of the puppies as being vivacious and aware, and having a good temperament. Not forgotten are their four "siblings," whom the article mentions as remaining on Spitsbergen.

Having safely returned from the Arctic and the Spitsbergen rescue expedition (with the addition of the two puppies), Lussi and Storm were reunited with their fellow South Pole expedition teammate Obersten, and with the Amundsen brothers Roald and Leon, who had awaited them, at Roald Amundsen's home "Uranienborg" in Svartskog, on the Bundefjord (Tahan, 2021, 322–323). There they would go on to do further work.

And, whether they had been fathered by Obersten or Storm, Lussi's four puppies who remained on Spitsbergen became the Spitsbergen rescuers dogs and the first of generations of rescuer dog teams who would continue the legacy of Lussi and Storm.

## The Results of the Staxrud Rescue Expedition on Spitsbergen

Now at home on September 3, and analyzing the outcome of his rescue expedition, Arve Staxrud shared his analysis with the newspaper that interviewed him on his first day back (*Aftenposten* 3 September 1913, evening: 4). In the article, a series of questions about the expedition are answered by Staxrud in what seems to be a candid process. In response to the interviewer's inquiry regarding the ultimate "result of the expedition," Staxrud replied that "the goal" had been entirely met, summarizing the following three points: First, that all (seven) surviving members of the Schröder-Stranz expedition had been returned safe and sound; second, that the manner of death of the three missing members had been determined; and, third, that a conclusion had been arrived at regarding the ship's sledging party comprising Herbert Schröder-Stranz and his three companions August Sandleben, Dr. Max Mayr, and Richard Schmidt, who, it was surmised, could not possibly have reached Northeast Land, but who evidently had perished while attempting to reach the land from the drift ice upon which they had disembarked.

When asked what experiences he had taken away from the rescue mission, Staxrud responded that, of the many significant experiences, principally and primarily was the solving of the question as to whether reindeer would be useful as draft animals in the Arctic regions: "They pulled absolutely excellently both on poor ice and in deep snow," he emphasized, immediately adding "But the 19 dogs, we had with us, also did good work," thus making sure not to omit mention of the sled dogs. It is notable that now, according to this newspaper interview, the sledge dogs, credited by Staxrud for working well on the expedition, numbered at 19. Thus, in Staxrud's eyes, based on a reading of this article, the importance of the reindeer and the dogs very much deserved to be stated, and the significance of their accomplishments was quite notable.

The second huge and significant experience gained by the expedition, as stated by Staxrud in the article, had been the opportunity to geographically reconnoiter and topographically survey several new areas along the way, which hitherto had been entirely unknown, as the expeditioners had made several crossings through completely unfamiliar areas during the entirety of the rescue expedition.

Staxrud concluded the interview with the assertion that there was still much to be answered and considered regarding Spitsbergen and the Spitsbergen expeditions—"pages" of information yet to come, he said, seemingly confidently affirming: "but they will come."

And indeed, following the rescue expedition, Staxrud would go on to prepare for his next topographical survey expedition on Spitsbergen (Hoel, 1933, 346). The following year would find him co-leading yet another surveying expedition on the at that time still not fully explored Arctic archipelago, that was Spitsbergen, and that is today called Svalbard.

## Notes on Original Material and Unpublished Sources

Hilmar Nøis's diary SVB is in the Historical Archive at the Svalbard Museum (SVB) in Svalbard. Hilmar Nøis's diary NPI is in the Library Archives at the Norwegian Polar Institute (NPI) in Tromsø. Arve Staxrud letters of correspondence, written from and to Arve Staxrud, are in the Manuscripts Collection at the National Library of Norway (NB) in Oslo. Roald Amundsen letters of correspondence, written from and to Roald Amundsen and Leon Amundsen, are in the Manuscripts Collection at the National Library of Norway (NB) in Oslo.

1. H. Nøis diary SVB, 1913, journal page 64, SVB-AP2
2. A. Staxrud to R. Amundsen, telegram, 11 August 1913, NB Brevs. 812:1
3. L. Amundsen to A. Staxrud, telegram, 13 August 1913, NB Brevs. 812:3:8
4. Spitsbergen Radio Personnel List, 1 July 1911–17 August 1914, National Archives Tromsø (Spi radio065 received from NPI).
5. L. Amundsen to Ø.W. Danielsen, telegram, 14 August 1913, NB Brevs. 812:3:8
6. L. Amundsen to Foreign Minister of Norway, letter, 14 August 1913, NB Brevs. 812:3:8
7. H. Nøis diary SVB, 1913, journal pages 65 and 87, SVB-AP2
8. H. Nøis diary NPI, notebook 3, 1913, pdf page 39, (NPI), D-307 / D00307_3_0001
9. F. G. Zapffe to L. Amundsen, telegram, 30 August 1913, NB Brevs. 812:1
10. Debbie Archeck DVM, Doctor of Veterinary Medicine, conversation with the author, Vancouver, BC, Canada, 6 March 2023; and written personal communication sent to the author, 18 June 2023.
11. Debbie Archeck DVM, Doctor of Veterinary Medicine, conversation with the author, Vancouver, BC, Canada, 6 March 2023.

12. Debbie Archeck DVM, Doctor of Veterinary Medicine, conversation with the author, during analysis of photograph of Lussi's puppies featured in *Morgenbladet* 15 September 1913 newspaper article "'Obersten's Barn.", Vancouver, BC, Canada, 6 March 2023; and written personal communication sent to the author, 8 March 2023.

13. L. Amundsen to Foreign Minister of Norway, letter, 14 August 1913, NB Brevs. 812:3:8

14. H. Nøis diary SVB, 1913, journal page 87, SVB-AP2

## Unpublished Sources and Original Material

Amundsen, L., 13 August 1913, Telegram from Leon Amundsen to Arve Staxrud, letter copybook NB Brevs. 812:3:8. Manuscripts Collection, National Library of Norway, Oslo. Received 9 June 2021.

Amundsen, L., 14 August 1913a, Telegram from Leon Amundsen to Øyvind Widding-Danielsen, NB Brevs. 812:3:8. Manuscripts Collection, National Library of Norway, Oslo. Received 9 June 2021.

Amundsen, L., 14 August 1913b, Letter from Leon Amundsen to Foreign Minister of Norway, NB Brevs. 812:3:8. Manuscripts Collection, National Library of Norway, Oslo. Received 23 June 2021.

Nøis, H., Diary NPI, handwritten journal of recollections recounting the events of the 1913 Staxrud expedition, Hilmar Nøis, "Minder og Fortelinger fra fangsttiden på Spitsbergen, Klade 3, 7 Mars 1970" [Hilmar Nøis, "Memories and Stories from the trapping time on Spitsbergen, Notebook 3, 7 March 1970"], dated 7 March 1970, D-307, ("Minder og fortelinger, Nøis, Hilmar, hefte 3"), D00307_3_0001. Library Archives, Norwegian Polar Institute, Tromsø, https://brage.npolar.no/npolar-xmlui/handle/11250/274077. Accessed 20 July 2021.

Nøis, H., Diary SVB, handwritten journal of recollections including an account of the 1913 Staxrud expedition, ("Hilmar Nøis. Fangstmann. 1909–1923, dagbok med erindringer fra fangstlivet" ["Hilmar Nøis. Trapper. 1909–1923, diary with recollections from the trapping life"], not dated (latest year referenced is 1942 in journal pages and 1950 in inserted sheet), SVB-AP2-Hilmar-Nøis-dagbok-1909–1923-compressed. Historical Archive, Svalbard Museum, Svalbard, https://svalbardmuseum.no/no/samlingene/historisk-arkiv/arkivprosjekt-fangst-og-annen-overvintringsvirksomhet-fra-perioden-1910-til-1970/. Accessed 6 June 2021.

Spitsbergen Radio, Personnel List, "Personalfortegnelse", handwritten Green Harbour telegraph station staff record 1 July 1911–17 August 1914, courtesy of Per Kyrre Reymert, from the National Archives in Tromsø (Arkivverket Statsarkivet i Tromsø), File Spi radio065, received from Petr Masat, Library, Norwegian Polar Institute, Tromsø (Bibliotek, Norsk Polarinstitutt). Received 31 March 2022.

Staxrud, A., 11 August 1913, Telegram from Arve Staxrud to Roald Amundsen, NB Brevs. 812:1. National Library of Norway, Oslo, online archives, https://www.nb.no. Retrieved 6 July 2021.

Zapffe, F. G., 30 August 1913, Telegram from Fritz G. Zapffe to Leon Amundsen, NB Brevs. 812:1. Manuscripts Collection, National Library of Norway, Oslo. Received 22 June 2021.

# References

Aftenposten. (1913, September 3). Kaptein Staxrud hjemme igjen. Udtaler sig til "Aftenposten" om redningsexpeditionen. [Captain Staxrud back home again. Speaks to "Aftenposten" about the rescue expedition.]. *Aftenposten* (Kristiania [Oslo]), 3 September 1913, evening edition, p. 4. National Library of Norway, Oslo, online archives. https://www.nb.no. Retrieved 4 April 2022.

Aftenposten. (1913, September 12). Postskøitens sidste tur til Spitsbergen. [The postal ship's last trip to Spitsbergen.]. *Aftenposten* (Kristiania [Oslo]), 12 September 1913, morning edition, p. 1. National Library of Norway, Oslo, online archives. https://www.nb.no. Retrieved 4 April 2022.

Aftenposten. (1913, September 15). Hjem fra Spitsbergen. Universitetsstipendiat Hoel. *Aftenposten* (Kristiania [Oslo]), 15 September 1913, evening edition, p. 1. National Library of Norway, Oslo, online archives. https://www.nb.no. Retrieved 2 April 2022.

Aftenposten. (1913, October 8). Lussi. *Aftenposten* (Kristiania [Oslo]), 8 October 1913, morning edition, p. 12. Reference archives, National Library of Norway, Oslo, online archives. www.nb.no. Author accessed 28 August 2012.

Amundsen, R. (1912). *The South Pole: An account of the Norwegian Antarctic expedition in the "Fram", 1910–1912*, 2 vols (A. G. Chater, Trans.). London: John Murray.

Fredrikshalds Avis. (1913, September 15). Til Spitsbergen. Dagboksnotiser og smaa reiseindtryk fra en tur sommeren 1913. *Fredrikshalds Avis* (Fredrikshald [Halden]), 15 September 1913, p. 2). Article by Jakob Moe. National Library of Norway, Oslo, online archives. https://www.nb.no. Retrieved 4 April 2022.

Hoel, A. (ed.). (1929). *Resultater av de Norske Statsunderstøttede Spitsbergenekspeditioner (Skrifter om Svalbard og Ishavet), Bind I, Nr. 1, Adolf Hoel: The Norwegian Svalbard Expeditions 1906–1926*. Utgitt på Den Norske Stats bekostning ved Spitsbergenkomiteen. Oslo: (I Kommisjon Hos) Jacob Dybwad. https://brage.npolar.no/npolar-xmlui/bitstream/handle/11250/173618/Skrifter001.pdf?sequence=1&isAllowed=y Accessed 19 September 2023.

Hoel, A. (1933). Arve Staxrud. *Norsk Geografisk Tidsskrift [Norwegian Journal of Geography]*, 4(6), 345–346 https://doi.org/10.1080/00291953308622018

Morgenbladet. (1913, September 10). Hjem fra Spitsbergen. [Home from Spitsbergen.]. *Morgenbladet* (Kristiania [Oslo]), 10 September 1913, Evening edition, p. 2. National Library of Norway, Oslo, online archives. https://www.nb.no. Retrieved 4 April 2022.

Morgenbladet. (1913, September 15). "Obersten"s Barn. [Children of "Obersten".]. *Morgenbladet* (Kristiania [Oslo]), 15 September 1913, evening edition, p. 1. Digital Archive, National Library of Norway, Oslo, received 30 March 2022; National Library of Norway, Oslo, online archives. https://www.nb.no. Retrieved 2 April 2022.

Nøis, D. (1929). Med kaptein Staxrud på leiting efter Schröder-Stranz og hans folk [With Captain Staxrud on the search for Schröder-Stranz and his people]. In *And-Ungen*, July 1929 (pp. 4–12). Andenes: Andøyposten. Library Archives, Norwegian Polar Institute, Tromsø. Received 20 July 2021.

Staxrud, A. (1914). Die Staxrudsche Hilfs-Expedition für Schröder-Stranz. In A. Miethe (Ed.), *Die Expedition zur Rettung von Schröder-Stranz und seinen Begleitern - geschildert von ihren Führern Hauptmann A. Staxrud und Dr. K. Wegener* (pp. 1–68). Berlin: Dietrich Reimer.

Tahan, M. R. (2019). *Roald Amundsen's sled dogs: The sledge dogs who helped discover the South Pole*. Cham: Springer International Publishing.

Tahan, M. R. (2021). *The return of the South Pole sled dogs: With Amundsen's and Mawson's Antarctic expeditions*. Cham: Springer International Publishing.

Tandberg, R. S. (1928). *Med hundespann på eftersøkning efter "Italia"-folkene*. Særtrykk av *Norsk Geografisk Tidsskrift*, Bind II, Hefte 3–4, 1928. Oslo: A. W. Brøggers Boktrykkeri A/S. Digital Archive, National Library of Norway, Oslo, received 8 June 2021. Also: Tandberg, R. S. (1928/1929). Med hundespann på eftersøkning efter "Italia"-folkene. In *Norsk Geografisk Tidsskrift*, Bind II –1928/1929, pp. 176–214. O. Holtedahl (ed.), 1929. Oslo: A. W. Brøggers Boktrykkeri A/S. National Library of Norway, Oslo, online archives. https://www.nb.no. Retrieved 6 February 2022 and 23 June 2023.

Trondhjems Adresseavis. (1913, September 6). Spitsbergen. Den norske ekspedition paa hjemtur. [Spitsbergen. The Norwegian expedition on its (return) trip home.]. *Trondhjems Adresseavis* (Trondheim), 6 September 1913, p. 1. National Library of Norway, Oslo, online archives. https://www.nb.no. Retrieved 5 April 2022.

Villinger, B. (1929). *Die Arktis ruft! Mit Hundeschlitten und Kamera durch Spitzbergen und Grönland*. Freiburg i. Br.: Herder.

# Part IV
# Progeny and Legacy

# Chapter 23
# Epilogue: Ensuing Expeditions, Subsequent Searches, and the Sled Dogs of Spitsbergen

## Honors and Heroes

Many ensuing events occurred following the conclusion of the Arve Staxrud Rescue Expedition that were a direct result of the Staxrud expedition as well as the Herbert Schröder-Stranz expedition.

Due to the experience with Schröder-Stranz's badly prepared expedition to Spitsbergen, a special committee was recommended in Germany to review further expedition plans (Anonym, 1914). The objective was that such a disaster should never happen again.

It was reported that, for his work on the rescue expedition to save the Schröder-Stranz expeditioners, Staxrud was awarded by the King of Saxony the Saxe-Ernestine House Order Commander's Cross First Class medal instituted in 1833 by the Saxon dukes of Altenburg, Coburg-Gotha, and Meiningen (Stenersen, 1988).

Staxrud continued his topographical and geographical survey work on Spitsbergen, co-leading his next expedition there in 1914 and again engaging in surveying expedition work on the archipelago in 1919 and 1920 (Hoel, 1933, 346). He went on to write and speak extensively about Spitsbergen and Svalbard, publishing his work as articles in newspapers and sending update letters of correspondence to Fridtjof Nansen (Staxrud 23 May 1919; Staxrud 22 October 1920; Staxrud 26 May 1925) expressing gratitude and appreciation to his mentor[1] and stating his approaches[2] and opinions[3] through speeches, reports, and articles.

In 1923, Staxrud informed Nansen that one of the Nansen sledges used for the rescue mission in search of the Schröder-Stranz expedition would be exhibited at the Ski Museum at Frognersætren (Frognerseteren), thus now effectively becoming an artifact of history (Staxrud 19 October 1923). Staxrud shared with Nansen the wording for the exhibition installation and catalogue, which would proclaim that the sledge had covered approximately 1500 km of distance during the Staxrud rescue expedition, traveling over inland ice that at times presented very challenging conditions for driving, including rocky and rough terrain, and yet still working well.[4]

© The Author(s), under exclusive license to Springer Nature Switzerland AG 2024    327
M. R. Tahan and C. Lüdecke, *Stranded at the Top of the World*,
https://doi.org/10.1007/978-3-031-56288-4_23

The sledge dogs who, together with the draft reindeer, had pulled the sledges during the rescue operation, also were part of the fabric of Spitsbergen's recent history as well as instrumental factors of its near future. The South Pole Expedition sled dogs Lussi and Storm were reunited with the South Pole dog Obersten at Roald Amundsen's home and went on to further work and adventures (Tahan, 2021). Lussi's four puppies who were left at Green Harbour, from her days working on the Staxrud rescue mission with fellow Antarctic expedition dog Storm, became the first generation of Spitsbergen Dogs who would help companies and expeditions conduct search-and-rescue missions on these Arctic islands (Hilmar Nøis Diary SVB a; Hilmar Nøis Diary NPI a) and who in the following years were dispersed from the Green Harbour telegraph station to all the coal mining companies on Spitsbergen; this included the mining company Store Norske when it started in 1916, which thus also benefitted from the descendants of the sled dogs of Roald Amundsen.[5] (Store Norske Spitsbergen Kulkompani A/S was the Norwegian-owned coal mining company that purchased the American-owned Arctic Coal Company in 1916 [Dole, 1922, vol. 2: 419–425]). The dogs' progeny became the Spitsbergen Rescuer Dogs,[6] a group of teams of sled dogs whose mission was to carry out life-saving expeditions across the Arctic archipelago throughout the coming decades.

Spitsbergen eventually became Svalbard: The Svalbard Treaty was signed in 1920 and came into force in 1925, giving Norway sovereignty over the entire archipelago (Berg, 2020). The new name encompassed the same region which was now considered West Spitsbergen and the other islands.

When the Italian aviator and explorer Umberto Nobile's airship *Italia* crashed on the ice in the north of Spitsbergen in late May 1928, it was Lussi's grandchildren and descendants, the Spitsbergen Rescuer Dogs, who traveled with Hilmar Nøis and Rolf S. Tandberg (ca. 1901–1978) in June 1928 to search for Nobile and his crew who might have survived the accident of the downed airship (Tandberg, 1928, 176). As reported by Tandberg, a team of 10 Greenland dogs—who, according to Tandberg, descended from the pair of dogs Lussi and Storm—worked as part of a complex, official, and international rescue effort comprising officers and personnel, sled dogs, ships, and airplanes. The trekking draft dogs came from Store Norske Spitsbergen Kulkompani A/S (Hilmar Nøis Diary NPI b) and, according to Hilmar Nøis, would spend 52 days with Nøis and Tandberg seeking Nobile's expeditioners.[7]

As documented in detail by Tandberg in a special journal article, the sled dogs were requested by the Norwegian government to work with the sledge driver and the local expert on an urgent rescue mission to find and secure any Nobile survivors (Tandberg, 1928, 176–214). They boarded the ship *Hobby* in Advent Bay on June 3 and circumnavigated the western, northern, and eastern coasts of Spitsbergen, conducting sledge drives on the ice during the journey, from Wahlenberg Bay and North Cape (near Beverly Sound) to as far as Scoresby Island, Cape Platen, Cape Bruun, and Esmarkøya, just before Cape Leigh Smith, and also providing essential assistance and support for the individual active airplane pilots Hjalmar Riiser-Larsen and Finn Lützow-Holm, including rescuing Lützow-Holm when he landed without fuel or food in Mossel Bay, for which Tandberg, Nøis, and the dogs made a 20 km drive from Verlegenhuk to Polhem to reach the stranded Lützow-Holm while hauling a filled petrol tank and provisions to allow Lützow-Holm to take off again and while simultaneously giving Riiser-Larsen a ride on the sledge to Lützow-Holm's location. Despite difficult conditions, dangerous terrain, and extreme weather, the sled dogs also allowed Tandberg and Nøis to establish depots for the survivors and searchers alike, pulling a load, intended for the depot-laying mission, that weighed approximately 300 kg, as well as on another occasion allowed the two men to search for the three individuals who had left Nobile's group, pulling a load of approximately 350 kg while navigating the far reaches of the islands and the north coast of Northeast Land. On the search excursion alone, the sledging party covered nearly 300 km over 13 days. The two men and 10 dogs finally returned in late July with the *Braganza*, with Nøis disembarking in Kings Bay, and Tandberg remaining with the dogs en route to Advent Bay and Green Harbour (Fig. 23.1).

Tandberg's entire dog team, Tandberg made it a point to say, returned safely to Advent Bay and Store Norske. Unfortunately, the other sledge team from Barentsburg (adjacent to Green Harbour), under the command of the Italian Alpine captain (Gennaro) Sora, with his Dutch colleague (Sjef) van Dongen—both of whom, according to Tandberg, had ventured out farther beyond the ice than had been advised, reaching as far northeast as Foynøya (Foyn Island)—was completely decimated. Sora and van Dongen had taken their dog sled team out on the risky ice—a move, upon discovering it, that Tandberg seems to have questioned. Tandberg, Nøis, and their dog sled team participated in the search for Captain Sora and van Dongen, who

**Fig. 23.1** Hilmar Nøis on board the Russian ship *Krassin* (also *Krasin*) during the rescue expedition to find Umberto Nobile's *Italia* survivors in 1928. (*Photographer/Byline* Adolf Hoel. *Source/Owner* Norwegian Polar Institute)

themselves had to be rescued by aircraft after they were seen by the ship *Krassin* (also *Krasin*). Tandberg reported that, on his way back toward Green Harbour with van Dongen, van Dongen himself stated that two of the dogs had had to be killed in order to be eaten as food, five of the dogs—most likely as a result of their critical conditions—had expired, and the remaining two dogs were left behind on Foynøya, along with the sledge and evidently the other items. The airplanes had landed on the water to pick them up, and, as the ice was now gathering together, Sora and van Dongen had to be quickly rescued. The indication, then, from Tandberg's account, is that, having been placed in a tenuous position, Sora's dogs had not fared well.

Tandberg and Nøis, as expressly relayed by Tandberg, had worked with their dog team to provide support, sustenance, and safety during and as part of their search-and-rescue efforts. The sled dog team working with Tandberg and Nøis included the dogs King, Jumbo, Pan, Boy, Abild, Fox, and Fliks (whose name Tandberg also evidently spelled as Flix), as well as three other dogs whose names Tandberg does not seem to disclose in his account, all of whom pulled and performed excellently and exhibited admirable commitment throughout the operation, as described by Tandberg. The Greenland sled dogs who worked with him were strong and superb, he reported. These sled dogs, affirmed Tandberg, were among Lussi's (and, according to Tandberg, Storm's) descendants who had by now been dispersed to all constructed sites and critical places throughout Svalbard.

Nobile and seven of his 16-man crew, as well as his Fox Terrier Titina, were found and rescued during the vast operation. The other eight people from the Nobile expedition sadly perished, most of them instantly lost following the crash. Roald Amundsen himself disappeared during his own attempted mission to rescue Nobile, on June 18, 1928, never to be seen again (Tahan, 2021, 448–450). Search efforts then turned their sights on Amundsen, and much of the rescue operation, including the aviation pilots Riiser-Larsen and Lützow-Holm as well as the ships *Hobby* and *Braganza*, set about searching for Amundsen and his crew (Tandberg, 1928, 196–212). Thus, in an affiliated role, some of the Spitsbergen Dogs were also among those teams supporting the search for Amundsen after the disappearance of the Latham 47 airboat in which he and his five-person crew were flying to desperately reach and save the Nobile expedition. Hilmar Nøis (Hilmar Nøis Diary SVB b) reported that the *Hobby*, after its rescue efforts with him and the sled dogs to find the Nobile expeditioners, went on to search for Amundsen and his lost Latham expedition members as part of the massive search effort incorporating ships and aircraft (and sled dog teams).[8] Tandberg reported that Riiser-Larsen and Lützow-Holm, who had been supported by the dog sledging excursions, flew from their base to find Amundsen (Tandberg, 1928, 196–212). Roald Amundsen's long-time Polar exploration colleague Oscar Wisting, who had been the sledge driver for Obersten's team in Antarctica, and who had ultimately taken Obersten home with him, also engaged in the search for Amundsen (Tahan, 2021, 55, 324, 449; Tahan, 2019, 349). It seems to be a certain sense of completed historical connection that these special Arctic rescuer dogs, the descendants of Lussi and either Storm or Obersten—the Greenland dogs who had accompanied Amundsen on the South Pole Expedition—were the ones who now also aided and supported in the search for the vanished Polar explorer—17 years after they had helped Amundsen reach the South Pole—now in their rescuer role here along this frozen archipelago near the North Pole.

Roald Amundsen and his Latham expedition mates were unfortunately presumed dead. Fridtjof Nansen died two years later. Arve Staxrud himself, after going on further to teach and mentor youth, died on April 4, 1933, and was eulogized by his surveying partner Adolf Hoel in a published obituary (Hoel, 1933). Christopher Rave also died that same year of 1933, and Leon Amundsen and Einar Rotvold died the following year of 1934.

## Echoes and Remains

Visits to Svalbard continued to take place by scientific and hunting expeditions, and, according to Daniel Nøis, it was reported that Wilhelm Eberhard's remains may have been found (Nøis, 1929, 12). Nøis reported in his account that an Arctic Ocean skipper by the name of Elev Elevsen Gratangen had become shipwrecked in Mossel Bay one summer prior to 1929 and thus was forced to travel over the challenging land route with his crew toward Isfjord, during which he apparently discovered Eberhard's remains. Based on the finding, it was surmised that Eberhard had probably lost his way and had finally given in to his profound tiredness and fatigue. The picture painted by the report in Nøis's account suggests that Eberhard had arranged rocks on either side of himself and had lain down between them, making this last effort, but ultimately succumbing to the severity of his situation and, sadly, expiring. As Nøis describes it, the lying-down location became Eberhard's final resting place, on this Christmas Eve of 1912, during which, simultaneously, his next of kin at home in Germany were thinking of him and missing him as they were gathering together to celebrate the Christmas holiday. Nøis's description of Eberhard as lost and dying and missed by his family members poignantly echoes the verbal illustration and drawing presented by Rave in his 1913 account, in which Rave imagined Eberhard all alone on the ice, and wrote that, after covering most of the distance to the ship, and having become lost, perhaps it was Eberhard's destiny to simply slip away from the world with a silent and unconscious farewell and with a "dream image" in his mind of "his dear relatives," thus his loved ones, all "gathered around the Christmas tree full of radiant lights" (Rave, 1913, 56).

Herbert Schröder-Stranz's camp on Northeast Land was evidently found in 1937, when remains were accidentally discovered by two fishermen on the shore of Northeast Land east of Dove Bay which could be assigned to a camp of the lost Schröder-Stranz expeditioners (Rüdiger, 1939) (Figs. 23.2 and 23.3).

The men had apparently reached the coast. But why they left, among other things, a kayak, a sleeping bag, clothing, cooking utensils, bandages, and cartridges, remained unclear, for no records were found.

Ernst Sorge, a member of Alfred Wegener's Greenland expedition (1930–1931), said that these were things "which an expedition likes to leave behind after an

**Fig. 23.2** Discovery of the last camp of Schröder-Stranz's sledge group on Northeast Land in 1937. (*Source* Schröder-Stranz estate, Leibniz Institute for Regional Geography, Leipzig, Germany)

**Fig. 23.3** Schröder-Stranz final camp site as seen later in 2006. (Photograph by Cornelia Lüdecke)

exhausting journey through the pack ice on the ascent to the inland ice, and presumably the Germans also found the ascent to the inland ice difficult and therefore laid down everything expendable. Since they perished, it is most probable to place the location of death in the interior or on the southwest half" (Sorge, 1937). The assumption was obvious from these items found that already, by the time they had made this camp, one of the four men was no longer alive. At the time of the camp's discovery, a new search expedition seemed to make sense only if it was carried out within the framework of a scientific expedition to Northeast Land.

This ensuing investigative scientific expedition to that site happened in 2005 and 2006, under the leadership of the biologist and wildlife filmmaker Hans Fricke, whose TV report *Lost in the Arctic* (German: *Verschollen in der Arktis*) offers speculations about the last camp (Fricke, 2009). Arved Fuchs also undertook an expedition in 2007 on the tracks of Schröder-Stranz, which also led to a TV documentary film entitled *Voyage Into the Unknown—Lost off Spitsbergen* (German: *Fahrt ins Ungewisse— Verschollen vor Spitzbergen*) being produced in Germany in 2008 (Fuchs, 2008).

## Antarctic Applications

Captain Alfred Ritscher went on to lead the third German Antarctic Expedition in 1938/1939, putting his experience in the north to use in the southern regions.

It seemed that Ritscher's experiences from the Schröder-Stranz expedition, and especially the grueling and adventurous march through the Polar night, qualified him to conduct this expedition intended to pave the way for whaling activities in the Antarctic. During this summer campaign in Antarctica, he and his expedition took aerial photogrammetric pictures of Neuschwabenland, which they had discovered using two Dornier Wal airplanes, in preparation for a future occupation of the area in order to secure the new German whaling enterprise (Lüdecke & Summerhayes, 2012).

## Invasions, Evacuations, and the Last of Lussi's Life-Saving Breed

Back in the Arctic, in Spitsbergen, according to Hilmar Nøis (Hilmar Nøis Diary SVB a), when World War II broke out, some of Lussi's Spitsbergen Dogs were taken to Iceland during the time when Nøis and all the humans on Svalbard were evacuated in 1941.[9]

Indeed, the evacuation of Spitsbergen, led by Canadian forces in combination with British forces and Norwegian troops, took place in late August and early September of 1941, during a stealth raid called Operation Gauntlet, instigated after the invasion of Russia by Germany in June 1941 (Germany had already invaded Norway in April

1940), and using Iceland as an operational base for refueling and for ships; the combined force operation resulted in the evacuation of all people on Spitsbergen, including nearly 800 Norwegian civilians who were taken to Scotland, and nearly 2000 Russians who were taken to Archangel (Archangelsk), as well as resulted in the destruction of the coal mines, coal stacks, oil and gasoline supplies, radio and wireless stations, and meteorological stations on the archipelago so as to keep these fuels, resources, and infrastructure out of the hands of the Germans (Dean & Lackenbauer, 2017, 1–46) (Fig. 23.4). The Director of the coal mining company Store Norske Spitsbergen Kulkompani A/S, Einar Sverdrup (1895–1942), was one of those people evacuated, although he departed from the archipelago quite unwillingly as well as protested the amount of destruction to his coal mines.[1]

Only one Norwegian hid during the evacuation and intentionally stayed behind on Spitsbergen (Schofield & Nesbit [1987] 2005, 62).

Apparently, it was reported by one of the Canadian army medical personnel, in their diary, that, as of August 25, there were 22 "huskies" at Longyearbyen (Advent Bay, where Store Norske was located) where much of the Norwegian community worked and resided (Dean & Lackenbauer, 2017, 17, 14–19). There were also 60 pigs, seven cows, and six horses kept by the Norwegians at Longyearbyen. This was in addition to the multitude of livestock located at the Russian settlement in Barentsburg (at Green Harbour), where 600 pigs, 60 cows, and 35 ponies resided. Another diary report is cited as indicating that the Russian population at Barentsburg was extremely dismayed when all the livestock were sadly ordered to be destroyed. Presumably this included the dogs, some of whom, a journalist's report is mentioned as documenting, could be heard howling during preparations for the evacuation.

Most likely, the 22 "huskies" mentioned in reference to the Canadian forces diary were possibly members of Lussi's Spitsbergen Dogs. And, evidently, some of the Spitsbergen Sled Dogs—those who were taken on to Iceland—were evacuated eventually.

Unfortunately, not all the dogs on Spitsbergen were taken off the islands during the evacuation. According to Hilmar Nøis, upon the arrival of the British forces, evacuation maneuvers were quickly undertaken in late August, and, with only a few days allotted for him to prepare, Hilmar Nøis's own nine dogs were forcibly shot, as the dogs were not allowed to be evacuated with the humans, and no one could remain behind (Hilmar Nøis Diary NPI c). The news was devastating to Hilmar Nøis, who considered his beloved sledge dogs to be his work mates and his dear, cherished friends, and who described many journeys together on Svalbard with his trusted and excellent draft dogs who had shared his work and struggles.[10] An expert dog teamer who, according to his diary, had never shot a dog or killed a puppy, Hilmar Nøis could not bring himself to shoot his dogs and asked two guards to do it instead, telling them to shoot all his dogs far away from him where he could neither hear nor

---

[1] It was reported that Einar Sverdrup was related to Otto Sverdrup (1854–1930), the Polar explorer, Arctic sailor, and first captain of the famous ship *Fram* (Dean & Lackenbauer, 2017; Schofield & Nesbit [1987] 2005, 63). (Roald Amundsen also later captained the *Fram*, which Fridtjof Nansen had designed for drifting with the ice, using it to travel to Antarctica to make his attempt to reach the South Pole (Tahan, 2019, 3, 6–8, 17–18].).

**Fig. 23.4** Destruction of the telegraph station radio masts in Green Harbour during the World War II evacuation of Spitsbergen. The photograph of the explosion originates from a British film shot during the operation on September 2, 1941. (*Photographer/Byline* unidentified. *Source/Owner* Norwegian Polar Institute)

see the deed or the demise of his dogs.[11] Hilmar Nøis himself and his wife Helfrid Nøis, along with all the remaining humans of Svalbard, were evacuated on September 3, taken—on board the liner, now troop transport ship, *Empress of Canada*—from Green Harbour to Glasgow, Scotland.[12] Daniel Nøis, too, was evacuated with the rest of the Norwegians to Scotland (*Andøya Avis* 6 October 1950, 1; Daniel Nøis Biography/Norwegian Polar Institute).

Of the Spitsbergen Rescuer Dogs who had been taken to Iceland, these remaining dogs, reported Hilmar Nøis (Hilmar Nøis Diary SVB a), went down with the sudden and violent sinking of the *Isbjørn* in 1942.[13]

To illustrate this statement, there is a connection between the dogs, the *Isbjørn*, and Sverdrup. It will be remembered that Hilmar Nøis stated that some of Lussi's Spitsbergen Dogs had been distributed to Store Norske in 1916, and that Rolf Tandberg had brought his rescue sled dog team to find Nobile in 1928 from Store Norske (see above). Hilmar Nøis (Hilmar Nøis Diary NPI c) also recalled running into Sverdrup, the director of Store Norsk, in Longyearbyen (Advent Bay) in 1941, on September 2, as the two men—evidently quite reluctantly—were readying themselves to be evacuated.[14] He also mentioned seeing the Norwegian icebreaker *Isbjørn* (which means "Polar Bear") at Spitsbergen in May of 1941.[15] Based on Nøis's reports, it appears that both Sverdrup and the sled dogs who had been evacuated in 1941 returned together with the Allied Forces and Norwegian troops in 1942, the dogs having been taken to the British base in Iceland after the evacuation and brought back again on the *Isbjørn* when the vessel was deployed on a mission to Spitsbergen.

Indeed, it was on May 14, 1942, that *Isbjørn* was sunk during a German surprise bombing attack by four Focke-Wulf Condor planes targeting the Norwegian icebreaker that had been painstakingly cutting a channel through the ice to the Green Harbour shore rather than use sledges to unload supplies (Schofield & Nesbit [1987] 2005, 94–114). The *Isbjørn* had returned to Spitsbergen as part of the Allied Forces' and Norwegian Navy's Operation Fritham, together with the sealing ship *Selis*, as well as with a landing force of approximately 60 Norwegians, most of whom had been part of the evacuation in September 1941, and all of whom were now under the command of Einar Sverdrup, previously the Director of the Store Norske Norwegian coal mining company and currently the Lieutenant Colonel leading the men's efforts to attempt to regain control of Spitsbergen. According to this report, the total number of participants was approximately 85 (other reports place it at 82 and 92). The coal mining director who had been made to evacuate Spitsbergen unwillingly had returned with the goal to reinstate Norwegian presence on Spitsbergen and ascertain the extent of the German presence there, including at the meteorological stations. Departing from Greenock on April 30, and traveling first to Akureyri, Iceland, where two officers joined and additional supplies were taken onboard the two ships, *Isbjørn* and *Selis* sailed from Iceland on May 8 toward Spitsbergen.

The forces, officers, and crew were not the only passengers on the ship. According to a report published in a Norway's War Sailors' Association magazine in 1982, based on a letter written to the magazine by Einar Oscar Holst, who had been the chief engineer onboard the icebreaker and who had survived the attack, Holst stated that, after leaving Greenock, Scotland on April 30, the *Isbjørn*, which had been based in Reykjavik, brought on board sled dogs in Iceland, picking them up at Akureyri, before proceeding on to Spitsbergen (Krigsseileren, 1982, 22; WarSailors). At Akureyri, bunker fuel also was loaded prior to departure, and the vessel arrived at Isfjorden (Isfjord) on the 13th of May and then was attacked on the following day of May 14.

It seems, then, that the Spitsbergen Rescuer Dogs—Lussi's descendants—had returned to Spitsbergen as members working with the Allied Forces. They were now war-effort dogs, helping the Norwegians reclaim their home, their place of work, and the place where they had built their lives.

It was reported that when the *Isbjørn* and the *Selis* arrived at Isfjorden on May 13, discovering that the ice was impassable to Advent Bay, they made their way to Green Harbour instead, which they also found to be completely covered with thick ice that would necessitate the icebreaker's cutting a lengthy channel to the dock (Schofield & Nesbit [1987] 2005, 94–114). Rather than begin sledging supplies to shore immediately, Sverdrup gave the men a night of rest and continued icebreaking activities early the next morning, during which time the two vessels were spotted by a German Ju88 flying above Isfjorden. According to the report, Sverdrup had already decided to proceed with cutting through the ice all the way to the dock rather than sledge the supplies off the ship to the shore, and *Isbjørn* toiled away for 15 h, making its way through the ice, and being surrounded by ice, when the four German planes, having been alerted by the Ju88, dropped the first of several bombs.

Einar Oscar Holst himself, who was working in the engine room at the time the first bomb was unleashed by the German airplanes, narrowly escaped the attack and barely survived the bombing, jumping onto the ice along with the remainder of the ship's crew members who were able to run from the shelling, and then being knocked into the freezing water by the explosion from a bomb that fell near him and through the ice, and then ultimately fortunately received aid from some of his fellow surviving crew members (Krigsseileren, 1982, 22; WarSailors). According to Holst, based on the report that quotes the letter he wrote, 17 men were killed and four men were wounded in the attack.[2]

And based on Hilmar Nøis's report (above), the dogs were evidently still on board when the bombing occurred.

The icebreaker *Isbjørn*, while it was diligently making its way through the ice, was unfortunately an easy target for the four German planes that dropped the bombs on the vessel and crew as well as on its accompanying ship, the sealer named *Selis* (Schofield & Nesbit [1987] 2005, 94–114). During the third bombing, *Isbjørn* was struck and instantly plunged to its demise. In a further bombing run, *Selis*, too, burned, and later sank. According to this report, the number of men sadly killed was 13, which included Einar Sverdrup, and the number of men wounded was nine, with two of those wounded later perishing (other reports place the number of deaths at 14 and 12).

Thus, tragically, Sverdrup, the men, and all the dogs, died.

As Hilmar Nøis described it (Hilmar Nøis Diary SVB a), this archipelago named Spitsbergen—the place where this special dog breed had arisen—was also the place where this rescuer dog breed came to its final watery rest.[16]

And so ended Lussi's Spitsbergen Dog Team breed. They had been rescuer sled dogs who had been a result of Arve Staxrud's rescue expedition that had saved members of Herbert Schröder-Stranz's expedition and that had exemplified compassionate cooperation between Norwegians and Germans and extraordinary effort to save every human life possible. Now the last of the dogs, and those humans attempting liberation, disappeared to the bottom of the sea in a tragic act of violence visited upon Norway's Svalbard from Nazi Germany—an act that indiscriminately ended lives and proved contradictory to the respect for life and the heroic acts of search and

---

[2] According to the *Krigsseileren* magazine report, at the time of its publishing in 1982, Einar Oscar Holst was 81 years of age and a veteran of both World War I and World War II. The report draws upon the letter Holst wrote to the magazine and quotes from some of his statements made in his letter.

rescue that had been exhibited on these very same ice fjords of Spitsbergen during the Staxrud expedition in the instrumental year of 1913.

## Notes on Original Material and Unpublished Sources

Hilmar Nøis's diaries SVB are in the Historical Archive at the Svalbard Museum (SVB) in Svalbard. Hilmar Nøis's diaries NPI are in the Library Archives at the Norwegian Polar Institute (NPI) in Tromsø. Arve Staxrud letters of correspondence, written from and to Arve Staxrud, are in the Manuscripts Collection at the National Library of Norway (NB) in Oslo.

1.  A. Staxrud to F. Nansen, letter, 23 May 1919, NB Brevs. 48
2.  A. Staxrud to F. Nansen, letter, 22 October 1920, NB Brevs. 48
3.  A. Staxrud to F. Nansen, letter, 26 May 1925, NB Brevs. 48
4.  A. Staxrud to F. Nansen, letter, 19 October 1923, NB Ms.fol. 1924:6:A:2.
5.  H. Nøis diary SVB a, 1913, journal pages 65 and 87, SVB-AP2
6.  H. Nøis diary NPI a, notebook 3, 1913, pdf page 39, (NPI), D-307 / D00307_3_0001
7.  H. Nøis diary NPI b, notebook No. 8, May–July 1928, pdf page 28, (NPI), D-307/D00307_8_0001
8.  H. Nøis diary SVB b, September 1928, pdf pages 39–42, SVB-AP3
9.  H. Nøis diary SVB a, 1913, journal pages 65 and 87, SVB-AP2
10. H. Nøis diary NPI c, notebook No. 11, Summer 1941, journal pages 81–84, (NPI), D-307/D00307_11_0001
11. H. Nøis diary NPI c, notebook No. 11, Summer 1941, journal pages 81–84, (NPI), D-307/D00307_11_0001
12. H. Nøis diary NPI c, notebook No. 11, Summer 1941 & 3 September 1941, journal pages 84–87, (NPI), D-307/D00307_11_0001
13. H. Nøis diary SVB a, 1913, journal page 65, SVB-AP2
14. H. Nøis diary NPI c, notebook No. 11, Summer 1941 & 2 September 1941, journal pages 84–86, (NPI), D-307/D00307_11_0001
15. H. Nøis diary NPI c, notebook No. 11, Spring 1941 & May 1941, journal page 60, (NPI), D-307/D00307_11_0001
16. H. Nøis diary SVB a, 1913, journal page 65, SVB-AP2

## Unpublished Sources and Original Material

Nøis, D. Daniel Nøis Biography/Norwegian Polar Institute, "Nøis_Daniel_Biografiarkiv_1", written by H. M. [possibly Haakon Aronsen, aka Haakon Magnus, per Gunnhild Holmen of the National Library of Norway, based on *Andøya Avis* 1964 article, according to communications from Odd Ivar Nøis Olsen of the Nøis Family and from Ivar Stokkeland and Petr Masat of the Norwegian Polar Institute, received 29 March 2022], Biography Archive/Library Archives, Norwegian Polar Institute, Tromsø. Received 20 July 2021.

Nøis, H., Diary NPI a, handwritten journal of recollections recounting the events of the 1913 Staxrud expedition, Hilmar Nøis, "Minder og Fortelinger fra fangsttiden på Spitsbergen, Klade 3, 7 Mars 1970" [Hilmar Nøis, "Memories and Stories from the trapping time on Spitsbergen, Notebook 3, 7 March 1970"], dated 7 March 1970, D-307, ("Minder og fortelinger, Nøis, Hilmar, hefte 3"), D00307_3_0001. Library Archives, Norwegian Polar Institute, Tromsø, https://brage.npolar.no/npolar-xmlui/handle/11250/274077. Accessed 20 July 2021.

Nøis, H., Diary NPI b, handwritten journal of recollections recounting events including the 1928 search for the Nobile expedition, Hilmar Nøis, "Minder og Fortelinger, No 8., Fra No 7 Den 23 Jan 1971" [Hilmar Nøis, "Memories and Stories, Number 8, From No. 7 on 23 January 1971"], D-307, ("Minder og fortelinger, Nøis, Hilmar, hefte 8"), D00307_8_0001. Library Archives, Norwegian Polar Institute, Tromsø, https://brage.npolar.no/npolar-xmlui/handle/11250/274077. Accessed 18 April 2022.

Nøis, H., Diary NPI c, handwritten journal of recollections recounting events including the evacuation from Svalbard in the summer and September of 1941, Hilmar Nøis, "1940 Evakuerengen 1941 slut, No. 11." [Hilmar Nøis, "1940 The Evacuation 1941 end, Number 11"], D-307, ("Minder og fortelinger, Nøis, Hilmar, hefte 11"), D00307_11_0001. Library Archives, Norwegian Polar Institute, Tromsø, https://brage.npolar.no/npolar-xmlui/handle/11250/274077 Accessed 18 April 2022.

Nøis, H., Diary SVB a, handwritten journal of recollections including an account of the 1913 Staxrud expedition, ("Hilmar Nøis. Fangstmann. 1909–1923, dagbok med erindringer fra fangstlivet") ["Hilmar Nøis. Trapper. 1909–1923, diary with recollections from the trapping life"], not dated (latest year referenced is 1942 in journal pages and 1950 in inserted sheet), SVB-AP2-Hilmar-Nøis-dagbok-1909–1923-compressed. Historical Archive, Svalbard Museum, Svalbard, https://svalbardmuseum.no/no/samlingene/historisk-arkiv/arkivprosjekt-fangst-og-annen-overvintringsvirksomhet-fra-perioden-1910-til-1970/ Accessed 6 June 2021.

Nøis, H., Diary SVB b, handwritten journal of events in 1928, Hilmar Nøis, "Dagbok for år 1928" [Hilmar Nøis, "Diary for year 1928"], ("Hilmar Nøis. Fangstmann. 1928, dagbok Roosneset/Woodfjorden"), SVB-AP3-Hilmar-Nøis-dagbok-1928-compressed. Historical Archive, Svalbard Museum, Svalbard, https://svalbardmuseum.no/no/samlingene/historisk-arkiv/arkivprosjekt-fangst-og-annen-overvintringsvirksomhet-fra-perioden-1910-til-1970/. Accessed 9 April 2022.

Schröder-Stranz estate, Leibniz Institute for Regional Geography, Leipzig, Germany.

Staxrud, A., 23 May 1919, Letter from Arve Staxrud to Fridtjof Nansen, NB Brevs. 48. National Library of Norway, Oslo, online archives, https://www.nb.no. Retrieved 6 July 2021.

Staxrud, A., 22 October 1920, Letter from Arve Staxrud to Fridtjof Nansen with three accompanying letters, NB Brevs. 48. National Library of Norway, Oslo, online archives, https://www.nb.no. Retrieved 6 July 2021.

Staxrud, A., 19 October 1923, Letter from Arve Staxrud to Fridtjof Nansen, NB Ms.fol. 1924:6:A:2. National Library of Norway, Oslo, online archives, https://www. nb.no. Retrieved 6 July 2021.

Staxrud, A., 26 May 1925, Letter from Arve Staxrud to Fridtjof Nansen with accompanying article, NB Brevs. 48. National Library of Norway, Oslo, online archives, https://www.nb.no. Retrieved 6 July 2021.

# References

Andøya Avis. (1950, October 6). Fangstmann og båtbygger Daniel Nøis runder 70 år. [Trapper and boat builder Daniel Nøis turns 70 years old.]. Written by H. M. [possibly Haakon Aronsen, aka Haakon Magnus, per Gunnhild Holmen of the National Library of Norway, based on *Andøya Avis* 1964 article, according to communications from Odd Ivar Nøis Olsen of the Nøis Family and from Ivar Stokkeland and Petr Masat of the Norwegian Polar Institute, received 29 March 2022]. *Andøya Avis* (Andøya), 6 October 1950, p. 1. Library Archives, Norwegian Polar Institute, Tromsø. Received 29 March 2022.
Anonym. (1914). Eine Organisation für Polar-und Forschungsexpeditionen. *Zeitschrift der Gesellschaft für Erdkunde zu Berlin* (pp. 223–224).
Berg, R. (2020). The Svalbard "Channel", 1920–2020—A geopolitical sketch. In J. Weber (Ed.), *Handbook on geopolitics and security in the Arctic*. Frontiers in International Relations (pp. 303–321). Cham: Springer. https://doi.org/10.1007/978-3-030-45005-2_18
Dean, R., & Lackenbauer, P. W. (2017). "A particularly spectacular piece of demolition": The Canadian-Led Raid on Spitzbergen, 1941. In A. Lajeunesse & P. W. Lackenbauer (Eds.), *Canadian armed forces Arctic operations, 1941–2015: Lessons learned, lost, and relearned* (pp. 1–46). Fredericton, NB: The Gregg Centre for the Study of War & Society/University of New Brunswick. https://www.unb.ca/fredericton/arts/centres/gregg/what/publications/CdnArcticOps2017.pdf. Retrieved 15 April 2022.
Dole, N. H. (1922). *America in Spitsbergen: The romance of an Arctic coal-mine*. In Two Volumes. Boston: Marshall Jones Company. Volume II. Facsimile reprint by Scholar Select.
Fricke, H. (2009). *Verschollen in der Arktis*. https://www.youtube.com/watch?v=TNScwgehIto. Visited 19 May 2021.
Fuchs, A. (2008). *Fahrt ins Ungewisse. Verschollen vor Spitzbergen*. https://www.spitzbergen.de/2008/04/20/schroder-stranz-expedition-filmmaterial-aufgetaucht.html. Visited 20 May 2021.
Hoel, A. (1933). Arve Staxrud. In *Norsk Geografisk Tidsskrift* [*Norwegian Journal of Geography*], *4*(6), 345–346. https://doi.org/10.1080/00291953308622018
Krigsseileren. (1982). Brev til redaksjonen: Fra en veterans erindringer [Letter to the editor: From a veteran's recollections]. In: *Krigsseileren*, Nr. 2–1982–11. årgang (p. 22). Norges Krigsseilerforbund. H. Wigestrand (ed.). Oslo: Norges Krigsseilerforbund. https://www.krigsseilerregisteret.no/no/tidsskriftet-krigsseileren/360382. Accessed 4 July 2023.

Lüdecke, C., & Summerhayes, C. (2012). *The third Reich in Antarctica: The German Antarctic expedition 1938–39.* Erskine Press and Bluntisham: Bluntisham Books.

Nøis, D. (1929). Med kaptein Staxrud på leiting efter Schrøder-Stranz og hans folk [With Captain Staxrud on the search for Schröder-Stranz and his people]. In *And-Ungen*, July 1929, (pp. 4–12). Andenes: Andøyposten. Library Archives, Norwegian Polar Institute, Tromsø. Received 20 July 2021.

Rave, C. (1913). *Tagebuch von der verunglückten Expedition Schröder-Stranz.* Schaffsteins Gründe Bändchen 49. Cöln: Schaffstein.

Rüdiger, H. (1939). Die Überreste der Schröder-Stranz-expedition. *Petermanns Geographische Mitteilungen, 85,* 158–162.

Schofield, E., & Nesbit, R. C. ([1987] 2005). *Arctic Airmen: The RAF in Spitsbergen and North Russia, 1942.* This edition published in 2005 in the UK. Staplehurst, Kent: Spellmount Limited.

Sorge, E. (1937). Das Nordostland von Spitzbergen. Karte. *Berliner-Lokal-Anzeiger.* Ritscher estate, Cornelia Lüdecke, München.

Stenersen, H. (editor/publisher). (1988). Tingelstad-brødrene som kartla Svalbard: Arve og Olav Staxrud har satt varige spor etter seg. In *Brandbu'stikka: Lokalhistorisk Kvartalskrift for Brandbu. 1. KV. 1988 Mars NR. 11* (March 1988). (Svalbard-forskerne Arve og Olav Staxrud.) (pp. 1–26). Biography Archive/Library Archives, Norwegian Polar Institute, Tromsø. Received 20 July 2021.

Tahan, M. R. (2019). *Roald Amundsen's sled dogs: The sledge dogs who helped discover the South Pole.* Cham: Springer International Publishing.

Tahan, M. R. (2021). *The return of the South Pole sled dogs: With Amundsen's and Mawson's Antarctic expeditions.* Cham: Springer International Publishing.

Tandberg, R. S. (1928). *Med hundespann på eftersøkning efter "Italia"-folkene.* Særtrykk av *Norsk Geografisk Tidsskrift*, Bind II, Hefte 3–4, 1928. Oslo: A. W. Brøggers Boktrykkeri A/S. Digital Archive, National Library of Norway, Oslo, received 8 June 2021. Also: Tandberg, R. S. (1928/1929). Med hundespann på eftersøkning efter "Italia"-folkene. In *Norsk Geografisk Tidsskrift*, Bind II –1928/1929, pp. 176–214. O. Holtedahl (ed.), 1929. Oslo: A. W. Brøggers Boktrykkeri A/S. National Library of Norway, Oslo, online archives, https://www.nb.no. Retrieved 6 February 2022 and 23 June 2023.

*WarSailors.com.* D/S Isbjørn. http://www.warsailors.com/singleships/isbjorn2.html. Accessed 16 April 2022.

# Chapter 24
# Conclusion: The Crucial Necessity for Experience, Preparation, and Animals

## The Aftermath of Unpreparedness and Inexperience

Due to the complete inexperience of the participants, Herbert Schröder-Stranz's expedition to Spitsbergen in 1912 ended in disaster, which was accordingly marketed in Germany in the sensational press. Out of 15 participants, only seven saw their homeland again. Apart from the Norwegian crewmembers Einar Rotvold, August Stenersen, Julius Jensen, and Jørgen Jensen—with the exception of the cook, Knut Stave, who sadly died during the overwintering on board the *Herzog Ernst*, leaving four Norwegian survivors who managed to save themselves—only three Germans survived: the two invalids Captain Alfred Ritscher and the geographer Hermann Rüdiger, and the marine painter Christopher Rave. The other seven participants remained missing: Herbert Schröder-Stranz, Richard Schmidt, August Sandleben, Dr. Max Mayr, Dr. Erwin Detmers, Dr. Walter Moeser, and Wilhelm Eberhard.

Even then, it touched one strangely when Rave luridly announced his film about *The Tragedy of the Schröder-Stranz Expedition* (German: *Die Tragödie der Schröder-Stranz-Expedition*) in 1913 with *The Death March of Ten Brave German Men through Night and Ice* (German: *Der Todeszug der zehn mutigen deutschen Männer durch Nacht und Eis*) and gave a lecture in the cinema of the Noellendorf Theater in Berlin after the performance (Rave 27 September 1913). Allgeier had also edited his Spitsbergen film and brought it to the cinema under the title *With the Camera in the Eternal Ice* (German: *Mit der Kamera im ewigen Eis*) in October 1913 (Allgeier, 1913). The premiere of the 90-min film took place together with a lecture by Theodor Lerner in Frankfurt (Allgeier, 1931, 40–41). The local press emphasized in the effusively positive review that this was probably the first documentary film of such an expedition.

In the Norwegian press, sympathy and empathy were expressed in articles such as the one titled "A Christmas-night tragedy on Spitsbergen" (*Aftenposten: Ugens Nyt* 2 September 1913, 2–3), which reported on the investigations taking place as of August 24, 1913, regarding the failed expedition, following the survivors' arrival

M. R. Tahan and C. Lüdecke, *Stranded at the Top of the World*, https://doi.org/10.1007/978-3-031-56288-4_24

343

in Tromsø, and which summarized Rotvold's and Stenersen's reported explanations about the events of the journey during which Moeser and Detmers had inexplicably parted ways with the rest of the group, Eberhard had sadly and mysteriously disappeared, and Ritscher had chosen to continue to Advent Bay solo. Calling Schröder-Stranz's disastrous excursion "one of the most ill-fated expeditions, that have traveled up there on the Arctic sea island," the article emphasized "the stirring account of that Christmas-night journey, when Eberhardt [sic] disappeared" (*Ugens Nyt* by *Aftenposten* 2 September 1913, 2).

The following year, the Austrian alpinist Karl Potpeschnigg strongly criticized the expedition. It would be common knowledge that, due to the rapidly changing ice conditions northeast of the main island of Spitsbergen, involuntary wintering with fatal ends could occur if one did not turn back in time or if one did not know the conditions there (Potpeschnigg, 1914, 9). It would have to be regarded as immense recklessness "if scientific expeditions without sufficient experience set off in late summer towards the east coast of Spitsbergen with the scarcest provisions and with a ship that is by no means suitable for wintering. Such a completely rash undertaking can only be described as blind bravado and must necessarily lead to disaster" (Potpeschnigg, 1914, 9–10).

The Norwegian rescue expedition's leader Arve Staxrud concluded his published report by noting that further search measures in this dangerous region were completely useless (Staxrud, 1914, 67–68).

Nevertheless, before the outbreak of World War I, Kaiser Wilhelm II donated a large sum of money from his private disposition fund for clues to finding the Schröder-Stranz expedition, and the Imperial Government asked the Norwegian government to "inform the population of the parts of the country under consideration of the suspension of the Imperial money awards" (Anonym, 1914, 643).

## A Successful Rescue Enterprise

Arve Staxrud's rescue expedition, in search of Herbert Schröder-Stranz's expedition to Spitsbergen, was successful in many respects. Two Germans from the Schröder-Stranz Expedition—Hermann Rüdiger and Christopher Rave—were found and rescued at Treurenberg Bay; two round-trip crossings of Spitsbergen—including over previously unknown areas—were successfully undertaken and completed; the frozen ship *Herzog Ernst* was released from the ice, re-floated, and returned intact; 13 men and nine dogs from Theodor Lerner's shipwrecked rescue expedition were encountered and saved; Staxrud's own human expeditioners returned safely and with no injuries; and his teams of sledge dogs also returned safe and healthy—this included the Antarctic Expedition dogs Lussi and Storm, as Staxrud had assured the Amundsens, Leon and Roald, and this also included Lussi's addition of six puppies, four of whom went on to become the first generation of Spitsbergen Rescuer Dogs who helped workers, expeditioners, scientists, and visitors in Spitsbergen for over a period of nearly three decades.

Seven remaining members of the Schröder-Stranz expedition unfortunately were not found, not for lack of effort, but due to extreme conditions and the late season—the Staxrud expeditioners had searched extensively near Mossel Bay and had made numerous attempts to cross the forbidding ice to Northeast Land.

Staxrud's conclusion that the four men of Schröder-Stranz's sledging group "had already perished on the drift ice" (Staxrud, 1914, 55) was wrong, however. For, as the discovery of Schröder-Stranz's camp in Dove Bay in 1937 seems to prove, it appears that at least three men had reached the coast of Northeast Land (Rüdiger, 1939). The rescuers, however, could not find them, but succeeded to find and save the others who were still stranded.

An organized and cooperative group under the leadership of an experienced officer, and comprising overwinterers, trappers/hunters, reindeer teamers, Arctic Ocean skippers/ice pilots, medical doctor, mining engineer, sledge dogs, and draft reindeer, had worked together to achieve life-saving results—with the loss of life being among the draft reindeer due to food supply depletion caused by delayed travel and consequently exhausion ending in euthanization.

Captain Arve Staxrud, deputy Daniel Nøis, Hilmar Nøis, Johan Nilsen Nøis, Per Hansen, Johannes Kemi, Samuel Klemmetsen, the physician Dr. P. W. K. Bøckmann, the mining engineer Jakob Ellingsen, Martin Pettersen Nøis, ice pilot Søren Zachariassen, and August Stenersen all contributed to the success of the mission—with Stenersen, who was the ice pilot on the *Herzog Ernst*, notably being a member of both the stranded expedition and the rescue expedition. The Norwegians brought their overwintering experience and professional skills to the effort. The Sámi reindeer walkers provided guidance and expertise appreciated by the other men. And the 20 draft reindeer and 18–19 sledge dogs contributed greatly, pulling heavy amounts of weight and escorting the men over miles of ice and through the rough and roller-coaster terrain.

## Thwarted Ambitions and Achieved Results

Staxrud had put his geographical/topographical surveying background to work to benefit the rescue mission and had employed an ice-worthy vessel, *Hertha*, with a capable crew, to reach his Polar destination and thus conduct the Spitsbergen rescue expedition.

Although it appears he did not enjoy a world-famous explorer's reputation, Staxrud brought together practical knowledge, pragmatic approach, careful planning, a band of expert and experienced expedition members who helped execute his plan, and even a consistent sense of camaraderie and effective leadership, to bear on his success. He also left room for well-thought-out improvisation when it was warranted. He seems to have kept his promises. And he attempted again and again to achieve the goal, without being unrealistic or overtaxing.

In contrast, the Schröder-Stranz expedition was woefully unprepared and overly ambitious, as well as liable to impromptu decisions. The other rescue expeditions that failed, also, did so either from lack of knowledge and pre-planning (such as Theodor Lerner) or lack of circumstances and suitable equipment (as in the unfortunate Arctic Coal Company-organized expedition), with one of the expeditions taking unmindful actions that jeopardized the lives of the very people they were hoping to rescue (as in the Kurt Wegener expedition).

Lerner's expedition, however, came the farthest north (Chap. 18) and explored the region between Walden Island and Cape Platen without finding any trace of Schröder-Stranz's people (Villinger, 1929, 42–46; Allgeier, 1931, 30–35). They came the closest to Schröder-Stranz's camp, but no one suspected that Schröder-Stranz and his group had reached Dove Bay east of Cape Platen. As Villinger reported, the Lerner expeditioners, too, had concluded that Schröder-Stranz and his companions did not reach the coast of Northeast Land. Villinger suspected that a storm had driven them northward, where they then perished in the pack ice.

Although Wegener took important supplies from the huts and did not replenish them, as well as was suspected to have burnt down the West Fjord hut (accidently), he was never called to account. Hugo Hergesell considered Wegener's expedition noteworthy for its boldness and energy shown, and in the end all Wegener participants were awarded a reward (Hergesell, 1913, 137). It is important to note that Hergesell was responsible for the observatory at Ebeltofthamna and, as Wegener's supervisor, had asked Wegener to overwinter there. Hergesell himself had no mountaineering experience and was not familiar with the habits and particular care taken in the Alps concerning the handling of firewood and provisions. Hence, he did not seem to oppose Wegener's unacceptable actions and perhaps did not realize the true detriment they posed. Wegener himself mentioned in his expedition report of 1914 that the Norwegians implied that his expedition had caused the fire in the hut, but he claimed that he and his group had left the hut eight days before the arrival of the Norwegians (Wegener, 1914, 96). He stated that he departed from the hut on March 25 and that the Norwegians arrived there on April 3 where it was found to be burnt down and still burning. He further commented, regarding their implying that his expedition was the cause, that "Maybe it should be a revenge for the sleeping bag affair" (Wegener, 1914, 96). Einar Rotvold, however, in his expedition diary written during the time of these events, documented that the Norwegians reached the burnt-down hut on March 30, not April 3 (Einar Rotvold Diary). Having seen smoke, they had left Cross Point and had gone to the burnt-down hut on March 30, where they saw it had already been burnt down and "was only burning in the ruins which caused the smoke,"[1] and so they then returned that same night to the Cross Point hut and stayed there for another few days, and later left Cross Point and came back to the burnt-down hut again on April 2 on their way trekking further south.[2] This was a shorter period of time after Wegener's departure than Wegener had stated, and it is conceivable that Wegener's group had unintentionally started a fire through carelessness while leaving on March 25, which later ignited, and that there was still smoke emanating from the ruins as of March 30 when the Norwegians arrived. Rotvold had surmised that the people who had visited the hut before the Norwegians arrived there "had been rather reckless with the fire, when they left the house."[3] Wegener and his people had preceded the Norwegians to the huts and seemed to be the only other group in the area, and the

Norwegians were being very careful in their own conduct (Chaps. 10 and 11). The Norwegians, after all, were struggling for their lives. By lengthening the timeline, perhaps Wegener was attempting to distance himself from his bad behavior. In the end, it seems there was no official criticism of Wegener, and no one was blamed. Perhaps the outbreak of WWI stopped any legal action, if any was attempted from the German or Norwegian side.

## The Mystery and Tragedy of the Disappearances

In writing about Schröder-Stranz's and his three companions' disappearance, Fridtjof Nansen, too, apparently later conjectured that perhaps they had gone out too far in the ice from the land with their boat and had unintentionally drifted out into the ocean, or that the boat had become crushed between the ice floes and their equipment was therefore lost (Nansen, 1920, 186). Nansen surmised that some kind of an accident must have befallen them, but even so, he seemed to imply that he would have thought they would have been able to make it back to the area where game was plentiful, and he made it a point to state that they had their dogs with them. Thus, he indicated, what had transpired was a tragic mystery.

Nansen attributed catastrophes such as the Schröder-Stranz expedition to a decrease in the enormous respect that is required for the Arctic sea's challenges, and an ascendancy of what he termed "the Wellman newspaper-spirit," which inspired inexperienced leaders and other untrained individuals to nonetheless venture out and thus subsequently exhibit carelessness which ended in misfortune (Nansen, 1920, 228).

It was a sad case of youth, bravery, and dedicated sacrifice completely wasted, lamented Nansen, and an especially bitter twist that, had fate allowed him to meet the Schröder-Stranz expeditioners on their way eastward, at the time that they must have passed him during his scientific cruise in Spitsbergen, he and his mates might have had a discussion with them about their expedition plans, and, just perhaps, their lives might not have been lost, the tragedy and suffering might not have occurred, and they might have gone on to accomplish good things, perhaps even joining Nansen in his oceanographic investigations (Nansen, 1920, 188).

And so it was that a rescue expedition, led by Captain Staxrud, and with draft reindeer and sledge dogs, drove from Green Harbour to search for the missing expeditioners, and was able to rescue the two stranded men Rave and Rüdiger (Nansen, 1920, 187–188).

# The Importance of Animals and Human-Animal Cooperation

The use of animals, both as reliable draft animals to pull the sledges carrying vital supplies and injured and exhausted men, and as hunted animals to provide required meat nutrition for the expeditioners, played an important part in the survival of both the rescue expeditioners and the stranded expeditioners. The sledge dogs and draft reindeer moved the rescue expedition's men and supplies across the ice, while draft dogs also guided the stranded men who took refuge in the huts or made a desperate march from the ship to civilization. The draft reindeer who were sacrificed due to lack of moss for food helped sustain the other expeditioners. The wild reindeer, Arctic foxes, birds, and polar bears trapped and killed were a source of sustenance and nourishment for both the stranded, injured, and sick expeditioners and the rescue expeditioners who attempted to save them. The capturing and captivity of polar bears, however, is not included in this equation, as that action was questionable at best and not necessary for survival.

The crucial role of the sled dogs and draft reindeer is made evident by the successful results of the search-and-rescue efforts that combined humans with working animals to locate and save stranded expeditioners. It is also evident through the ensuing creation of a network of rescuer sled dog teams—the Spitsbergen Rescuer Dogs—which in turn enabled improved search-and-rescue capabilities for Spitsbergen and for the future of the Arctic archipelago.

The Spitsbergen Rescuer Dogs' demise was a tragic one—both during the evacuation in 1941 and the bombing attack in 1942 (Chap. 23). Their ancestors had returned from the South Pole Expedition with Roald Amundsen, who, with the help of his Greenland dogs, had reached the Pole prior to Robert Falcon Scott and who was subsequently lauded in London in 1912 for the achievement. Lord Curzon's words "I almost wish that in our tribute of admiration we could include those wonderful good-tempered, fascinating dogs, the true friends of man, without whom Captain Amundsen would never have got to the Pole. I ask you to signify your assent by your applause," delivered after Amundsen's lecture at the Royal Geographical Society, as reported the following year in 1913 in their official journal, were later represented by Amundsen in his 1927 autobiography as a proposal of "three cheers for the dogs" intended as "satirical and derogatory" toward him (Tahan, 2021, 253–256). Amundsen's interpretation of Earl Curzon of Kedleston's words was contested by the RGS, and controversy ensued. Nonetheless, the remark from the RGS President was a tribute by the English to the sled dogs. It is a bitter irony, then, that years later, during WWII, the British-Canadian evacuation orders (Hilmar Nøis Diary NPI) would result in most of the dogs on Spitsbergen—some of whom were descended from those two dogs who were members of Amundsen's South Pole expedition and thus had been applauded by the British—now being summarily destroyed.[4] These Spitsbergen Rescuer Dogs' direct ancestor Lussi, along with Storm, had been part of Staxrud's successful rescue expedition.

Indeed, credit must be given to the animals who were instrumental actors in saving human lives.

## Polar Arctic Ambitions and Effective Interspecies Efforts to Save Lives

In the case of the Staxrud rescue expedition, humans and animals bravely worked in concert to save the lives of those who were marooned on the frozen islands of Spitsbergen. Cooperation was created to combat the chaos of the situation; calm planning was implemented to methodically reach those who were stranded; and safety measures pre-empted unnecessary risks. This is ultimately an exemplary tale of interspecies and international cooperation. It is an example of dedicated determination and perseverance, courageous life-saving strategy, and quick action.

The Staxrud rescue expedition on Spitsbergen was an admirable endeavor to rescue and save expeditioners left stranded and injured due to the overly high ambitions and lack of experience of their leaders, during an unseasonable time of year, in the beckoning yet extremely challenging Polar region that is the Arctic.

(Figs. 24.1, 24.2, 24.3, 24.4, 24.5, 24.6, 24.7, 24.8, and 24.9).

**Fig. 24.1** Unfamiliar areas were traveled to save stranded expeditioners. (*Photographer/Byline* Arve Staxrud. *Source/Owner* Norwegian Polar Institute)

**Fig. 24.2** Glaciers and ice fjords were crossed to search for missing expedition members. (*Photographer/Byline* Arve Staxrud. *Source/Owner* Norwegian Polar Institute)

**Fig. 24.3** Distances were diligently traversed, over ice and over sea. (*Source* Staxrud, 1914, Tf. 11)

**Fig. 24.4** Draft reindeer were instrumental in pulling provisions and supplies as well as transporting the injured and exhausted expeditioners who were rescued. (Photograph slightly cropped from the original.) (*Photographer/Byline* Ernest Mansfield. *Source/Owner* Norwegian Polar Institute)

**Fig. 24.5** Sámi reindeer walkers were crucial guides and important rescue expedition members. (Photograph slightly cropped from the original.) (*Photographer/Byline* Ernest Mansfield. *Source/ Owner* Norwegian Polar Institute)

**Fig. 24.6** The family of overwintering experts, the Arctic and seafaring professionals, the mining engineer and expedition physician, and the experienced sled dogs including Lussi and Storm and Nøis's draft dogs, all worked closely together throughout the rescue mission. (*Photographer/Byline* Arve Staxrud. *Source/Owner* Norwegian Polar Institute)

**Fig. 24.7** Two round-trip crossings of Spitsbergen were made by the men and by dogs like Storm and Lussi and the other sledge dogs. (*Photographer/Byline* Arve Staxrud. *Source/Owner* Norwegian Polar Institute)

**Fig. 24.8** The Spitsbergen Rescuer Dog breed was created, courtesy of Lussi, resulting in future generations of search-and-rescue sled dogs helping workers, visitors, and explorers on Svalbard. (*Photographer/Byline* Arve Staxrud. *Source/Owner* Norwegian Polar Institute)

**Fig. 24.9** Captain Arve Staxrud led the rescue expedition that searched for Lieutenant Herbert Schröder-Stranz's expedition to Spitsbergen. Amid the tragedy and loss that were a consequence of Schröder-Stranz's ambitious yet ill-planned expedition, Staxrud's coordinated, interspecies rescue mission resulted in the saving of lives. (Close-up cropped image excerpted from original portrait photograph by Anders Beer Wilse). (*Photographer* Anders Beer Wilse. *Source/Owner* Norsk Folkemuseum/National Library of Norway)

## Notes on Original Material and Unpublished Sources

Einar Rotvold's diary is in the Library Archives at the Norwegian Polar Institute (NPI) in Tromsø. Hilmar Nøis's diary NPI is in the Library Archives at the Norwegian Polar Institute (NPI) in Tromsø.

1. E. Rotvold diary, 30 March 1913, (NPI), D00125
2. E. Rotvold diary, 30 March 1913, 31 March 1913, 1 April 1913, and 2 April 1913, (NPI), D00125
3. E. Rotvold diary, 30 March 1913, (NPI), D00125
4. H. Nøis diary NPI, notebook No. 11, Summer 1941, journal pages 81–84, (NPI), D-307/D00307_11_0001

## Unpublished Sources and Original Material

Nøis, H., Diary NPI, handwritten journal of recollections recounting events including the evacuation from Svalbard in the summer and September of 1941, Hilmar Nøis, "1940 Evakuerengen 1941 slut, No. 11." [Hilmar Nøis, "1940 The Evacuation 1941 end, Number 11"], D-307, ("Minder og fortelinger, Nøis, Hilmar, hefte 11"), D00307_11_0001. Library Archives, Norwegian Polar Institute, Tromsø, https://brage.npolar.no/npolar-xmlui/handle/11250/274077. Accessed 18 April 2022.

Rotvold, E., Diary, Einar Rotvold's expedition diary written during the Schröder-Stranz expedition of 1913, "Dagbok for August Stenersen og Einar Rotvold, Tromsø: Treurenberg Bay fra 1.januar – 7.juni 1913: Schröder-Stranz-ekspedisjonen 1912–13", "Einar Rotvold har skrevet dagboken", ["Diary for August Stenersen and Einar Rotvold, Tromsø: Treurenberg Bay from 1 January–7 June 1913: The Schröder-Stranz Expedition 1912–13", "Einar Rotvold has written the diary"], dated 12 January 1913 through 7 June 1913, D00125. Library Archives, Norwegian Polar Institute, Tromsø, https://brage.npolar.no/npolar-xmlui/handle/11250/2426394. Accessed 24 July 2021.

## References

Aftenposten: Ugens Nyt. (2 September 1913). En julenats-tragedie paa Spitsbergen. [A Christmas-night tragedy on Spitsbergen.]. *Ugens Nyt* published by *Aftenposten* (Kristiania [Oslo]), 2 September 1913, (pp. 2–3). National Library of Norway, Oslo, online archives. https://www.nb.no. Retrieved 2 April 2022.

Allgeier, S. (1913). *Mit der Kamera im ewigen Eis*. https://de.wikipedia.org/wiki/Mit_der_Kamera_im_ewigen_Eis. Visited 19.5.2021.

Allgeier, S. (1931). *Die Jagd nach dem Bild. 18 Jahre Kameramann in Arktis und Hochgebirge*. Stuttgart: Engelhorn.

Anonym. (1914). Nord-Polargegenden. *Geographische Zeitschrift* XX, (p. 643).

Hergesell, H. (1913). Die Deutsche Wissenschaftliche Station auf Spitzbergen und die Schröder-Stranz-Expedition. *Petermanns Geographische Mitteilungen* (Vol. 59, II, pp. 137).

Nansen, F. (1920). *En ferd til Spitsbergen*. Kristiania: Jacob Dybwads Forlag. Digital Archive, National Library of Norway, Oslo, online archives. https://www.nb.no. Accessed 28 September 2023.

Potpeschnigg, K. (1914). Verlauf und Ausrüstung der Expedition. In H. Philipp (ed.) Ergebnisse der W. Filchnerschen Vorexpedition nach Spitzbergen 1910. *Petermanns Geographische Mitteilungen*, Ergänzungsheft (Vol. 179, pp. 1–13).

Rave, C. (1913, September 27). Die Tragödie der Schröder-Stranz-Expedition. Der Todeszug der zehn mutigen deutschen Männer durch Nacht und Eis. Berlin: *Berliner Tageblatt*, 27 September 1913.

Rüdiger, H. (1939). Die Überreste der Schröder-Stranz-Expedition. *Petermanns Geographische Mitteilungen, 85*, 158–162.

Staxrud, A. (1914). Die Staxrudsche Hilfs-Expedition für Schröder-Stranz. In A. Miethe (ed.), *Die Expedition zur Rettung von Schröder-Stranz und seinen Begleitern - geschildert von ihren Führern Hauptmann A. Staxrud und Dr. K. Wegener* (pp. 1–68). Berlin: Dietrich Reimer.

Tahan, M. R. (2021). *The Return of the South Pole Sled Dogs: With Amundsen's and Mawson's Antarctic Expeditions.* Cham: Springer International Publishing.

Villinger, B. (1929). *Die Arktis ruft! Mit Hundeschlitten und Kamera durch Spitzbergen und Grönland.* Freiburg i. Br.: Herder.

Wegener, K. (1914). Die Hilfsexpedition von Cross-und Kings Bay, 21.II.-31.III.1913. In A. Miethe (Ed.), *Die Expeditionen zur Rettung von Schröder-Stranz und seinen Begleitern geschildert von ihren Führern Hauptmann A. Staxrud und Dr. K. Wegener.* Berlin: Dietrich Reimer, pp. 69–101.

# Bibliography

## *Unpublished Sources*

Allgemeiner Plan. (1912). Allgemeiner Plan der Deutschen Arktischen Expedition durch die Nordostpassage (Taimyr-Halbinsel) und durch den Stillen Ozean. Schröder-Stranz estate, Kasten 26, Nr. 13, Leibniz Institute for Regional Geography, Leipzig, Germany.

Amundsen Letters of Correspondence, Manuscripts Collection, National Library of Norway, Oslo.

Amundsen, L. (1913, 13 August). Telegram from Leon Amundsen to Arve Staxrud, letter copybook NB Brevs. 812:3:8. Manuscripts Collection, National Library of Norway, Oslo. Received 9 June 2021.

Amundsen, L. (1913a, 14 August). Telegram from Leon Amundsen to Øyvind Widding-Danielsen, NB Brevs. 812:3:8. Manuscripts Collection, National Library of Norway, Oslo. Received 9 June 2021.

Amundsen, L. (1913b, 14 August). Letter from Leon Amundsen to Foreign Minister of Norway, NB Brevs. 812:3:8. Manuscripts Collection, National Library of Norway, Oslo. Received 23 June 2021.

Committee Letter to Norske Geografiske Selskap, (1913, 18 March). Letter from Committee to Norske Geografiske Selskap board, 1913_03_18_Brev_til_Norske_Geografiske_Selskap, Library Archives, Norwegian Polar Institute, Tromsø. Received 20 July 2021.

Gazert, H. (1913, 24 February). Letter to Drygalski. Gazert estate, Volkert Gazert, Partenkirchen.

Henriksen, O. (1911–1913). Diary, Olaf Henriksen's diary written while he was the Green Harbour Telegraph Station Manager, "Dagbok for Olaf Henriksen, 1911–13" ["Diary for Olaf Henriksen, 1911–13"], dated 14 August 1911 to 2 May 1913, D-305, D00305_0001. Library Archives, Norwegian Polar Institute, Tromsø. https://brage.npolar.no/npolar-xmlui/han dle/11250/2600678. Accessed 24 July 2021.

Hoel, A. (1913, 29 April). Letter from Adolf Hoel to The Royal Ministry of Foreign Affairs, 1913_ Brev fra Hoel til Utenriksdepartementet. Library Archives, Norwegian Polar Institute, Tromsø. Received 20 July 2021.

Nøis, D. Daniel Nøis Biography / Norwegian Polar Institute, "Nøis_Daniel_Biografiarkiv_1", written by H. M. [possibly Haakon Aronsen, aka Haakon Magnus, per Gunnhild Holmen of the National Library of Norway, based on *Andøya Avis* 1964 article, according to communications from Odd Ivar Nøis Olsen of the Nøis Family and from Ivar Stokkeland and Petr Masat of the Norwegian Polar Institute, received 29 March 2022], Biography Archive/Library Archives, Norwegian Polar Institute, Tromsø. Received 20 July 2021.

Nøis, H., Diary NPI a, a handwritten journal of recollections recounting the events of the 1913 Staxrud expedition, Hilmar Nøis, "Minder og Fortelinger fra fangsttiden på Spitsbergen, Klade 3, 7 Mars

© The Editor(s) (if applicable) and The Author(s), under exclusive license to Springer Nature Switzerland AG 2024
M. R. Tahan and C. Lüdecke, *Stranded at the Top of the World*,
https://doi.org/10.1007/978-3-031-56288-4

1970" [Hilmar Nøis, "Memories and Stories from the trapping time on Spitsbergen, Notebook 3, 7 March 1970"], dated 7 March 1970, D-307, ("Minder og fortelinger, Nøis, Hilmar, hefte 3"), D00307_3_0001. ("Opskrifter fra Dagbøker Hilmar Nøis.") Library Archives, Norwegian Polar Institute, Tromsø. https://brage.npolar.no/npolar-xmlui/handle/11250/274077. Accessed 20 July 2021.

Nøis, H., Diary NPI b, handwritten journal of recollections recounting events including the 1928 search for the Nobile expedition, Hilmar Nøis, "Minder og Fortelinger, No 8., Fra No 7 Den 23 Jan 1971" [Hilmar Nøis, "Memories and Stories, Number 8, From No. 7 on 23 January 1971"], D-307, ("Minder og fortelinger, Nøis, Hilmar, hefte 8"), D00307_8_0001. ("Opskrifter fra Dagbøker Hilmar Nøis.") Library Archives, Norwegian Polar Institute, Tromsø. https://brage. npolar.no/npolar-xmlui/handle/11250/274077. Accessed 18 April 2022.

Nøis, H., Diary NPI c, handwritten journal of recollections recounting events including the evacuation from Svalbard in the summer and September of 1941, Hilmar Nøis, "1940 Evakuerengen 1941 slut, No. 11." [Hilmar Nøis, "1940 The Evacuation 1941 end, Number 11"], D-307, ("Minder og fortelinger, Nøis, Hilmar, hefte 11"), D00307_11_0001. ("Opskrifter fra Dagbøker Hilmar Nøis.") Library Archives, Norwegian Polar Institute, Tromsø. https://brage.npolar.no/ npolar-xmlui/handle/11250/274077. Accessed 18 April 2022.

Nøis, H., Diary SVB a, handwritten journal of recollections including an account of the 1913 Staxrud expedition, ("Hilmar Nøis. Fangstmann. 1909–1923, dagbok med erindringer fra fangstlivet") ["Hilmar Nøis. Trapper. 1909–1923, diary with recollections from the trapping life"], not dated (latest year referenced is 1942 in journal pages and 1950 in inserted sheet), SVB-AP2-Hilmar-Nøis-dagbok-1909–1923-compressed. Historical Archive, Svalbard Museum, Svalbard. https://svalbardmuseum.no/no/samlingene/historisk-arkiv/arkivprosjekt-fangst-og-annen-overvintringsvirksomhet-fra-perioden-1910-til-1970/. Accessed 6 June 2021.

Nøis, H., Diary SVB b, handwritten journal of events in 1928, Hilmar Nøis, "Dagbok for år 1928" [Hilmar Nøis, "Diary for year 1928"], ("Hilmar Nøis. Fangstmann. 1928, dagbok Roosneset/ Woodfjorden"), SVB-AP3-Hilmar-Nøis-dagbok-1928-compressed. Historical Archive, Svalbard Museum, Svalbard. https://svalbardmuseum.no/no/samlingene/historisk-arkiv/arkivpros jekt-fangst-og-annen-overvintringsvirksomhet-fra-perioden-1910-til-1970/. Accessed 9 April 2022.

Norwegian Polar Institute, Tromsø. Olaf Henriksen correspondence in Library Archives. Approximate date-of-birth and date-of-death dates per Fred I. Presteng in memo dated 7 February 2008. Received 22 July 2021.

Notruf. (1913). Notruf zur Rettung der Schroeder-Stranz'schen Spitzbergen Expedition. Schröder-Stranz estate, Kasten 26, Nr. 76, Leibniz Institute for Regional Geography, Leipzig, Germany.

Oertz, M., (1912). Schröder-Stranz estate, Box 26, Leibniz Institute for Regional Geography, Leipzig, Germany.

Rotvold, E., Diary, Einar Rotvold's expedition diary written during the Schröder-Stranz expedition of 1913, "Dagbok for August Stenersen og Einar Rotvold, Tromsø: Treurenberg Bay fra 1.januar – 7.juni 1913: Schröder-Stranz-ekspedisjonen 1912–13", "Einar Rotvold har skrevet dagboken", ["Diary for August Stenersen and Einar Rotvold, Tromsø: Treurenberg Bay from 1 January – 7 June 1913: The Schröder-Stranz Expedition 1912–13", "Einar Rotvold has written the diary"], dated 12 January 1913 through 7 June 1913, D00125. Library Archives, Norwegian Polar Institute, Tromsø. https://brage.npolar.no/npolar-xmlui/handle/11250/2426394. Accessed 24 July 2021.

Schröder-Stranz estate, Leibniz Institute for Regional Geography, Leipzig, Germany.

Spitsbergen Radio, Personnel List, "Personalfortegnelse", handwritten Green Harbour telegraph station staff record 1 July 1911–17 August 1914, courtesy of Per Kyrre Reymert, from the National Archives in Tromsø (Arkivverket Statsarkivet i Tromsø), File Spi radio065, received from Petr Masat, Library, Norwegian Polar Institute, Tromsø (Bibliotek, Norsk Polarinstitutt). Received 31 March 2022.

Staxrud, A., Letters of correspondence, Arve Staxrud letters of correspondence, written from and to Arve Staxrud, are in the Library Archives at the Norwegian Polar Institute (NPI) in Tromsø.

Staxrud, A., Letters of correspondence, Arve Staxrud letters of correspondence, written from and to Arve Staxrud, are in the Manuscripts Collection at the National Library of Norway (NB) in Oslo.

Staxrud, A. (1913, 17 February). Letter from Captain Arve Staxrud to Norway's Minister of Foreign Affairs Nils Claus Ihlen regarding plans for rescue expedition after Schröder-Stranz, with notes, "Redningsekspeditionen for Schröder-Stranz 1913" ["The Rescue Expedition for Schröder-Stranz 1913"], copybook and notes, 1913_02_17_Redningsexpeditionen_Schroder_St[r]anz_1913_Hoels_og_Staxruds_Spitsbergenexpedisjoner, Library Archives, Norwegian Polar Institute, Tromsø. Received 20 July 2021.

Staxrud, A. (1913a, 4 March). Letter from Arve Staxrud to Leon Amundsen, NB Brevs. 812:1. National Library of Norway, Oslo, online archives. https://www.nb.no Retrieved 6 July 2021.

Staxrud, A. (1913b, 4 March). Letter from Arve Staxrud to Alex Nansen, NB Brevs. 812:1. National Library of Norway, Oslo, online archives. https://www.nb.no Retrieved 6 July 2021.

Staxrud, A. (1913, 11 March). Letter from Arve Staxrud to Fridtjof Nansen, NB Ms.fol. 1924:6:A:2. National Library of Norway, Oslo, online archives. https://www.nb.no. Retrieved 6 July 2021.

Staxrud, A. (1913, 12 March). Telegram from Arve Staxrud to Fridtjof Nansen, NB Ms.fol. 1924. National Library of Norway, Oslo, online archives. https://www.nb.no. Retrieved 6 July 2021.

Staxrud, A. (1913, 13 March). Telegram from Arve Staxrud to Adolf Hoel, 1913_mai_Telegrams_fra_Staxrud_Part3, Library Archives, Norwegian Polar Institute, Tromsø. Received 20 July 2021.

Staxrud, A. (1913, 17 March). Letter from Arve Staxrud to Fridtjof Nansen, with two newspaper article clippings enclosed featuring handwritten notation (*Nord Norge* [Tromsø], 21 February 1913, pages 1–2, and *Dagsposten*, Trondhjem [Trondheim], 6 March 1913), NB Ms.fol. 1924:6:A:2. National Library of Norway, Oslo, online archives. https://www.nb.no. Retrieved 6 July 2021.

Staxrud, A. (1913, 24 March). Letter from Arve Staxrud (to Adolf Hoel), 1913_mai_Telegrams_fra_Staxrud_Part2, Library Archives, Norwegian Polar Institute, Tromsø. Received 20 July 2021.

Staxrud, A. (1913, 25 March). Telegram from Arve Staxrud to Fridtjof Nansen, NB Ms.fol. 1924. National Library of Norway, Oslo, online archives. https://www.nb.no. Retrieved 6 July 2021.

Staxrud, A. (1913, 14 May). Telegram from Arve Staxrud to Fridtjof Nansen, NB Ms.fol. 1924. National Library of Norway, Oslo, online archives. https://www.nb.no . Retrieved 6 July 2021.

Staxrud, A. (1913, 25 May). Letter from Arve Staxrud to Adolf Hoel, 1913_05_25_Brev fra Staxrud til Hoel. Library Archives, Norwegian Polar Institute, Tromsø. Received 20 July 2021.

Staxrud, A., (1913, 26 May and 30 May 1913). (the latter "delayed due to line fault" and re-sent by the telegraph office on 6 June 1913), Telegrams from Arve Staxrud to Adolf Hoel, 1913_mai_Telegrams_fra_Staxrud. Library Archives, Norwegian Polar Institute, Tromsø. Received 20 July 2021.

Staxrud, A. (1913, 11 August). Telegram from Arve Staxrud to Roald Amundsen, NB Brevs. 812:1. National Library of Norway, Oslo, online archives. https://www.nb.no. Retrieved 6 July 2021.

Staxrud, A. (1919, 23 May). Letter from Arve Staxrud to Fridtjof Nansen, NB Brevs. 48. National Library of Norway, Oslo, online archives. https://www.nb.no. Retrieved 6 July 2021.

Staxrud, A. (1920, 22 October). Letter from Arve Staxrud to Fridtjof Nansen with three accompanying letters, NB Brevs. 48. National Library of Norway, Oslo, online archives. https://www.nb.no. Retrieved 6 July 2021.

Staxrud, A. (1923, 19 October). Letter from Arve Staxrud to Fridtjof Nansen, NB Ms.fol. 1924:6:A:2. National Library of Norway, Oslo, online archives. https://www.nb.no. Retrieved 6 July 2021.

Staxrud, A. (1925, 26 May). Letter from Arve Staxrud to Fridtjof Nansen with accompanying article, NB Brevs. 48. National Library of Norway, Oslo, online archives. https://www.nb.no. Retrieved 6 July 2021.

Staxrud, A. and Kemi, J. (1913, 31 May). Contract between Arve Staxrud and Johannes Kemi signed in Green Harbour, Johannes_Kemi_Kontrakt_Green_Harbour_31_mai_1913. Library Archives, Norwegian Polar Institute, Tromsø. Received 20 July 2021.

Tessem, E., Jenssen, I., Rognli, J. & Pedersen, E. (1913). (Translated from Norwegian to German by Volkert Gazert and later translated to English by Cornelia Lüdecke.) Ritscher estate, Cornelia Lüdecke, München.

Zapffe, F. G. (1913, 30 August). Telegram from Fritz G. Zapffe to Leon Amundsen, NB Brevs. 812:1. Manuscripts Collection, National Library of Norway, Oslo. Received 22 June 2021.

# References

Allgeier, S. (1913). *Mit der Kamera im ewigen Eis.* https://de.wikipedia.org/wiki/Mit_der_Kam era_im_ewigen_Eis. Visited 19.5.2021.

Allgeier, S. (1931). *Die Jagd nach dem Bild. 18 Jahre Kameramann in Arktis und Hochgebirge.* Stuttgart: Engelhorn.

Amundsen, R. (1912). *The South Pole: An account of the Norwegian Antarctic Expedition in the "Fram", 1910–1912,* 2 vols (A. G. Chater, Trans.). London: John Murray.

Andree, R. (1893). *Andrees allgemeiner Handatlas.* 3., völlig neubearb. und verm. Aufl. Bielefeld: Velhagen & Klasing.

Angenheister, G. G. (1974). Geschichte des Samoa-Observatoriums von 1902–1921. In: H. Birett, K. Helbig, W. Kertz und U. Schmucker: *Zur Geschichte der Geophysik. Festschrift zur 50-jährigen Wiederkehr der Gründung der Deutschen Geophysikalischen Gesellschaft.* Berlin, Heidelberg, New York: Springer, pp. 43–66.

Anonym. (1913). Nord-Polargegenden. *Geographische Zeitschrift,* XIX, pp. 408–409.

Anonym. (1914). Eine Organisation für Polar- und Forschungsexpeditionen. *Zeitschrift der Gesellschaft für Erdkunde zu Berlin,* pp. 223–224.

Anonym. (1914). Nord-Polargegenden. *Geographische Zeitschrift* XX, p. 643.

Behm. (1912). Die deutsche Arktische Expedition Schröder-Stranz. *Annalen der Hydrographie und Maritimen Meteorologie, 40,* 449.

Berg, R. (2020). The Svalbard "Channel", 1920–2020—A geopolitical sketch. In: Weber, J. (Eds.), *Handbook on geopolitics and security in the Arctic.* Frontiers in international relations, pp. 303–321. Cham: Springer. https://doi.org/10.1007/978-3-030-45005-2_18

Brunner, K. and C. Lüdecke. (2012). Übung für die Antarktis: Wilhelm Filchners Vorexpedition nach Spitzbergen im Jahr 1910. Ein Beitrag zur Expeditionskartographie. In: C. Lüdecke, & K. Brunner (Eds.), Von A(ltenburg) bis Z(eppelin). Deutsche Forschung auf Spitzbergen bis 1914. 100 Jahre Expedition des Herzogs Ernst II. von Sachsen-Altenburg. *Schriftenreihe des Instituts für Geodäsie der Universität der Bundeswehr München,* Neubiberg, Heft 88, pp. 69–76.

Capelotti, P. J. (1999). *By airship to the North Pole: An archaeology of human exploration.* New Brunswick, New Jersey, and London: Rutgers University Press.

Capelotti, P. J. (2002). The train has left the station: archaeological formation processes and the abandoned coal mining village of Pyramiden, Svalbard. In Bouzney, E. (Ed.), *International Scientific Cooperation in the Arctic: Proceedings of a Conference Devoted to the Centenary of the Swedish-Russian Arc-of-Meridian Expedition to Spitsbergen and 125 Years of V.A. Rusanov's Birth.* Moscow: Scientific World, pp. 116–127.

Capelotti, P. J. (2013). *Shipwreck at Cape Flora: The expeditions of Benjamin Leigh Smith, England's forgotten Arctic explorer.* Calgary: University of Calgary Press.

Dean, R., & Lackenbauer, P. W. (2017). "A particularly spectacular piece of demolition": The Canadian-Led Raid on Spitzbergen, 1941. In A. Lajeunesse & P. W. Lackenbauer (Eds.), *Canadian armed forces Arctic operations, 1941–2015: Lessons learned, lost, and relearned* (pp. 1–46). Fredericton, NB: The Gregg Centre for the Study of War & Society/University of New Brunswick. https://www.unb.ca/fredericton/arts/centres/gregg/what/publications/CdnArc ticOps2017.pdf. Retrieved 15 April 2022.

Dole, N. H. (1922). *America in Spitsbergen: The Romance of an Arctic Coal-Mine.* In Two Volumes. Boston: Marshall Jones Company. Volume II. Facsimile reprint by Scholar Select.

Filchner, W. and Seelheim, H. (1911). *Quer durch Spitzbergen. Eine deutsche Übungsexpedition im Zentralgebiet östlich des Eisfjords*. Berlin: E.S. Mittler und Sohn.

Filchner, W. (1922). *Zum sechsten Erdteil*. Berlin: Ullstein.

Filchner, W. (1994). *To the Sixth Continent: The Second German South Polar Expedition, 1911–1913*. Translated and edited by W. Barr. Eccles: Erskine Press and Bluntisham: Bluntisham Books.

Fricke, H. (2009). *Verschollen in der Arktis*. https://www.youtube.com/watch?v=TNScwgehIto. Visited 19.5.2021

Fuchs, A. (2008). *Fahrt ins Ungewisse. Verschollen vor Spitzbergen*. https://www.spitzbergen.de/2008/04/20/schroder-stranz-expedition-filmmaterialaufgetaucht.html. Visited 20.5.2021

*Hanseater.no*. Etterkommere av Hanseatene i Bergen: Geolog Jakob Andreas Martin Ellingsen. The Hanseatic Museum, and Slekt og Data Hordaland, "Hansa" Project. http://hanseater.no/tng/getperson.php?personID=I14579&tree=hansa. Accessed 19 September 2023.

Heidbrink, I. (2015). Vom Scheitern vor dem Beginnen. Überlegungen zu den Motiven der gescheiterten Schroeder-Stranz-Expedition nach Spitzbergen. *Jahrbuch für Europäische Überseegeschichte* 13. Wiesbaden: Harrassowitz, pp. 147–166.

Hergesell, H. (1911a). Das arktische Luftschiffunternehmen und der Zweck unserer Studienreise. In: A. Miethe and H. Hergesell (Eds.) *Mit Zeppelin nach Spitzbergen*. Berlin: Deutsches Verlagshaus Bong & Co., pp. 4–16.

Hergesell, H. (1911b). Die Fahrten des „Fönix". In: A. Miethe, & H. Hergesell (Eds.) *Mit Zeppelin nach Spitzbergen*. Berlin: Deutsches Verlagshaus Bong & Co., pp. 227–282.

Hergesell, H. (1913). Die Deutsche Wissenschaftliche Station auf Spitzbergen und die Schröder-Stranz-Expedition. *Petermanns Geographische Mitteilungen* 59, II, p. 137.

Hergesell, H. (1914). Die Deutsche wissenschaftliche Station in Spitzbergen. In: H. Hergesell (Ed.): Das Deutsche Observatorium in Spitzbergen. Beobachtungen und Ergebnisse I, *Schriften der Wissenschaftlichen Gesellschaft in Straßburg* 21, pp. 1–5.

Hergesell, H. (1933). Der erste Registrierballonaufstieg in hohen arktischen Breiten. *Beiträge zur Physik der Atmosphäre*, *20*, 261–268.

Herzog Ernst. (1943). Beschreibung der Spitzbergenexpedition des „Herzog Ernst" II. von Sachsen-Altenburg (1943) In: U. Gillmeister. (2009): *Vom Thron auf den Hund*. Borna: Südraum, 3. Auflage, pp. 240–269.

Hoel, A. (Ed.). (1929). *Resultater av de Norske Statsunderstøttede Spitsbergenekspeditioner (Skrifter om Svalbard og Ishavet), Bind I, Nr. 1, Adolf Hoel: The Norwegian Svalbard Expeditions 1906–1926*. Utgitt på Den Norske Stats bekostning ved Spitsbergenkomiteen. Oslo: (I Kommisjon Hos) Jacob Dybwad. https://brage.npolar.no/npolar-xmlui/bitstream/handle/11250/173618/Skrifter001.pdf?sequence=1&isAllowed=y. Accessed 19 September 2023.

Hoel, A. (1933). Arve Staxrud. In: *Norsk Geografisk Tidsskrift [Norwegian Journal of Geography]*, *4*(6), 345–346. https://doi.org/10.1080/00291953308622018

Kneitz, A. (2016). German Water Infrastructure in China: Colonial Qingdao 1898–1914). *Zeitschrift für Geschichte der Wissenschaften, Technik und Medizin* (N.T.M.) 24, pp. 421–450.

Koldewey, K. and Petermann, A. (1871). Die erste Deutsche Nordpolar-Expedition im Jahre 1868. *Petermanns Geographische Mitteilungen*. Ergänzungsheft 28.

Krigsseileren. (1982). Brev til redaksjonen: Fra en veterans erindringer [Letter to the editor: From a veteran's recollections]. In: *Krigsseileren*, Nr. 2 – 1982 – 11. årgang (p. 22). Norges Krigsseilerforbund. H. Wigestrand (Ed.). Oslo: Norges Krigsseilerforbund. https://www.krigsseilerregisteret.no/no/tidsskriftet-krigsseileren/360382. Accessed 4 July 2023.

Lerner, T. (2005). *Polarfahrer: Im Banne der Arktis*. F. Berger (Ed.). Zürich: Oesch/Kontrapunkt.

*Longyear Museum*. John Munro Longyear. https://www.longyear.org/learn/pioneer-index/longyear-john-m/ and https://www.longyear.org/learn/research-archive/john-longyear-landlooker-from-michigan/. Accessed April 13, 2022.

Lüdecke, C. (1995). Die deutsche Polarforschung seit der Jahrhundertwende und der Einfluß Erich von Drygalskis. Ph.D. Thesis. *Berichte zur Polarforschung* 158. Bremerhaven: Alfred-Wegener-Institut für Polar- und Meeresforschung.

Lüdecke, C. (2012). Die Zeppelin-Studienexpedition nach Spitzbergen (1910). In: C. Lüdecke, & K. Brunner (Eds.), Von A(ltenburg) bis Z(eppelin). Deutsche Forschung auf Spitzbergen bis 1914. 100 Jahre Expedition des Herzogs Ernst II. von Sachsen-Altenburg. *Schriftenreihe des Instituts für Geodäsie der Universität der Bundeswehr München*, Neubiberg, Heft 88, pp. 99–107.

Lüdecke, C., and Summerhayes, C. (2012). *The Third Reich in Antarctica: The German Antarctic Expedition 1938–39*. Eccles: Erskine Press and Bluntisham: Bluntisham Books.

Marschall, B. (1991). Reisen und Regieren. Die Nordlandfahrten Kaiser Wilhelms II. *Schriften des Deutschen Schiffahrtsmuseums* 27. Bremen: Schiffahrtsmuseum, Hamburg: Ernst Kabel.

Martens, F. (1675). *Friderich Martens vom Hamburg Spitzbergische oder Groenlandische Reise Beschreibung gethan 1671. Aus eigner Erfahrung beschrieben / die dazu erforderte Figuren nach dem Leben selbst abgerissen / (so hierbey in Kupffer zu sehen) und jetzo durch den Druck mitgetheilet.* Hamburg.

Miethe, A. (1911). Die Reise der „Mainz". In: A. Miethe & H. Hergesell (Eds.) *Mit Zeppelin nach Spitzbergen.* Berlin: Deutsches Verlagshaus Bong & Co., pp. 17–164.

Miethe, A., and Hergesell, H. (Eds.). (1911). *Mit Zeppelin nach Spitzbergen.* Berlin: Deutsches Verlagshaus Bong & Co.

Miethe, A. (Ed.). (1914). *Die Expedition zur Rettung von Schröder-Stranz und seinen Begleitern - geschildert von ihren Führern Hauptmann A. Staxrud und Dr. K. Wegener.* Berlin: Dietrich Reimer.

Nansen, F. (1914). *Through Siberia, the land of the future,* (A. G. Chater, Trans.). New York: Frederick A. Stokes Company; London: William Heinemann. Digitized by Google. https://books.google.com. Accessed 4 August 2014.

Nansen, F. (1920). *En ferd til Spitsbergen.* Kristiania: Jacob Dybwads Forlag. Digital Archive, National Library of Norway, Oslo, online archives, https://www.nb.no. Accessed 28 September 2023.

*National Mining Hall of Fame & Museum.* John Munro Longyear. https://www.mininghalloffame. org/hall-of-fame/john-munro-longyear. Accessed 13 April 2022.

Nøis, D. (1929). Med kaptein Staxrud på leiting etter Schrøder-Stranz og hans folk [With Captain Staxrud on the search for Schröder-Stranz and his people]. In: *And-Ungen*, July 1929, pp. 4–12. Andenes: Andøyposten. Library Archives, Norwegian Polar Institute, Tromsø. Received 20 July 2021.

*Nøis (slekt) – lokalhistoriewiki.no.* Nøis Family. https://lokalhistoriewiki.no/wiki/Nøis_(slekt) Accessed 29 March 2022 and 20 June 2022.

Norddeutscher Lloyd. (1911). *Polarfahrt 1911 mit dem großen Schraubendampfer „Großer Kurfürst".* Bremen: Norddeutscher Lloyd.

Nordenskjöld, A. E. (1880). *Die Nordpolarreisen Adolf Erik Nordenskjöld's 1858 bis 1879.* Leipzig: F.A. Brockhaus.

*Norsk Polarhistorie – Polarhistorie.no.* Søren Zachariassen. "Søren Zachariassen: 1837–1915, Ishavsskipper fra Tromsø som innledet kulldriften på Spitsbergen." (Norwegian Polar Institute, University of Tromsø, Troms County Municipality). https://polarhistorie.no/personer/Zac hariassen,%20Soren.html. Accessed 12 May 2022.

*Norsk Polarhistorie – Polarhistorie.no.* Spitsbergen Radio. "Etablering av Spitsbergen Radio i 1911." Oddvar Ulvang. (Norwegian Polar Institute, University of Tromsø, Troms County Municipality). https://www.polarhistorie.no/artikler/2014/etablering-spitsbergen-radio. html. Accessed 12 May 2022.

*Norsk Polarhistorie – Polarhistorie.no.* Staxrud's Rescue Expedition. "Staxruds unnsetningsekspedisjon etter Schröder-Stranz, 1913." ["Staxrud's rescue expedition after Schrøder-Stranz, 1913"]. (Norwegian Polar Institute, University of Tromsø, Troms County Municipality). http://www.polarhistorie.no/ekspedisjoner/Staxruds%20unnsetningseksp.html. Accessed 22 August 2012, and 19 September 2023.

*Norwegian Polar Institute.* Hertha. https://data.npolar.no/vessel/hertha. Accessed 24 July 2021.

*Norwegian Polar Institute.* The History of the Norwegian Polar Institute. https://www.npolar.no/en/history/. Accessed 21 July 2021.

Norwegian Polar Institute. (2003). *The Place Names of Svalbard*. Rapportserie nr. 122. G. S. Jaklin, technical editor. (O. Orheim, director; A. Urset, committee chair.) Peder Norbye grafisk. Tromsø: Norwegian Polar Institute. https://brage.npolar.no/npolar-xmlui/bitstream/han dle/11250/173470/Rapport122.pdf?sequence=1&isAllowed=y. Accessed 19 September 2023.

Oesau, W. (1937). *Schleswig-Holsteins Grönlandfahrt auf Walfischfang und Robbenschlag vom 17. - 19. Jahrhundert*. Glückstadt: Augustin.

Oesau, W. (1955). *Hamburgs Grönlandfahrt: auf Walfischfang und Robbenschlag vom 17.-19. Jahrhundert*. Glückstadt: Augustin.

Penck, A. (1912). Polargebiete. *Zeitschrift der Gesellschaft für Erdkunde zu Berlin*, pp. 91–93.

Philipp, H. (Ed.). (1914). Ergebnisse der W. Filchnerschen Vorexpedition nach Spitzbergen 1910. *Petermanns Geographische Mitteilungen*, Ergänzungsheft 179.

Piepjohn, K. (2021). *English version of the German map, first published in black and white*. In: K. Piepjohn (2012), p. 61, (updated 2023). From: Piepjohn, K. (2012). Weg-Zeit-Diagramm der Schröder-Stranz-Expedition und der norwegischen und deutschen Rettungsexpeditionen 1912/1913. In: Lüdecke, C. und K. Brunner (Eds.), Von A(ltenburg) bis Z(eppelin). Deutsche Forschung auf Spitzbergen bis 1914. 100 Jahre Expedition des Herzogs Ernst II. von Sachsen-Altenburg. *Schriftenreihe des Instituts für Geodäsie der Universität der Bundeswehr München*, Neubiberg, Heft 88, pp. 59–68.

Pluntke, M. (2008). Expedition des Herzogs Ernst II. von Sachsen-Altenburg im Sommer 1911 nach Spitzbergen. *Altenburger Geschichts- und Hauskalender für den Kreis Altenburger Land* 17 NF, pp. 93–98.

Potpeschnigg, K. (1914). Verlauf und Ausrüstung der Expedition. In: H. Philipp (Ed.) Ergebnisse der W. Filchnerschen Vorexpedition nach Spitzbergen 1910. *Petermanns Geographische Mitteilungen*, Ergänzungsheft 179, pp. 1–13.

Przigoda, S. (2012). Bergbau auf der Bäreninsel? Deutsche Rohstoffinteressen und die Erkundung Svalbards (1871–1914). In: C. Lüdecke & K. Brunner (Eds.), Von A(ltenburg) bis Z(eppelin). Deutsche Forschung auf Spitzbergen bis 1914. 100 Jahre Expedition des Herzogs Ernst II. von Sachsen-Altenburg. *Schriftenreihe des Instituts für Geodäsie der Universität der Bundeswehr München*, Neubiberg, Heft 88, pp. 77–91.

Rabot, C. (1919). The Norwegians in Spitsbergen. *The Geographical Review, 8*(4/5), (Oct.-Nov. 1919), pp. 209–226. https://doi.org/10.2307/207837. Accessed 21 July 2021.

Rave, C. (1913). *Tagebuch von der verunglückten Expedition Schröder-Stranz*. Schaffsteins Gründe Bändchen 49. Cöln: Schaffstein.

Rempp, G., and Wagner, A. (1914). Die Station in der Adventbai. In: H. Hergesell (Ed.), Das Deutsche Observatorium in Spitzbergen. Beobachtungen und Ergebnisse I. *Schriften der Wissenschaftlichen Gesellschaft in Straßburg* 21, pp. 6–20.

Ritscher, A. (1916). Wanderung in Spitzbergen im Winter 1912. *Zeitschrift der Gesellschaft für Erdkunde zu Berlin*, pp. 16–34.

Rüdiger, H. (1913). *Die Sorge Bay. Aus den Schicksalstagen der Schröder-Stranz-Expedition*. Berlin: Georg Reimer.

Rüdiger, H. (1939). Die Überreste der Schröder-Stranz-Expedition. *Petermanns Geographische Mitteilungen, 85*, 158–162.

Rudmose Brown, R. N. (1919). Spitsbergen, Terra Nullius. *The Geographical Review, 7*(5), (May 1919), 311–321. https://doi.org/10.2307/207588. Accessed 29 March 2022.

Schmidt, C. (1913). Übersichtskarte von Spitzbergen zur Veranschaulichung des Verlaufs der Expedition Schröder-Stranz und der Hilfsexpeditioin zu deren Rettung. *Petermanns Geographische Nachrichten* 59, II, table 29.

Schofield, E. and R. C. Nesbit. ([1987] 2005). *Arctic Airmen: The RAF in Spitsbergen and North Russia, 1942*. This edition published in 2005 in the UK. Staplehurst, Kent: Spellmount Limited.

Schröder Stranz, H. (1910). *"Süd-West". Kriegs- und Jagdfahrten*. Berlin: Wilhelm Süsserott.

Schröder Stranz, H. (1911). Quer durch Russisch-Lappland. *Die Woche, Nr. 26*, 1090–1095.

Sieberg, A. (1913). *Im Nordpolareis des nordwestlichen Spitzbergen*. Straßburg: M. DuMont.

Sorge, E. (1937). Das Nordostland von Spitzbergen. Karte. *Berliner-Lokal-Anzeiger*. Ritscher estate, Cornelia Lüdecke, München.

Staxrud, A. (1914). Die Staxrudsche Hilfs-Expedition für Schröder-Stranz. In: A. Miethe (Ed.), *Die Expedition zur Rettung von Schröder-Stranz und seinen Begleitern - geschildert von ihren Führern Hauptmann A. Staxrud und Dr. K. Wegener* (pp. 1–68). Berlin: Dietrich Reimer.

Steinhagen, H. (2008). *Max Robitzsch. Polarforscher und Meteorologe*. Jakobsdorf: Die Furt.

Stenersen, H. (editor/publisher). (1988). Tingelstad-brødrene som kartla Svalbard: Arve og Olav Staxrud har satt varige spor etter seg. In: *Brandbu'stikka: Lokalhistorisk Kvartalskrift for Brandbu. 1. KV. 1988 Mars NR. 11* (March 1988). (Svalbard-forskerne Arve og Olav Staxrud.) pp. 1–26. Biography Archive/Library Archives, Norwegian Polar Institute, Tromsø. Received 20 July 2021.

*Svalbard Museum* a. https://www.arkivportalen.no/contributor/no-SVAM_arkiv_000000000112. Accessed 3 March 2023.

*Svalbard Museum* b. https://www.arkivportalen.no/contributor/no-SVAM_arkiv_000000000361? ins=SVAM and https://bildearkiv.svalbardmuseum.no/fotoweb/archives/. Accessed 6 June 2021.

*Svalbard Museum* c. https://www.arkivportalen.no/entity/no-SVAM_arkiv_000000000022? ins=SVAM and https://www.arkivportalen.no/contributor/no-SVAM_arkiv_000000000032. Accessed 7 June 2021.

*Svalbard Museum*. Mining Communities. Gerd Johanne Valen/Bjørg Evjen. https://svalbardm useum.no/en/kultur-og-historie/gruvesamfunn/. Accessed 26 January 2022.

*Svalbard Museum*. The Pomors Enter the Area. Per Kyrre Reymert/Christian Lydersen/Kit Kovacs. https://svalbardmuseum.no/en/kultur-og-historie/pomorene/. Accessed 26 January 2022.

*Svalbard Museum*. Trapper Life/Hunting Life. Marit Anne Hauan/Gerd Johanne Valen. https://sva lbardmuseum.no/en/kultur-og-historie/pelsjegerliv/. Accessed 26 January 2022.

*Svalbard Museum*. Whaling in the Arctic. Kristin Prestvold. https://svalbardmuseum.no/en/kultur-og-historie/hvalfangst/. Accessed 26 January 2022.

*Svalbard Museum*. Willem Barentsz and the Discovery of Svalbard. Thor Bjørn Arlov. https://sva lbardmuseum.no/en/kultur-og-historie/oppdagelsen/. Accessed 26 January 2022.

Tahan, M. R. (2019). *Roald Amundsen's Sled Dogs: The Sledge Dogs Who Helped Discover the South Pole*. Cham: Springer International Publishing.

Tahan, M. R. (2021). *The Return of the South Pole Sled Dogs: With Amundsen's and Mawson's Antarctic Expeditions*. Cham: Springer International Publishing.

Tandberg, R. S. (1928). *Med hundespann på eftersøkning efter "Italia"-folkene*. Særtrykk av *Norsk Geografisk Tidsskrift*, Bind II, Hefte 3–4, 1928. Oslo: A. W. Brøggers Boktrykkeri A/S. Digital Archive, National Library of Norway, Oslo, received 8 June 2021. Also: Tandberg, R. S. (1928/ 1929). Med hundespann på eftersøkning efter "Italia"-folkene. In *Norsk Geografisk Tidsskrift*, Bind II –1928/1929, pp. 176–214. O. Holtedahl (Ed.), 1929. Oslo: A. W. Brøggers Boktrykkeri A/S. National Library of Norway, Oslo, online archives, https://www.nb.no. Retrieved 6 February 2022 and 23 June 2023.

Villinger, B. (1928). Erlebnisse auf der deutschen Arktischen Hilfsexpedition für Schröder-Stranz. In: *Akademischer Skiclub Freiburg im Breisgau 1908–1928*. C.A. Wagner, Freiburg, i. B. pp. 20–43.

Villinger, B. (1929). *Die Arktis ruft! Mit Hundeschlitten und Kamera durch Spitzbergen und Grönland*. Freiburg i. Br.: Herder.

*WarSailors.com*. D/S Isbjørn. http://www.warsailors.com/singleships/isbjorn2.html. Accessed 16 April 2022.

Wedemeyer, Dr. (1914). Die Spitzbergen-Expedition des Leutnant Schröder-Stranz. In: A. Miethe (Ed.), *Die Expeditionen zur Rettung von Schröder-Stranz und seinen Begleitern geschildert von ihren Führern Hauptmann A. Staxrud und Dr. K. Wegener*. Berlin: Dietrich Reimer, pp. VI-XII.

Wegener, G. (1897). *Zum Ewigen Eise. Eine Sommerfahrt in nördliche Polarmeer und Begegnung mit Andrée und Nansen*. Berlin: Allgemeiner Verein für Deutsche Litteratur.

Wegener, K. (1913). Die Hilfsexpedition von Cross- und Kings-Bay nach Wijde-Bay. *Petermanns Geographische Mitteilungen* 59, II, pp. 137–140.

Wegener, K. (1914a). Das Observatorium in der Crossbai 1912/13. In: H. Hergesell (Ed.). Das Deutsche Observatorium in Spitzbergen, Beobachtungen und Ergebnisse I. *Schriften der Wissenschaftlichen Gesellschaft in Straßburg, 21*, pp. 21–29.

Wegener, K. (1914b). Die Hilfsexpedition von Cross- und Kings Bay, 21.II.- 31.III.1913. In: A. Miethe (Ed.), *Die Expeditionen zur Rettung von Schröder-Stranz und seinen Begleitern geschildert von ihren Führern Hauptmann A. Staxrud und Dr. K. Wegener* (pp. 69–101). Berlin: Dietrich Reimer.

Westphal, W. (1984). *Geschichte der deutschen Kolonien*. München: C. Bertelsmann.

Wichmann, H. (1911). Plan einer deutschen Expedition nach der Taimyrhalbinsel. *Petermanns Geographische Mitteilungen, 57*(2), 26.

Wichmann, H. (1912a). Deutsche Nordostdurchfahrt. *Petermanns Geographische Mitteilungen, 58*, II (7), p. 34.

Wichmann, H. (1912b). Deutsche Nordostdurchfahrt. *Petermanns Geographische Mitteilungen, 58*, II (7), p. 94.

*Wikipedia* / Sepp Allgeier. https://de.wikipedia.org/wiki/Sepp_Allgeier. Visited 12 August 2021

*Wikipedia* / Wilhelm Bade. https://de.wikipedia.org/wiki/Wilhelm_Bade. Accessed 6.10.2021

Zeppelin, F. Graf von. (1911). Hat unsere Expedition die Zweckmäßigkeit der Verwendung meiner Luftschiffe zur Erforschung der Arktis ergeben? In: A. Miethe & H. Hergesell (Eds.): *Mit Zeppelin nach Spitzbergen*. Berlin: Deutsches Verlagshaus Bong & Co., pp. 283–291

Zorgdrager, C. G. (1723 [1975]). *Alte und neue Grönländische Fischerei und Wallfischfang*. Leipzig: P. C. Monath. Reprint: Kassel: H. Hamecher.

# *Newspapers*

Aftenposten. (25 February 1913). Undsætnings-expeditionen til Spitsbergen. Fartøiet og deltagerne. Slæde-expeditionens udrustning. [The rescue expedition to Spitsbergen. The vessel and the participants. The sled expedition's equipment.]. *Aftenposten* (Kristiania [Oslo]), 25 February 1913, evening edition, p. 1. Library Archives, Norwegian Polar Institute, Tromsø. Received 22 July 2021. (Also in the National Library of Norway, Oslo, online archives. www.nb.no. Accessed 7 April 2022.)

Aftenposten. (15 May 1913). Hjælpeexpeditionen paa Spitsbergen. Kaptein Staxruds beretning til "Aftenposten". [The rescue expedition on Spitsbergen. Captain Staxrud's account to "Aftenposten".]. *Aftenposten* (Kristiania [Oslo]), 15 May 1913, evening edition, pp. 1–2. National Library of Norway, Oslo, online archives. www.nb.no. Accessed 28 August 2012.

Aftenposten. (3 September 1913). Kaptein Staxrud hjemme igjen. Udtaler sig til "Aftenposten" om redningsexpeditionen. [Captain Staxrud back home again. Speaks to "Aftenposten" about the rescue expedition.]. *Aftenposten* (Kristiania [Oslo]), 3 September 1913, evening edition, p. 4. National Library of Norway, Oslo, online archives. https://www.nb.no. Retrieved 4 April 2022.

Aftenposten. (12 September 1913). Postskøitens sidste tur til Spitsbergen. [The postal ship's last trip to Spitsbergen.]. *Aftenposten* (Kristiania [Oslo]), 12 September 1913, morning edition, p. 1. National Library of Norway, Oslo, online archives. https://www.nb.no. Retrieved 4 April 2022.

Aftenposten. (15 September 1913). Hjem fra Spitsbergen. Universitetsstipendiat Hoel. *Aftenposten* (Kristiania [Oslo]), 15 September 1913, evening edition, p. 1. National Library of Norway, Oslo, online archives. https://www.nb.no. Retrieved 2 April 2022.

Aftenposten. (8 October 1913). Lussi. *Aftenposten* (Kristiania [Oslo]), 8 October 1913, morning edition, p. 12. Reference archives, National Library of Norway, Oslo, online archives. www.nb.no. Author accessed 28 August 2012.

Aftenposten: Ugens Nyt. (1913, 27 February). Undsætnings-expeditionen til Spitsbergen. *Ugens Nyt* published by *Aftenposten* (Kristiania [Oslo]), 27 February 1913, pp. 1–2. National Library of Norway, Oslo, online archives. https://www.nb.no. Retrieved 7 April 2022.

Aftenposten: Ugens Nyt. (1913, 6 March). Hjelpeexpeditionen til Spitsbergen. [The rescue expedition to Spitsbergen.]. *Ugens Nyt* published by *Aftenposten* (Kristiania [Oslo]), 6 March 1913, p. 1. National Library of Norway, Oslo, online archives. https://www.nb.no. Retrieved 4 April 2022.

Aftenposten: Ugens Nyt. (1913, 2 September). En julenats-tragedie paa Spitsbergen. [A Christmasnight tragedy on Spitsbergen.]. *Ugens Nyt* published by *Aftenposten* (Kristiania [Oslo]), 2 September 1913, pp. 2–3. National Library of Norway, Oslo, online archives. https://www. nb.no. Retrieved 2 April 2022.

Andøya Avis. (1950, 6 October). Fangstmann og båtbygger Daniel Nøis runder 70 år. [Trapper and boat builder Daniel Nøis turns 70 years old.]. Written by H. M. [possibly Haakon Aronsen, aka Haakon Magnus, per Gunnhild Holmen of the National Library of Norway, based on *Andøya Avis* 1964 article, according to communications from Odd Ivar Nøis Olsen of the Nøis Family and from Ivar Stokkeland and Petr Masat of the Norwegian Polar Institute, received 29 March 2022]. *Andøya Avis* (Andøya), 6 October 1950, p. 1. Library Archives, Norwegian Polar Institute, Tromsø. Received 29 March 2022.

Dagsposten. (1913, 6 March). Hos Kaptein Staxrud. Han udtaler sig om Planerne. – Man faar høre fra Undsætningsekspeditionen i Slutten af Marts. [At Captain Staxrud's. He speaks about the Plans. – One will hear from the Rescue-expedition at the End of March.]. Written by Th. K. *Dagsposten* (Trondhjem [Trondheim]), 6 March 1913. (Newspaper clipping featuring handwritten notation by Staxrud, attached to letter from Arve Staxrud to Fridtjof Nansen dated 17 March 1913, NB Ms.fol. 1924:6:A:2). National Library of Norway, Oslo, online archives. https://www.nb.no. Retrieved 6 July 2021.

Der Zeitspiegel Berlin im Frühjahr. (1913). *Newspaper clipping.* Ritscher Estate, Cornelia Lüdecke, München.

Dominion. (1913, 31 October). Arctic Tragedy. *Dominion*, Volume 7, Issue 1894, 31 October 1913, p. 5. National Library of New Zealand. https://paperspast.natlib.govt.nz/newspapers/DOM191 31031.2.34. Accessed 4 August 2014.

Fredrikshalds Avis. (1913, 15 September). Til Spitsbergen. Dagboksnotiser og smaa reiseindtryk fra en tur sommeren 1913. *Fredrikshalds Avis* (Fredrikshald [Halden]), 15 September 1913, p. 2). Article by Jakob Moe. National Library of Norway, Oslo, online archives. https://www. nb.no. Retrieved 4 April 2022.

Hamburger Nachrichten. (1913a). Professor Miethe über die Hilfsexpedition nach Spitzbergen. *Hamburger Nachrichten*, 25 February 1913.

Hamburger Nachrichten. (1913b). Notruf für die Spitzbergen-Rettungs-Expedition. *Hamburger Nachrichten*, 11 March 1913.

Morgenbladet. (1913, 10 September). Hjem fra Spitsbergen. [Home from Spitsbergen.]. *Morgenbladet* (Kristiania [Oslo]), 10 September 1913, Evening edition, p. 2. National Library of Norway, Oslo, online archives. https://www.nb.no. Retrieved 4 April 2022.

Morgenbladet. (1913, 15 September). "Oversten"s Barn. [Children of "Obersten".]. *Morgenbladet* (Kristiania [Oslo]), 15 September 1913, evening edition, p. 1. Digital Archive, National Library of Norway, Oslo, received 30 March 2022; National Library of Norway, Oslo, online archives. https://www.nb.no, retrieved 2 April 2022.

Nidaros. (1913, 26 February). Undsætningen til Spitsbergen. Amundsens Sydpols-hunder blir med. [The rescue to Spitsbergen. Amundsen's South Pole dogs join.]. *Nidaros* (Trondheim), 26 February 1913, p. 1. National Library of Norway, Oslo, online archives. https://www.nb.no. Retrieved 7 April 2022.

Nidaros: Trøndelagen. (1913, 1 March). Undsætningen til Spitsbergen. Amundsens Sydpols-hunder blir med. [The rescue to Spitsbergen. Amundsen's South Pole dogs join.]. *Nidaros: Trøndelagen* (Trondheim), 1 March 1913, p. 3. Digital Archives, National Library of Norway, Oslo, online archives. https://www.nb.no. Received 8 June 2021; retrieved 7 April 2022.

Nord Norge. (1913a, 21 February). Spitsbergen-ekspeditionen. En oversigt. [The Spitsbergen expedition. An overview.]. *Nord Norge* (Tromsø), 21 February 1913, p. 1. (Newspaper clipping attached to letter from Arve Staxrud to Fridtjof Nansen dated 17 March 1913, NB Ms.fol.

1924:6:A:2). National Library of Norway, Oslo, online archives. https://www.nb.no. Retrieved 6 July 2021.

Nord Norge. (1913b, 21 February). Undsætnings-ekspeditionen. Nansen anbefaler rensdyr. [The Rescue expedition. Nansen recommends reindeer.]. *Nord Norge* (Tromsø), 21 February 1913, p. 2. (Newspaper clipping attached to letter from Arve Staxrud to Fridtjof Nansen dated 17 March 1913, NB Ms.fol. 1924:6:A:2). National Library of Norway, Oslo, online archives. https://www.nb.no. Retrieved 6 July 2021.

Nord Norge. (1913c, 21 February). Mye skrik og litet uld sa manden da han klipte grisen. Editorial by H. C. Johannesen. *Nord Norge* (Tromsø), 21 February 1913, p. 2. (Newspaper clipping attached to letter from Arve Staxrud to Fridtjof Nansen dated 17 March 1913, NB Ms.fol. 1924:6:A:2). National Library of Norway, Oslo, online archives. https://www.nb.no. Retrieved 6 July 2021.

Rave, C. (1913, 27 September). Die Tragödie der Schröder-Stranz-Expedition. Der Todeszug der zehn mutigen deutschen Männer durch Nacht und Eis. Berlin: *Berliner Tageblatt,* 27 September 1913.

Social-Demokraten. (1913, 27 February). Hjælpeekspeditionen til de indefrosne tyskere. *Social-Demokraten* (Kristiania [Oslo]), 27 February 1913, p. 2. National Library of Norway, Oslo, online archives. www.nb.no. Retrieved 7 April 2022.

Tromsø Stiftstidende. (1913a, 11 March). "Hertha". *Tromsø Stiftstidende* (Tromsø), 11 March 1913, p. 2. National Library of Norway, Oslo, online archives. https://www.nb.no. Retrieved 4 April 2022.

Tromsø Stiftstidende. (1913b, 11 March). Staxruds redningsekspedition. [Staxrud's rescue expedition.]. *Tromsø Stiftstidende* (Tromsø), 11 March 1913, p. 2. National Library of Norway, Oslo, online archives. https://www.nb.no. Retrieved 4 April 2022.

Trondhjems Adresseavis. (1913, 26 February). Undsætningsekspeditionen. Tre av Amundsens hunde blir med. [The Rescue expedition. Three of Amundsen's dogs join.]. *Trondhjems Adresseavis* (Trondheim), 26 February 1913, p. 1. National Library of Norway, Oslo, online archives. https://www.nb.no. Retrieved 7 April 2022.

Trondhjems Adresseavis. (1913, 6 September). Spitsbergen. Den norske ekspedition paa hjemtur. [Spitsbergen. The Norwegian expedition on its (return) trip home.]. *Trondhjems Adresseavis* (Trondheim), 6 September 1913, p. 1. National Library of Norway, Oslo, online archives. https://www.nb.no. Retrieved 5 April 2022.

Trondhjems Folkeblad. (1913a, 1 March). Undsætnings-ekspeditionen. Professor Miethe kritiserer Schröder Stranz. [The Rescue Expedition. Professor Miethe criticizes Schröder Stranz (sic).]. *Trondhjems Folkeblad* (Trondheim), 1 March 1913, p. 2. National Library of Norway, Oslo, online archives. https://www.nb.no. Retrieved 4 April 2022.

Trondhjems Folkeblad. (1913b, 1 March). Undsætningsekspeditionen. Tre av Amundsens hunde blir med. [The Rescue expedition. Three of Amundsen's dogs join.]. *Trondhjems Folkeblad* (Trondheim), 1 March 1913, p. 3. National Library of Norway, Oslo, online archives. https://www.nb.no. Retrieved 4 April 2022.

## *Photographs*

Drygalski estate, Mörder, Feldkirchen-Westerham
Herzog Ernst II estate, Mauritianum, Altenburg
Lüdecke, Cornelia
National Library of Norway, Oslo
Norwegian Polar Institute, Tromsø
Norsk Folkemuseum, Oslo
Ritscher estate, Leibniz Institute for Regional Geography, Leipzig, Germany
Ritscher estate, Lüdecke, Munich
Svalbard Museum, Svalbard
Tahan, Mary R.